Complex Oxides

An Introduction

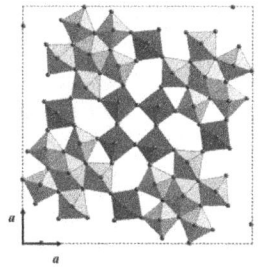

Complex Oxides

An Introduction

Editors

Thomas Vogt
University of South Carolina, USA

Douglas J Buttrey
University of Delaware, USA

World Scientific

NEW JERSEY · LONDON · SINGAPORE · BEIJING · SHANGHAI · HONG KONG · TAIPEI · CHENNAI · TOKYO

Published by

World Scientific Publishing Co. Pte. Ltd.

5 Toh Tuck Link, Singapore 596224

USA office: 27 Warren Street, Suite 401-402, Hackensack, NJ 07601

UK office: 57 Shelton Street, Covent Garden, London WC2H 9HE

Library of Congress Cataloging-in-Publication Data

Names: Vogt, Thomas (Professor of Chemistry & Biochemistry), editor. | Buttrey, Douglas J., editor.

Title: Complex oxides : an introduction / edited by Thomas Vogt (University of South Carolina, USA),
    Douglas J. Buttrey (University of Delaware, USA).

Description: New Jersey : World Scientific, 2019. | Includes bibliographical references and index.

Identifiers: LCCN 2018049402 | ISBN 9789813278578 (hardcover : alk. paper)

Subjects: LCSH: Metallic oxides.

Classification: LCC QD509.M46 C66 2019 | DDC 546.3--dc23

LC record available at https://lccn.loc.gov/2018049402

British Library Cataloguing-in-Publication Data

A catalogue record for this book is available from the British Library.

For any available supplementary material, please visit
https://www.worldscientific.com/worldscibooks/10.1142/11220#t=suppl

Typeset by Stallion Press

Email: enquiries@stallionpress.com

Contents

Complex Oxides — An Introduction

The following seven chapters provide a selection of research reviews on complex oxides involving solid-state battery research, the chemistry of transparent conductors, ternary uranium oxides, magnetic perovskites, nonlinear optical materials, and complex molybdenum–vanadium bronzes used as selective oxidation catalysts. While this represents a heterogeneous and wide array of physical and chemical properties, two threads can be followed through the book: (i) the materials are complex oxides and (ii) guiding design principles are outlined for obtaining and optimizing desired properties, many of them heuristic. We hope that this collection of chapters find use as topical reviews and current assessments of research directions for research scientists and graduate students.

In the contribution by Helena Braga and John B. Goodenough, 'Dielectric Amorphous-Oxide Electrolytes,' we are reminded that amorphous oxides remain an underappreciated class of materials which often reveal unusual and surprising properties due to their high degree of disorder. The authors present us with an all-solid-state rechargeable battery where a dielectric amorphous oxide supports fast-moving cations and electric dipoles which permit self-charge and self-cycling, bringing an all-battery-powered electrical vehicle within reach.

In metal oxides, the understanding and subsequent 'engineering' of competing interactions allows us to combine what appear to be, on first sight, mutually exclusive properties such as transparency to light *and* electrical conductivity. Transparent conductors, as the paradigmatic indium-tin oxide are ubiquitous and vitally important in consumer electronics, energy harvesting in solar cells, substrates in solar-driven water splitting, and organic light-emitting devices. Karl Rickart, Steven Flynn, and Kenneth R. Poeppelmeier introduce us to the basic materials, concepts, and opportunities associated with transparent conducting oxides, which are the basis of a multibillion dollar industry.

Uranium oxides are the foundation of the nuclear fuel cycle, and their long-term geological disposal requires a thorough understanding of their chemical reactivity, phase transitions, and physical and chemical properties. Surprisingly, the structural chemistry of ternary monouranate compounds AUO_{4-x} is not well known, and even simple materials such as $SrUO_{4-x}$ and $CaUO_{4-x}$ surprise us with unusual oxygen transport behavior at higher temperatures. In an up-to-date review, Gabriel L. Murphy, Zhaoming Zhang, and Brendan J. Kennedy present the solid-state chemistry of this important class of ternary uranium oxides.

In an authoritative review on the 'Magnetic Properties of Perovskite Structure Oxides', John E. Greedan provides us with an overview of the most pertinent magnetic phases in the most studied class of oxides. The ongoing discovery of new phases with perovskite structure, notoriously tolerant of substitutions, reveals an increasing array of unusual and useful dielectric, piezoelectric, superconducting, and magnetic properties and shows no sign of abating. The author guides us through the key magnetic structures of perovskite phases and perovskite-related variants, and develops main concepts using paradigmatic examples. This represents a true *tour-de-force* which we believe will serve as a valuable reference for research scientists in condensed matter physics and solid-state chemistry.

The discovery of oxides and oxide-related materials that allow the generation of ultraviolet (UV) and deep ultraviolet coherent radiation in oxides using nonlinear optical effects such as second harmonic generation (SHG) are the underpinnings of atto-second pulse generation, new laser systems, photolithography, and semiconductor manufacturing. P. Shiv Halasyamani describes the current state-of-the-art materials chemistry of these materials after systematically laying out the materials' design requirements for coherent emitters of UV or deep UV light and introducing refractive index and phase-matching conditions.

Selective oxidation and ammoxidation reactions account for roughly 25% of critical industrial organic products and intermediates needed for the manufacturing of important industrial products and consumer goods. Douglas J. Buttrey, Douglas A. Blom, and Thomas Vogt outline the structural characterization of a structurally and compositionally complex molybdenum vanadium bronze catalyst using X-ray and neutron powder diffraction in combination with electron microscopy. The emergence of a structural model for a catalyst that has the potential to use propane instead of propene as feedstock was only possible by combining novel instrumental approaches and pushing data analysis and simulation techniques to new limits.

In 'New crystalline complex metal oxide catalysts with porous, acidic, and redox properties' Satoshi Ishikawa, Zhenxin Zhang, Toru Murayama, Masahiro Sadakane, and Wataru Ueda highlight recent advances in synthesis of 3D pore structures of complex oxides with group 5 and 6 metals in order to create new heterogeneous catalysts with optimized morphologies for complex chemical reactions such as partial oxidation.

One of us (T.V.) spent part of 2018 as Fellow at Durham University's Institute of Advances Study working on the completion of this book and is very grateful to Linda Crowe and the rest of the IAS team as well as David Wilkinson, principal at St. John's College, for the warm welcome and providing an environment so conducive to scholarly work.

Dielectric Amorphous-Oxide Electrolytes

M. Helena Braga* and John B. Goodenough†

*The University of Porto, Portugal
†The University of Texas at Austin, TX 78712, USA

Introduction

Complex oxides are essential components of the electrochemical cells that are poised to revolutionize the energy economy of the world. Modern society now runs on the energy stored in a fossil fuel. This dependence is not sustainable. A fossil fuel, once burned, is not recyclable, and the exhaust gases of fossil-fuel combustion not only contribute to global warming, they are also already choking large populations in cities like Beijing in China and New Delhi in India. Since the nuclear option has proven to be a problematic energy source, it has become essential to find a way to return to dependence on the energy that reaches Earth daily from the sun. Converting the sun's radiant energy and the mechanical energy from wind into electric power is feasible, and the electric power can be transported to distributed collection points, but this power cannot be used unless it is stored. The rechargeable battery stores electric power. Therefore, a priority technical target worldwide is the development of large-scale rechargeable batteries consisting of stacks of identical electrochemical cells that are safe and cost competitive with the energy stored in a fossil fuel. This chapter describes dielectric amorphous-oxide alkali-metal electrolytes that may enable safe, low-cost, large-scale rechargeable batteries of sufficient volumetric density of stored electric power for replacing the air-polluting internal combustion engine powering today's road vehicles. Such a battery could also be used for distributed storage of electric power harvested from solar and wind energy to supplement the grid. Dielectric amorphous-oxide electrolytes developed at the University of Porto, Portugal have been used at the University of Texas at Austin to develop novel designs for a safe, low-cost rechargeable battery with long charge–discharge cycle life targeted for over 150,000 miles driving and a large volumetric density of stored electric power targeted for a driving range greater than 300 miles with one charge. Fast-moving A^+ cations (A = Li, Na) and slower-moving electric dipoles coexist in the dielectric glass electrolytes; this coexistence gives rise to two novel phenomena: self-charge and self-cycling of a rechargeable electrochemical cell. The complex oxides provide a basis for optimism that an all-electric road vehicle powered by batteries charged by solar and wind energy can soon begin to replace the road vehicles now powered by an internal combustion engine running on the chemical energy stored in a fossil fuel.

Rechargeable Batteries

A battery may consist of a single electrochemical cell or contain many identical cells. On discharge, a battery cell delivers electric power $P = IV$ as an electric current I at a voltage V for a time Δt in an external circuit. The cells of a multicell battery are connected in series to give a desired output voltage V and in parallel to give a desired output current I; the stack of cells must be managed so as to give similar power outputs from each cell. Not only the materials of the individual cells but also the cell design determine the cell performance and the cost of multicell management of a large-scale rechargeable battery.

A rechargeable battery cell stores electric power as chemical energy in its two electrodes, a negative electrode (anode) and a positive electrode (cathode) that are separated by an electrolyte, as is illustrated schematically in Figure 1. A liquid electrolyte is absorbed in a porous, flexible separator that keeps the two electrodes apart without being either reduced by the anode reductant or oxidized by the cathode oxidant. The chemical reaction of a cell is normally between the two electrodes and is reversible in a rechargeable battery; the reaction has an electronic and an ionic component. The electrolyte conducts the ionic component inside the cell and forces the electronic component to traverse the external circuit as a current I at a voltage V until the chemical reaction is completed after a time Δt. When the external circuit is open, the cell voltage is $V_{oc} = (\mu_A - \mu_C)/e$, where μ_A and μ_C are, respectively, the electrochemical potentials (Fermi levels) of the anode and the cathode, and e is the magnitude of the electron charge. With a fixed charge and discharge current $I_{dis} = I_{ch}$, the voltage of the applied charging power $P_{ch} = I_{ch}V_{ch}$ is $V_{ch} > V_{dis}$, which makes the efficiency of power storage $100\ P_{dis}/P_{ch} < 100\%$. The voltages are as follows:

$$V_{ch} = V_{oc} + \eta_{ch}I_{ch} \quad \text{and} \quad V_{dis} = V_{oc} - \eta_{dis}I_{dis}$$

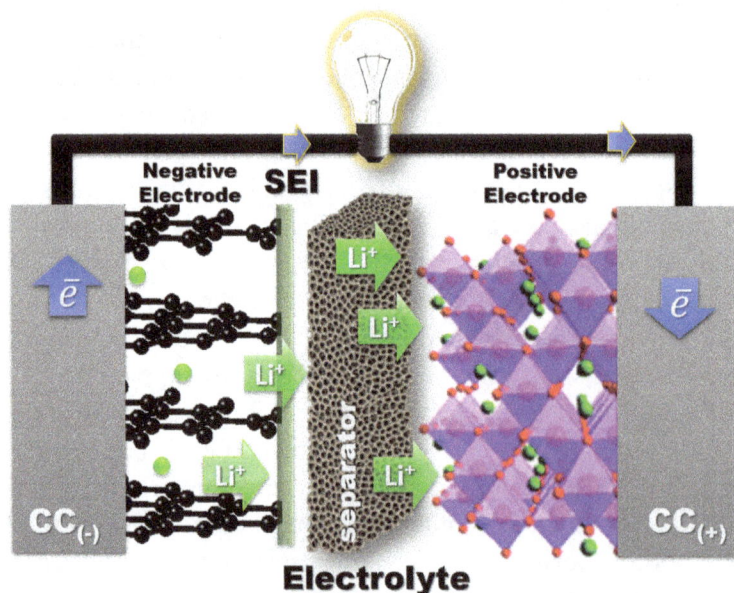

Figure 1. Battery cell with a liquid or gel electrolyte; the cell contains a negative and a positive electrode, a separator, and a current collector for each electrode. A solid–electrolyte interphase forms on the surface of the negative electrode in contact with the electrolyte where the anode Fermi level is above the electrolyte LUMO; see Figure 2.

where η_{ch} and η_{dis} are the resistances to the ionic transfer between the anode and the cathode. The $\eta_{ch}I_{ch}$ is called the overvoltage, the $\eta_{dis}I_{dis}$ the polarization. The charge delivered at a fixed current $I = dq/dt$ in the time Δt is the capacity per unit weight or volume

$$Q(I) = \int_0^{Q(I)} dq = \int_0^{\Delta t} I dt$$

and the total gravimetric or volumetric density of power stored as chemical energy is

$$\Delta E = \int_0^{\Delta t} P dt = \langle V(q) \rangle \cdot Q(I)$$

where q is the state of charge. The density of stored power ΔE at a desired current I is an important parameter.

During charge and discharge, cations are transferred between the electrode and the electrolyte, so the electrodes change volume. A liquid electrolyte is plastic and, therefore, can retain good electrode–electrolyte contact during the electrode volume changes. Moreover, the resistance to cation diffusion in the electrolyte is $R_b = l/\sigma_i A$, where l/A is the ratio of the thickness to surface area of the electrolyte and σ_i is the conductivity of the working cation transported by the electrolyte. A $\sigma_i \approx 10^{-2}$ S cm^{-1} at 25°C requires, for an acceptably fast charge–discharge, an electrolyte thickness $l < 40$ μm, where l includes the electrolyte penetration into a thick, porous cathode required for a large $Q(I)$. Therefore, today's batteries with a large energy density use a liquid electrolyte with a $\sigma_i \geq 10^{-2}$ S cm^{-1}.

The open-circuit voltage V_{oc} of a cell with a liquid electrolyte is limited by the energy gap $E_g = $ (LUMO – HOMO) between its lowest unoccupied and highest occupied molecular orbital. As is illustrated in Figure 2, if the Fermi level of the anode is above the LUMO energy, the electrolyte is reduced by the anode, and if the Fermi level of the cathode is below the HOMO energy, the electrolyte is oxidized by the cathode. A traditional battery cell uses a strongly alkaline or acidic aqueous electrolyte capable of fast H$^+$ transport $(\sigma_H > 10^{-1}$ S cm$^{-1})$, but an $E_g = 1.23$ eV limits the open-circuit voltage to $V_{oc} \leq 1.5$ V in a cell with a stable shelf life when charged. This limitation keeps the energy density of the above-mentioned equation too low and the cost too high for a large-scale battery that can compete with the internal combustion engine.

The Li-ion battery of the wireless revolution uses a flammable organic liquid electrolyte with a $\sigma_i \approx 10^{-2}$ S cm^{-1}. Although fabrication of a cell with an electrolyte thickness $l \lesssim 30$ μm is feasible, there are three problems

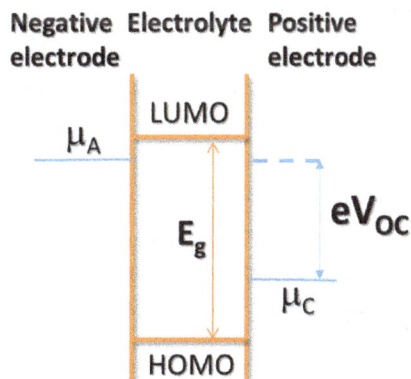

Figure 2.   Negative and positive electrode electrochemical potentials or Fermi levels in comparison with the liquid electrolyte's LUMO and HOMO to avoid formation of SEI layers.

with the liquid-carbonate electrolyte commonly used: (1) it has $E_g \approx 3.0$ eV, (2) the energy of its LUMO is 1.2 eV below the Fermi level of lithium, and (3) plating of a lithium anode during charge results in anode dendrite formation and, on repeated charges, growth across the thin electrolyte to the cathode, which creates an internal short-circuit that ignites the electrolyte. Although ethylene carbonate as an electrolyte additive can create a solid–electrolyte interphase (SEI) layer that passivates reduction of the electrolyte by an anode with a Fermi level above the electrolyte LUMO, the SEI limits the charge–discharge cycle life and low-temperature operation. Unless the Fermi levels of the two electrodes are within the electrolyte E_g, safety concerns and the cost of too frequent battery replacement make problematic large-scale batteries competitive with the internal combustion engine. Moreover, an air cathode is not feasible for a road vehicle and a sulfur S_8 cathode has soluble Li_2S_x intermediates in an organic liquid electrolyte that has prevented commercialization. To date, an inorganic liquid electrolyte that is inflammable with $E_g > 5$ eV and $\sigma_i \geq 10^{-2}$ S cm^{-1} has not been identified.

These problems with a liquid electrolyte have motivated reconsideration of an all-solid-state rechargeable battery that has been restricted previously to thin-film cells. New solid electrolytes and battery designs that can provide cells of higher energy density are needed. A solid electrolyte is also a separator; it may be a polymer, a crystalline ceramic, an amorphous ceramic, or a polymer–ceramic composite. It may also be a two-stage tandem electrolyte with a ceramic electrolyte contacting the anode and a polymer or plasticizer contacting the cathode. An elastic polymer may accommodate the electrode volume changes during charge and discharge. A solid electrolyte may have an $E_g = (E_c - E_v) > 5$ eV, where E_c and E_V are, respectively, the bottom of the conduction band and the top of the valence band. Nevertheless, a solid electrolyte faces formidable challenges. The two foremost challenges are (1) the cation conductivity and (2) the retention of a good solid–solid electrode–electrolyte contact during repeated charge–discharge cycling. Simple blocking of anode dendrites by a solid electrolyte will not retain a good anode–electrolyte contact. However, a dendrite-free alkali-metal anode can be plated on and stripped from a solid if the alkali-metal bonding to the solid is stronger than the alkali-metal bonding with itself, i.e. if the alkali metal wets the solid. Moreover, strong anode–electrolyte bonding can be expected to constrain the anode volume changes during charge–discharge cycling to be perpendicular to the interface, a change that can be accommodated by cell design. On the other hand, most solid electrolytes have a cation conductivity $\sigma_i < 5 \times 10^{-3}$ S cm^{-1} at 25°C, which requires fabrication as a mechanically robust, thin ($l < 30$ μm) membrane and operation at a temperature $T_{op} > 50$°C. This chapter introduces a remarkable complex oxide that can provide new designs for an all-solid-state rechargeable battery.

The Dielectric Amorphous-Oxide Electrolytes

The dielectric amorphous oxides contain electric dipoles coexisting with mobile A$^+$ cations (A = Li, Na, or K). Solid electrolytes have traditionally been crystalline materials. The nominal crystalline antiperovskite Li_3OCl was being investigated as a crystalline Li$^+$ electrolyte, but when it was found at the University of Porto, Portugal, that the nominal crystalline phase that formed was $Li_{3-x}H_xOCl$, $0 < x \leq 1$, the precursors $Li_{3-2x-y}A_xH_yOCl$ (A = Mg, Ca, Ba; $0 < y \leq 1$) (obtained by wet synthesis using LiCl, Li(OH), and Ba(OH)$_2 \cdot$ 8H$_2$O as reagents) were investigated and found to form dielectric amorphous-oxide Li$^+$ electrolytes if heated to dry without crystallization or formation of a crystalline second phase.[1] The value of x was adjusted to tune the lower glass transition temperature T_g to be near 45°C. During the drying process, a certain amount of Cl$^-$ was evaporated as Cl$_2$ and, perhaps, as HCl, along with water to leave an amorphous product containing LiO$^-$ and Li$_2$O electric dipoles as well as mobile Li$^+$ and Cl$^-$ ions in a neutral electrolyte. A similar product was obtained on substitution of Na for Li in the precursor material. Hereinafter, we refer to these products as the Li$^+$-glass and Na$^+$-glass electrolytes. Figure 3 illustrates Arrhenius plots of the Li$^+$ conductivity and permittivity of the Li$^+$ glass; comparable behavior is exhibited by the Na^+ glass. When the Li^+-glass data of Figure 3 was brought to the University of Texas at Austin, the existence of a Li^+ conductivity $\sigma_{Li} \simeq 2 \times 10^{-2}$ S cm^{-1} at 25°C immediately turned attention to its use in an all-solid-state battery since dendrite-free plating of lithium due to wetting of a solid electrolyte had just been demonstrated there.[2]

The Li^+-glass was shown to have a large E_g; it was not reduced on contact with a metallic lithium anode and it was not oxidized by an oxide cathode charged by a $V_{ch} = 8$ V.[3] It was also shown that the high conductivity and permittivity seen in Figure 3 could be obtained in minutes at 100°C, whereas aging of the glass takes days at room temperature.[4] The next step was to determine whether a metallic lithium anode can be plated/stripped dendrite-free at low impedance in a symmetric Li/Li-glass/Li cell. Figure 4 shows a voltage oscillating about 0 V at 3 mA cm^{-2} with a negligible plating/stripping impedance for over 5,000 charge–discharge cycles (still cycling). This experiment demonstrates that lithium bonds strongly to

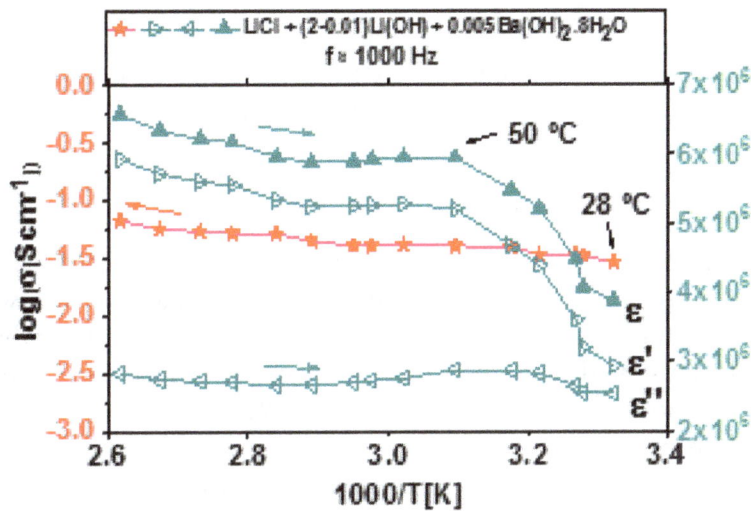

Figure 3.   Permittivity and ionic conductivity of a Li^+-glass electrolyte vs. temperature.

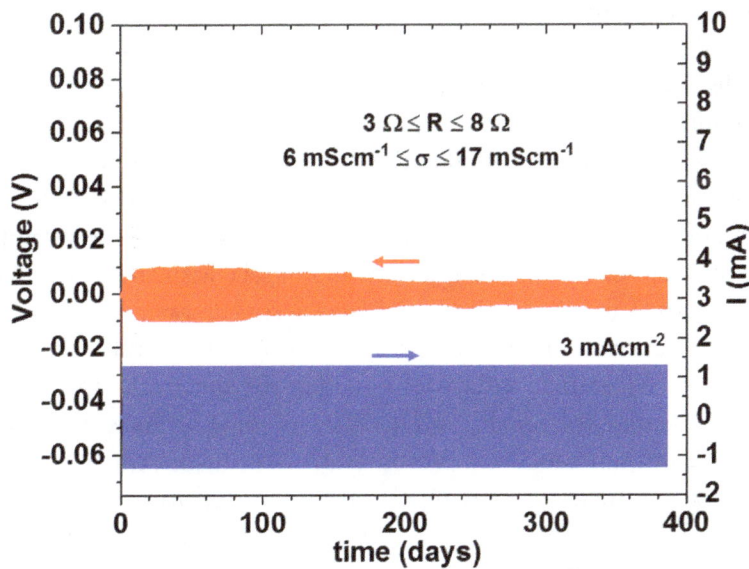

Figure 4.   Charge–discharge cycles of a Li/Li^+-glass/Li symmetric cell.

the Li$^+$-glass, i.e. wets it, and that the volume changes of the lithium electrode are accommodated in the cell as anticipated for 1D lithium growth perpendicular to the interface. This demonstration showed that the glass electrolyte solves the anode problem that has plagued the flammable liquid electrolyte; cells with a lithium or sodium anode can be safe with a fast charge and do not prevent the realization of a cell with a long cycle life.

The next problem was to learn how to fabricate a ceramic that is strongly hygroscopic into a mechanically robust and flexible large-area membrane with a thickness $l \leq 40$ μm that can be easily applied over a large area in a dry room rather than in a glove box. The glass bulk product was first broken up and ground under ethanol to form a slurry that can be immobilized by absorption in a porous cellulose paper sheet. By mixing into the slurry a lithium (or sodium) polyacrylate, a flexible, thin electrolyte membrane could be pressed into a thin membrane in a dry room for securing to an electrode.[5]

The Cathode Problem

The final, more challenging problem is to create a cathode and cell design that has an acceptable discharge voltage and a high volumetric capacity $Q(I)$ at an acceptably high I_{dis} and I_{ch}. We have developed three solutions.

Solution I

Encouraged by the low-impedance plating/stripping of the symmetric alkali-metal cells, we first investigated an asymmetric cell in which the Fermi level of the cathode was at a significantly lower energy than that of the anode so that we could transfer lithium from the anode and plate it on the cathode at a finite voltage. This strategy, although demonstrated experimentally,[6] has been challenged to be spurious as it was claimed to violate the First Law of Thermodynamics.[7] This challenge reflects a common failure to appreciate the role of the two electrolyte–electrode interfaces as heterojunctions. A heterojunction is an interface between two dissimilar materials in contact with one another. At the junction, the Fermi levels of the two materials are equalized by charge transfers within or between the two materials to create a capacitor that provides an electric field across the interface that is just large enough to equalize the Fermi levels of the two materials. In a battery cell at open circuit, charge transfer within the electrolyte creates an electric double-layer capacitor at each electrolyte–electrode interface. The work done in moving electrolyte charges fulfills the requirements of the First Law of Thermodynamics.

In our asymmetric cells, the electrolyte interfaces the anode on one side and, on the other side, an insulator relay particle that contacts both the electrolyte and the cathode current collector; the relay sets the Fermi level of the cathode. Work done by the cell in the formation of EDLCs at the two electrolyte interfaces creates a voltage difference across an asymmetric cell even with the same alkali metal on each electrode since the Fermi level of the cathode alkali metal is determined by that of a cathode relay insulator, provided the relay does not cover the majority of the cathode surface in contact with the electrolyte.

Figures 5 and 6 illustrate the variation of the Fermi level across an asymmetric Li/Li$^+$-glass/S–C–Cu cell before and after cell assembly at open-circuit; S represents an S$_8$-particle relay. The EDLCs at the lithium–electrolyte and the electrolyte–S$_8$-particle interfaces equalize the electrode and electrolyte Fermi levels at each interface to create a measured voltage $V_{me} = [\mu_A(Li) - \mu_C(S_8)]/e$. As is illustrated in Figure 6, during discharge, electrons from the lithium anode are prevented from reducing the S$_8$ particles by negative-charge buildup in the S$_8$ surface states; instead the arriving electrons plate Li$^+$ from the electrolyte onto the cathode metallic surface at a voltage $2.34 \leq V_{dis} \leq 2.65$ V, Figure 7. The voltage at which the S$_8$ particles are known to begin to be reduced to Li$_2$S$_8$ in a liquid electrolyte is $V_{dis} = 2.34$ V. If the rate of electron arrival from the

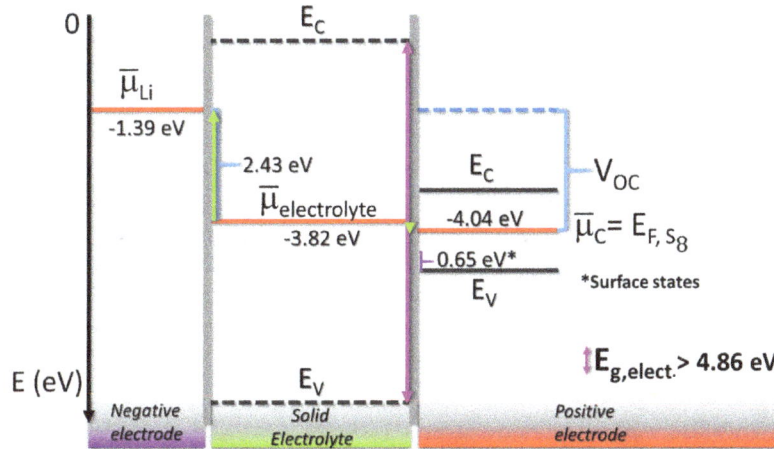

Figure 5.   Chemical potentials of an open circuit Li/Li$^+$-glass/S cell prior to alignment of the electrochemical potentials or Fermi levels by the formation of EDLC at the interfaces.

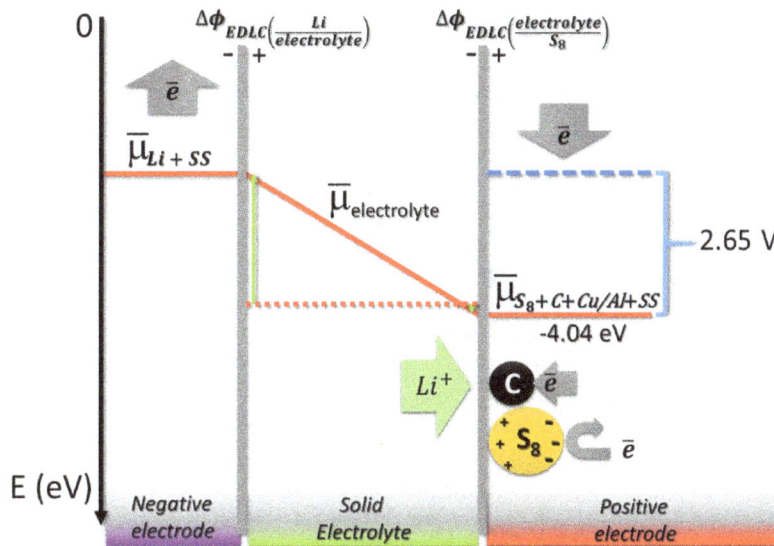

Figure 6.   Electrochemical potentials during discharge of a Li$^+$SS/Li$^+$-glass/S8+C+Cu/Al+SS cell. SS is stainless steel.

anode is faster than the rate of plating of lithium on the cathode, the S_8 surface states become more and more filled, raising the Fermi level of the relay and, therefore, lowering V_{dis}; at $V_{dis} = 2.34$ V, reduction of the S_8 particles makes V_{dis} fall more rapidly with time until the chemical reaction is completed once most of the S_8 particles are reduced to Li_2S_2 or Li_2S. Optimization of the $Q(I_{dis})$ of such a cell for a given I_{dis} awaits cell development.

Since plating of a dendrite-free alkali metal on an electrode only requires a solid electrolyte having a surface that is wet by the alkali metal, a rechargeable asymmetric cell that transports the alkali-metal cation A^+ (A = Li, Na, or K) reversibly between the two electrodes at a finite voltage can be fabricated with any solid A^+ electrolyte that is wet, but not reduced, by the alkali metal. On the other hand, a $T_{op} = 25°C$ requires a $\sigma_i \simeq 10^{-2}$ S cm^{-1} at 25°C. The amorphous ceramic electrolyte meets this criterion and even allows operation to –20°C.

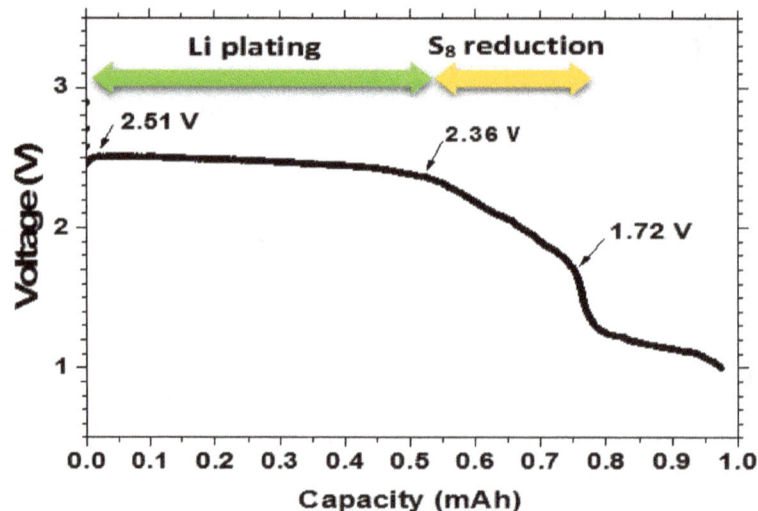

Figure 7.   Voltage vs capacity of an asymmetric Li/Li$^+$-glass/S–C–Cu cell.

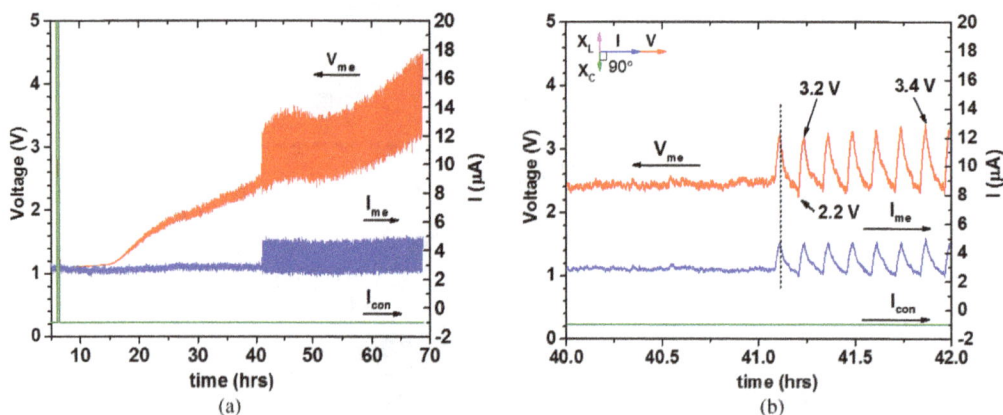

Figure 8.   Electrochemical cycling of Al/90 wt.% Li$^+$-glass 10 wt.% Li$_2$S/Cu cell; phasors description: X_C is the impedance of the capacitor, X_L is the impedance of the inductor, $V = V_{me}$, $I = I_{me}$. (a) Time variation of the measured voltage V_{me} and of the measured and control currents I_{me} and I_{con} after an initial charge–discharge cycle; (b) expanded time scale in an interval where the onset of the self-cycling occurs.

Solution II

The glass electrolytes also contain electric dipoles, and this coexistence produces two remarkable new phenomena, *self-charge* and an associated *self-cycling*.[8] These phenomena have only been found in a solid electrolyte in which fast-moving A$^+$ cations (A = Li or Na) coexist with slower-moving electric dipoles. In the following experiments that illustrate the phenomena, a control current I_{con} is the current specified by the potential difference between the two electrodes that is controlled by the load in a potentiostat whereas the measured current I_{me} is the actual measured current, which includes both the I_{con} and a current resulting from a self-charge. An $I_{con} \neq I_{me}$ distinguishes a cell with a self-charge from a traditional cell.

These phenomena are illustrated by the cell shown in Figure 8, a cell containing a Li$^+$-glass electrolyte to which 10 wt.% Li$_2$S was added to increase the concentration of electric dipoles in the glass. Neither electrode contained any detectable atom of Li or any other alkali metal in the initial cell fabrication. The anode consisted of an aluminum current collector coated with non-bonding carbon, Al–C, and the cathode was a copper current collector with a non-bonding carbon layer containing S_8 particles, S–C–Cu, to facilitate Li plating on discharge. Some Li^+-glass was mixed into the S–C cathode coat. The non-bonding layers proved to be ineffective as coatings to protect the electrode from reacting chemically with the electrolyte, and they do not fix the Fermi levels of the electrodes. Figure 8(a) shows the evolution with time of the measured voltage V_{me} and of the measured and control currents after an initial charge–discharge cycle of 6–7 h during which the cell was heated to 148°C and cooled back down to room temperature immediately after discharge. The heating was done in order to age the electrolyte for high Li^+ conductivity and permittivity.[4] The initial charge followed by discharge is represented by the vertical line on the left side of Figure 8(a). With an I_{con} = −1 μA, the steady increase in the V_{me} over a period of 41 h without an external charge represents a self-charge of the cell. After 41 h still with I_{con} = −1 μA, the cell exhibited a charge–discharge self-cycling for 27 h with a cycling period of about 8 min. A cycling I_{me} accompanied in-phase the cycling V_{me}, indicating a plating/stripping of a metallic lithium anode, i.e. a charge–discharge self-cycling phenomenon. Figure 8(b) shows an extended time scale graph of the time between 40 and 42 h of Figure 8(a) when the change from a self-charge to a self-cycling occurred. Figure 8(b) shows that the plating of the self-charge is faster than the stripping during a self-cycling period, which can give an increase in capacity with cycling as illustrated in Figure 8a.

An important feature of Figures 8a and 8b is the voltage 2.4 V at which self-cycling is initiated. The difference in the electrode chemical potentials (Fermi levels) is $[\mu_A(Al) - \mu_C(Cu)]$ = 2.0 eV for Cu(I) and 2.2 eV for Cu(II), and the cycling is over the voltage range $2.0 < V < 3.4 – 3.6$ V, where $[\mu_A(Li) - \mu_C(Cu)]$ = 3.5 V. This observation shows that the discharge corresponds to stripping of Li to a $\mu_A(Al)$ whereas the charge voltage increases to that of a layer of metallic lithium at a $\mu_A(Li)$. The self-cycling can be modeled as that of a *relaxation oscillator* and represented by the introduction of an inductor in the equivalent circuit of the cell as shown in Figure 9. Calculation shows that the self-cycling can be sustained by ambient heat.

The self-charge and self-cycling phenomena can be understood to be the result of the EDLCs that form at the electrode–electrolyte interfaces and the different time scales of the A^+ cation and electric-dipole contributions to the electrolyte charge at the anode EDLC. Equilibration of the Fermi levels is accomplished first by the motion of the fast-moving cations. The addition of more electrolyte charges at the interface by slower-moving electric dipoles requires a reequilibration of the Fermi levels by a reduction of the initial charge by the fast-moving cations. This can be done by increasing the width of the EDLC; but at the anode side, it can also be done by plating the electrolyte cations on the anode current collector. Plating of electrolyte Li^+, as in the example of Figure 8, represents a self-charge. However, removal of Li^+ from the electrolyte without the ability to replenish Li^+ to the electrolyte from the cathode creates a negative charge in the

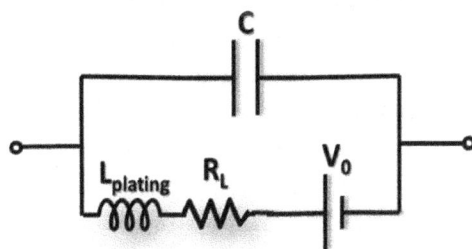

Figure 9. Equivalent circuit for a metal anode/A^+-glass/metal cathode cell that self-chargs and self-cycles as in a *relaxation-oscillator*. C is the equivalent capacitance for the association of EDLCs at the internal interfaces, $L_{plating}$ is the self-inductance, usually just called inductance, of the inductor that results from the accumulation of negative charge in the electrolyte due to plating without replenishing of the A^+-ions, R_L is the resistance to ionic movement due to accumulation of negative charges at the interface, and $V_0 = (\mu A - \mu C)/e$ in which μA and μC are the electrochemical potentials of the anode and cathode and e is the magnitude of the electron charge.

electrolyte that can strip the Li^+ back to the electrolyte. Once the V_{me} reaches the maximum faradaic voltage, i.e. where V_{me} plateaus at $(\mu_A - \mu_C)/e$, the self-cycling begins. The different rates of plating/stripping determine the period of the self-cycling. At longer times of self-charge, the linear increase in $V_{me}(t)$ reflects a capacitive component of stored electric power. An initial external charging pulse plates Al from the Al_2O_3 surface layer and Li from the electrolyte back on the Al current collector. The subsequent slower discharge not only returns most, if not all, of the plated Li as Li^+ to the electrolyte but also plates back on the cathode Cu sheet any copper cations in copper compounds, e.g. $CuCl_2$ at the copper surface. This process provides fresh Li and Cu surfaces for the self-charge and self-cycling phenomena.

Solution III

The phenomenon of self-charge with the dielectric amorphous-oxide electrolyte invited the development of an alternative cell design that can eliminate the need to transfer A^+ cations across the electrolyte.[9] An electrochemical capacitor stores electric power as a capacitive energy in the EDLCs at the electrode–electrolyte interfaces; charge–discharge cycling is fast without a capacity fade, but the capacity $Q(I)$ is relatively small and $V(q)$ changes strongly with q. A rechargeable battery cell with a two-phase product on A^+ insertion (q) into a cathode host has a $V(q)$ that is independent of q over a two-phase solid-solution range; it stores electric power as chemical (faradaic) energy with a $Q(I)$ that depends on the solid-solution range of the A^+ ion in the host at the discharge current I_{dis}. The cation conduction in the host may be fast enough to give a large $Q(I)$ at a desired I_{dis}. The challenge is to create a rechargeable cell that stores electric power as both capacitive and faradaic energy while keeping the cation displacements close to the electrode–electrolyte interfaces. The cathode oxide host providing the faradaic component should give a high V_{dis}. We chose as cathode a spinel host, $Li[Ni_{0.5+x}Mn_{1.5-x}]O_{y-2x}F_{2x}$ ($0 \leq 0.5$), hereafter LNMO, as the Li^+ diffusion is fast in this spinel and it gives a $V_{dis} \lesssim 4.8$ V.

Our solution to this cathode challenge began with the realization that a solid-solution cathode host that changes volume in three dimensions on A^+ insertion–extraction in a discharge–charge cycle needs to contact a plastic solid if the contact between the two solids is to be maintained over many discharge–charge cycles. We could choose an elastic polymer or a plasticizer to contact the high-voltage cathode oxide, but it would need to transport charge, which we originally assumed would be a mobile A^+ cation. We chose the plasticizer succinonitrile ($N \equiv C - CH_2 - CH_2 - C \equiv N$) with the mistaken assumption that the $C \equiv N$ terminal groups would allow a greater A^+ mobility. In fact, it did the opposite; the added lithium salt did not have a good solubility in this plasticizer; it was a poor Li^+ conductor. However, the plasticizer had a high dielectric constant because the $N \equiv C - CH_2 - CH_2 - C \equiv N$ molecules are electric dipoles that condense into a solid as a result of only weak dipole–dipole interactions, which is why it is plastic.[10] Charge is transported in the plasticizer by dipole rotations rather than by cation diffusion. Therefore, on fabrication of an all-solid-state rechargeable cell with a Li^+-glass electrolyte contacting a lithium anode and the plasticizer contacting the cathode oxide spinel particles, no Li^+ is transported across the plasticizer–Li^+-glass interface and the Li^+ extracted from the spinel cathode during charge are confined to the cathode surface; the Li^+ plated on the anode from the cathode are not replenished to the glass electrolyte from the cathode as occurs in a conventional rechargeable cell in which the working cations move between the electrodes in a charge and discharge. The cell has a faradaic component of electric power storage in the cathode spinel and a capacitive component not only at the electrode–electrolyte interfaces, but also at the electrolyte–plasticizer interface. The resulting charge behavior illustrated in Figure 10 shows an increase in capacity at a high $I_{dis} = I_{ch}$ rather than a capacity fade on cycling until the voltage plateaus and a self-cycling is initiated.

In conclusion, the dielectric amorphous oxide containing the coexistence of fast-moving cations and electric dipoles is a complex oxide that has not only displayed new electrochemical phenomena of self-charge

(a)

(b)

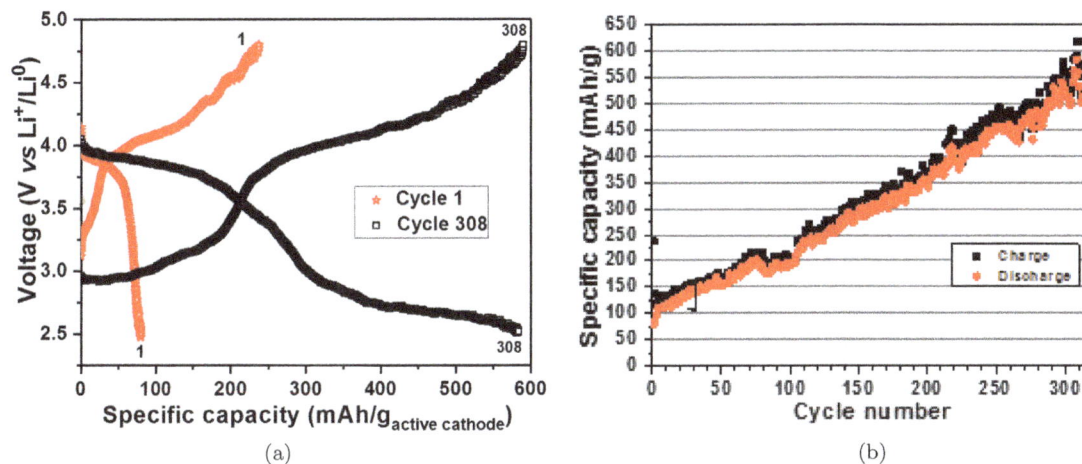

Figure 10. Charge–discharge cycles of a Li/Li$^+$-glass/plasticizer/LNMO high-voltage cell showing that the capacity increased seven times with cycling due to a faradic and especially to a capacitive component; (a) cycles 1 and 308; (b) specific capacity as a function of cycle number.

and self-cycling but also motivated the development of new strategies for the design of rechargeable batteries. The dielectric amorphous-oxide electrolytes may provide safe, low-cost, rechargeable batteries charged by solar and/or wind power and a battery-powered all-electric road vehicle that competes with vehicles powered by the air-polluting internal combustion engine that runs on the energy stored in a fossil fuel.

Acknowledgments

This work was supported by the COMPETE2020 and FCT project, EU and Portugal, PTDC-CTM-ENE-2391-2014. Professor John B. Goodenough thanks the Robert Welch Foundation Grant F-1066.

References

1. Braga, M. H.; Stockhausen, V.; Ferreira, J. A.; Oliveira, J. C. E.; El-Azab, A., Novel Li3ClO Based Glasses with Superionic Properties for Lithium Batteries, *J. Mater. Chem. A* 2014, 2, 5470–5480.

2. Yu, B.-C.; Park, K.; Jang, J.-H.; Goodenough, J. B., Cellulose-Based Porous Membrane for Suppressing Li Dendrite Formation in Lithium-Sulfur Battery, *ACS Energy Lett.* 2016, 1, 633–637.

3. Braga, M. H.; Murchison, A. J.; Ferreira, J. A.; Singh, P. Goodenough, J. B., Glass-Amorphous Alkali-Ion Solid Electrolytes and Their Performance in Symmetrical Cells. *Energy Environ. Sci.* 2016, 9(3), 948–954.

4. Braga, M. H.; Jorge A. Ferreira, Murchison, A. J.; Goodenough, J. B.; Electric Dipoles and Ionic Conductivity in a Na$^+$ Glass Electrolyte, *J. Electrochem. Soc.* 2017, 164 (2), A1–A7.

5. John B. Goodenough, M. Helena Braga, Andrew Murchison Heat Energy-Powered Electrochemical Cells, U.S. Patent Application No. 15/782, 443.

6. Braga, M. H.; Grundish, N. S.; Murchison, A. J.; Goodenough, J. B., Alternative Strategy for a Safe Rechargeable Battery. *Energy Environ. Sci.* 2017, 10, 331–336.

7. Steingart and Viswanathan, *Energy Environ. Sci.,* 2018 doi: 10.1039/C7EE01318C

8. Braga, M. H.; Grundish, N. S.; Murchison, A. J.; Goodenough, J. B., Reply to comment on Alternative Strategy for a Safe Rechargeable Battery, submitted.

9. Braga, M. H.; Murchison, A. J.; Oliveira, J. E.; Ferreira, J. A.; Goodenough, J. B.; Self-Charging and Self-Cycling of All-Solid-State Electrochemical Cells, submitted.

10. John B. Goodenough, M. Helena Braga, Andrew Murchison, Joana E. Oliveira, Jorge A. Ferreira, Self-charging and/or self-cycling electrochemical cells. U.S. Patent Application No. 15/478, 099.

2

Magnetic Properties of Perovskite Structure Oxides

J. E. Greedan

Department of Chemistry and Chemical Biology and the Brockhouse Institute for Materials Research, McMaster University, Hamilton, Canada

Introduction

Oxides with the perovskite or a related structure are likely the most intensely studied class of oxide materials. A database search for 'perovskite oxides' can yield $\gg 10^4$ references. In part, this is due to the central role occupied by perovskite structure oxides in areas such as dielectrics, piezo-electrics, fuel cells, high-temperature superconductors, catalysis, and many others. Also, the perovskite structure can accommodate almost 80% of the stable chemical elements, enabling the synthesis of literally thousands of chemically distinct materials. In this chapter, our concern will be the magnetic properties of perovskite oxides. Some basic knowledge of magnetic phenomena will be assumed, at a level similar to that of existing texts and monographs. A recent chapter, 'Magnetic Oxides' by the present author can be found in the *Encyclopedia of Inorganic and Bioinorganic Chemistry*.[1] A much more detailed but still excellent source is the classic 'Magnetism and the Chemical Bond' by Goodenough.[2]

Structural Preliminaries

The chemical formula for the simplest perovskite oxide is ABO_3, where A is a large radius cation and B a smaller radius cation. In its highest symmetry manifestation, sometimes called the aristotype, perovskite can be described in space group (*Pm-3m*). There are two useful settings within *Pm-3m* as indicated in Figures 1(a) and 1(b), along with the most common polyhedral representation, Figure 1(c). Figure 1(a) emphasizes the ideal 12-fold coordination of the A-site cation , Figure 1(b) the six-fold octahedral coordination of the B cation, and Figure 1(c) shows the 3D array of corner-sharing BO_3 octahedra characteristic of perovskite. From either Figure 1(a) or 1(b), the ideal relationship of the A and B cation radii can be deduced, as was done by Goldschmidt many years ago, i.e. $[r(A) + r(O)]/[r(B) + r(O)]^{1/2} = t$.[3] This expression, with t substituted for unity, is known as

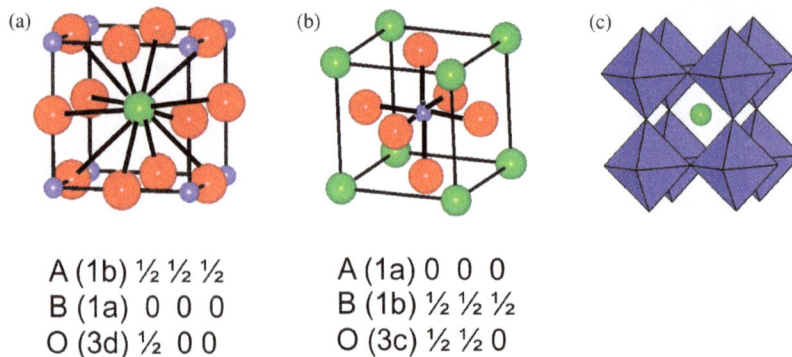

A (1b) ½ ½ ½
B (1a) 0 0 0
O (3d) ½ 0 0

A (1a) 0 0 0
B (1b) ½ ½ ½
O (3c) ½ ½ 0

Figure 1. (a) and (b) Two settings for the perovskite structure, ABO₃, with A (green), B (blue), and O (red) in *Pm-3m*. (a) Shows the 12-fold coordination of the larger A cation, (b) The six-fold coordination of the B cation. (c) Shows the corner-sharing connectivity of the BO₃ octahedra.

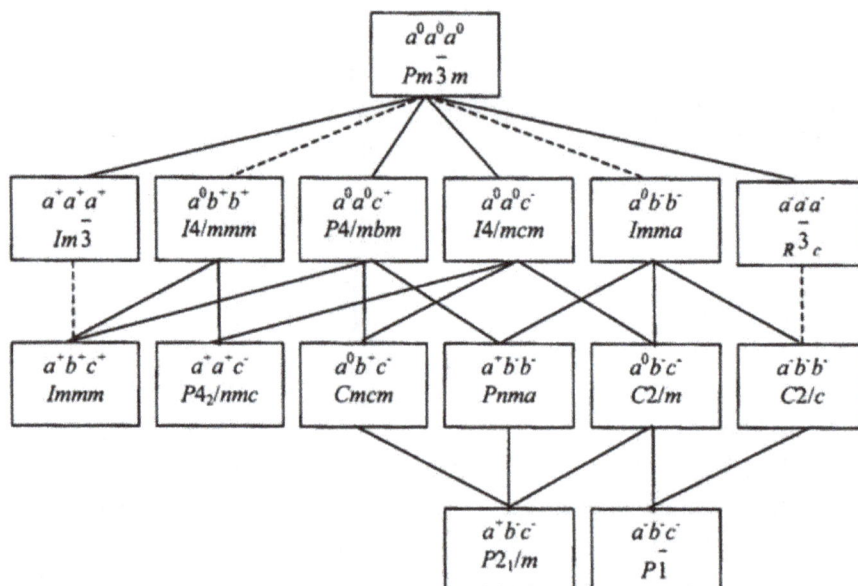

Figure 2. The group theory 'family tree' for perovskites with either a single B-site (A-site) ion or random occupation of the B-site (A-site).[4]

the Goldschmidt Tolerance Factor (GTF). The octahedra of Figure 1(c) can execute various rotations, called tilt systems, about the three axes, often to accommodate A cations which are too small for 12-fold coordination, giving rise to many other symmetries. Glazer was the first to recognize and codify the tilt systems, and the result is a group theory family tree as shown in Figure 2.[4]

The frequency with which the possible tilt systems (space groups) occur among known ABO₃ perovskites, as determined by Lufaso and Woodward, is shown in Figure 3.[5] Note that more than half crystallize in the orthorhombic space group *Pnma*, Number. 62, ($a^-b^+a^-$ or $a^+b^-b^-$ which are equivalent to a permutation of the orthorhombic axes). Rhombohedral, *R-3c* ($a^-a^-a^-$), I-centered cubic, *Im-3* ($a^+a^+a^+$), two tetragonal space groups, *P4/mbm* ($a^0a^0c^+$) and *I4/mcm* ($a^0a^0c^+$), and the aristo type *Pm-3m* ($a^0a^0a^0$) account for most of the remaining cases. As already noted, most of these lower symmetries arise from octahedral tilts to accommodate smaller A cations. The descent in symmetry roughly follows the departure of the GTF from unity, as *t* can range from 1 to about 0.9 or so, depending on the choice of ionic radii. Symmetry lowering

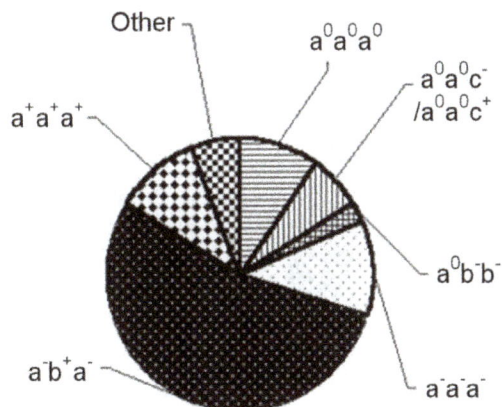

Figure 3.   Occurrence of the various tilt systems among known ABO_3 perovskites.[5]

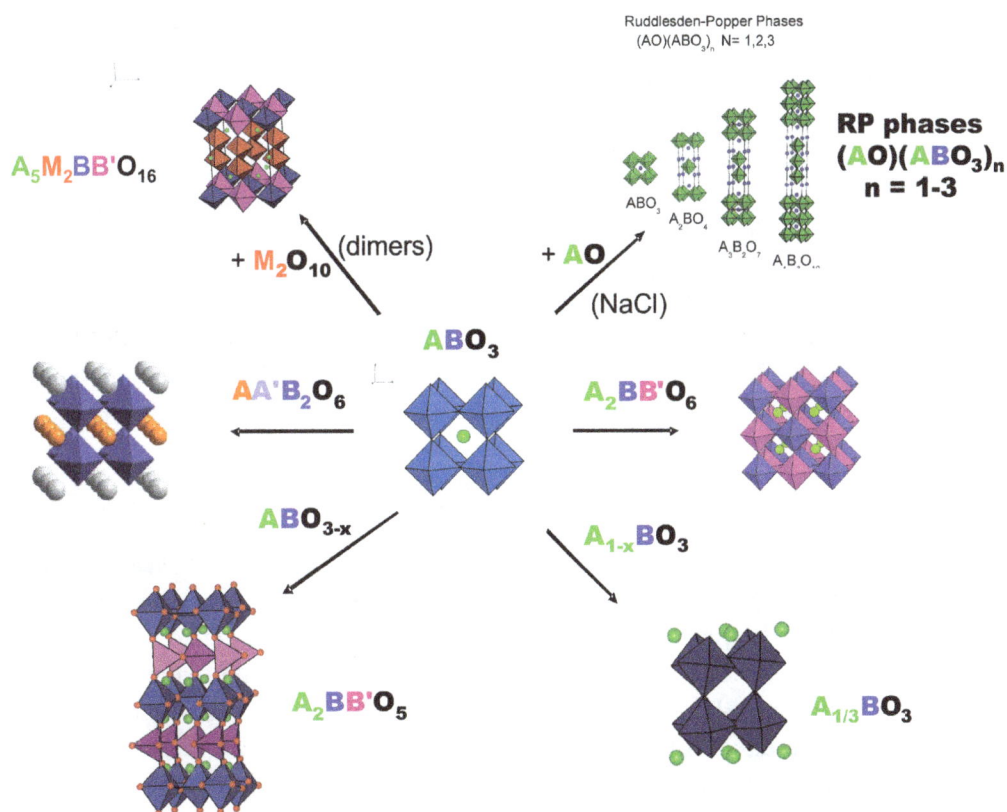

Figure 4.   Selected members of the family of perovskite related structures. Reproduced with permission from John Wiley and Sons.

has profound implications for magnetic properties in terms of both the B–O–B bond angle, which is critical for the strength (and sometimes the sign) of the superexchange interaction, and the local crystal field symmetry at the B-site, which determines the single ion ground state.

The Perovskite Family of Structures

There exist a large number of ways to modify the basic ABO_3 perovskite structure resulting in an extensive family of related phases. Selected members of this family are illustrated in Figure 4.

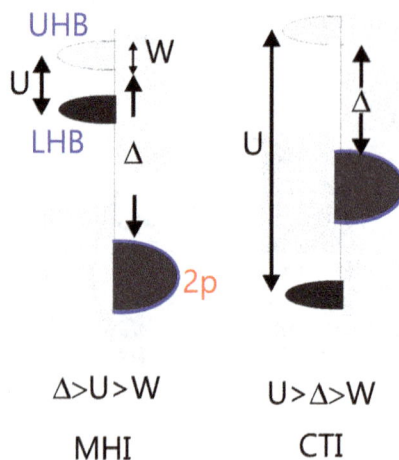

Figure 5.   Two insulating ground states in the ZSA scheme, the Mott-Hubbard insulator and the CTI.

The remainder of this chapter will be structured on Figure 4. The structure types to be considered are as follows:

(1) The ABO_3 perovskites.
(2) The B-site ordered double perovskites, $A_2BB'O_6$.
(3) The A-site ordered perovskites, $AA'B_2O_6$.
(4) The O-site deficient perovskites, ABO_{3-x}.
(5) Intergrowths with dimeric M_2O_{10} pillaring units, the pillared perovskites, $A_5M_2BB'O_{16}$.
(6) Intergrowths of ABO_3 with Rock Salt, AO, the Ruddlesden–Popper (RP) phases $(AO)(ABO_3)_n$, $n = 1–3$.
(7) Products of hydride reductions of perovskites and RP phases.

Electronic Structure of Transition Metal Oxides — A Comment

Before beginning a survey of magnetic perovskite oxides, some brief, general remarks about the electronic structure are in order. According to the model of Zaneen, Sawatsky, and Allen (ZSA), the electronic structure of transition metal oxides, of which the perovskite oxides are a special class, is determined by the relative magnitudes of three parameters, namely Δ — the energy difference between the top of the oxide $2p$ band and the lowest unoccupied d-levels, U — the Mott–Hubbard correlation energy, and w — the width of the Mott-Hubbard d bands.[6] Two situations which result in insulating ground states, which pertain for the vast majority of perovskite oxides, are illustrated in Figure 5, namely the Mott–Hubbard insulator, for which $\Delta > U > w$ and the charge-transfer insulator (CTI) defined by $U > \Delta > w$. Metallic ground states can also result in two cases, $w > U$ or $w > \Delta$. The following trends can be anticipated. For B = $3d$ elements, U is generally large and w is small. U increases across the row, and thus we might expect metallic behavior to be possible at the beginning of the series, followed by the Mott–Hubbard insulator behavior and, near the end of the series, CTI behavior. As w increases down a group and U decreases, metallic states should be more common among the $4d$ and $5d$ elements. For oxides with symmetries lower than cubic, there will be a strong correlation between w and the B–O–B bond angle, with w decreasing as the angle deviates from 180°; thus, metallic states should be more rare.

Perovskite Magnetic Structures

A system for describing the magnetic structures of perovskite oxides is well established and is illustrated in Figure 6 for a *Pm-3m* chemical cell.[2,7] While there are a few more exotic structures, see Ref. 2 , these,

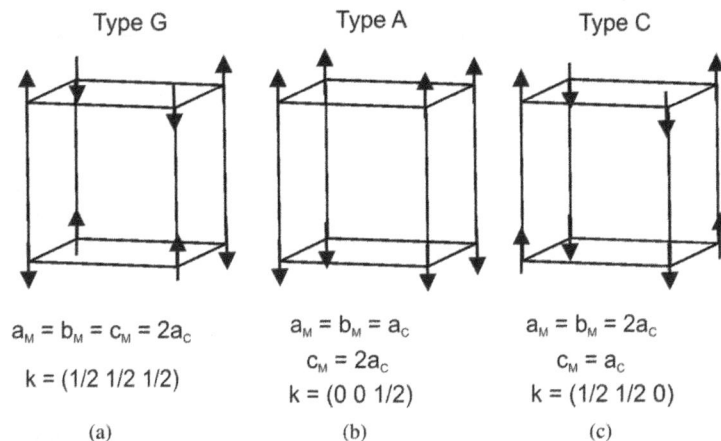

Figure 6. Three common magnetic structures, Types G, A, and C, for AF order in perovskite oxides illustrated on a *Pm-3m* chemical cell. The relationships between the magnetic and chemical cell axes and the ordering wave vectors, k, are indicated.

Types G, A, and C, are by far the most common for antiferromagnetic perovskites. Increasingly, magnetic structures are being described in terms of group theory representation analysis and, of course, magnetic space groups. These schemes will be introduced in a later section.

As the nearest neighbor (nn) exchange in an ABO_3 perovskite (Figure 1(a)) involves a B–O–B superexchange pathway for which the angle is often 180° (*Pm-3m*) or nearly so, the sign of the exchange can be predicted from the Goodenough–Kanamori (G–K) rules.[2,8] In most cases, this is AF, and if the nn exchange dominates, Type G AF order would be expected. There are only a very few cases of F order among perovskite oxides due strictly to superexchange.

ABO_3

There are two formal charge combinations which are relevant here, $A^{2+}B^{4+}O_3$ and $A^{3+}B^{3+}O_3$. There are few, if any, examples of $A^+B^{5+}O_3$ perovskites which contain magnetic ions.

$A^{2+}B^{4+}O_3$

$$B = V^{4+} (3d^1, t_{2g}^{~1}, S = 1/2), A = \text{Ca, Sr, Ba}$$

All V^{4+} perovskites are metallic and show Pauli paramagnetism, indicating a low level of correlation.[9-11] $SrVO_3$ is cubic (*Pm-3m*) and the V–O–V angle is of course 180°, optimum for a large d-band width (w), while $CaVO_3$ is orthorhombic (*Pnma*) with V–O–V ~160°. Evidently, the smaller w for the Ca phase is still sufficient for the metallic state, $w > U$. $BaVO_3$ is not a perovskite under ambient conditions, but cubic $BaVO_3$ can be prepared under high-pressure conditions and is also metallic and Pauli paramagnetic.

$$B = Cr^{4+}(3d^2, t_{2g}^{~2}, S = 1), A = \text{Ca, Sr, Pb}$$

As the Cr^{4+} state is difficult to stabilize under ambient conditions, all perovskites require high-pressure synthesis. Recent, quite thorough, studies of these three perovskites indicate that $CaCrO_3$ and $SrCrO_3$ are

Table 1. Summary of magnetic properties of the $A^{2+}Cr^{4+}O_3$ perovskites.[12]

A	I or M	Space group	T_N (K)	Magn. struct.	μ (Cr^{4+}) μ_B
Ca	M	*Pnma*	~ 90	C-Type (b)	1.09(4)
Sr	M	*P4/mmm*	~ 40	C-Type	0.8(3)
Pb	I	*Pm-3m*	~ 160	G-Type	2.1(3)

correlated metals, while $PbCrO_3$ is an insulator.[12] Unlike the case of metallic AVO_3 phases above, the correlation level is sufficiently strong to induce magnetic ordering. $CrCrO_3$ is *Pnma* at all temperatures, while $SrCrO_3$ shows phase separation in the form of a minority tetragonal phase (*P4/mmm*) at low temperature. Interestingly, *P4/mmm* is not a direct subgroup of *Pm-3m*. In this case, only the tetragonal phase shows long-range magnetic order. $PbCrO_3$ appears to be *Pm-3m* at all temperatures. Table 1 summarizes the known behavior of the $A^{2+}Cr^{4+}O_3$ perovskites. Note that the ordered moments for A = Ca and Sr are well below that expected for d^2, $S = 1$, $gS = 2$ μ_B, while that for A = Pb is within error.

$$B = Mn^{4+} (3d^3, t_{2g}^{\ 3}, S = 3/2), A = Ca, Sr$$

In contrast to the B = V and Cr perovskites, the Mn^{4+} materials are insulating, indicating $U > w$. $CaMnO_3$ (*Pnma*) is the stable form and has been studied for some time.[13,14] Perovskite $SrMnO_3$ (*Pm-3m*) is actually a high-temperature polymorph, the low-temperature form being hexagonal. The cubic form can be stabilized by low-temperature or high-pressure synthesis.[15] $CaMnO_3$ is a canted moment AF (sometimes called a weak ferromagnet) with $T_N = 123$ K.[13] Spin canting is allowed in *Pnma*. T_N for $SrMnO_3$ is predictably higher, $T_N = 233$ K, and no spin canting is observed as it is forbidden in *Pm-3m*.[15] The magnetic structure for both is Type G with ordered moments of 2.75 (Ca) and 2.6(2)μ_B(Sr).[15–17]

$$B = Fe^{4+} (3d^4, t_{2g}^{\ 3}e_g^{\ 1}, S = 2), A = Sr.$$

$SrFeO_{3-\delta}$ with $\delta \sim 0$ can be synthesized only using high O_2 pressures of ~ 6 GPa.[18] For small δ, it is cubic (*Pm-3m*) and metallic, likely due to $w > \Delta$, i.e. the overlap of the $3d$ band with the O $2p$ band. $SrFeO_3$ orders in a complex AF spiral structure below $T_N = 134$ K.[19] The origin of the spiral structure has been attributed to competing nn F and nnn AF interactions. The ordered moment is 2.7(4) μ_B, but is difficult to interpret due to the metallic ground electronic state. In any case, it is significantly smaller than the nearly 4 μ_B expected for an $S = 2$ high-spin state.

$$B = Co^{4+} (3d^5, t_{2g}^{\ 5}, S = 1/2)$$

Again, this is a high-pressure phase, although it can be prepared using ambient-pressure electrochemical methods.[20–22] All samples are *Pm-3m*, metallic and ferromagnetic, but the details vary widely as shown in Table 2. The single-crystal sample shows the highest T_c and Co ordered moment. In all cases, the ordered moment is significantly larger than that expected for the low spin configuration, ~1 μ_B. This has been interpreted in favor of an intermediate spin state for Co^{4+} ($t_{2g}^{\ 4}e_g^{\ 1}$).

To summarize the magnetic properties of the $A^{2+}B^{4+}O_3$ perovskites involving $3d$ ions , most are cubic and all but B = Mn^{4+} are metallic. AF order dominates for B = Cr and Mn, giving way to a complex spiral state for B = Fe and full F order for B = Co. The difficulty in stabilizing the +4 state for $3d$ elements limits the number of stable perovskites of this type. The +4 state is generally more accessible for $4d$ and $5d$ ions.

Table 2. Cell constants (Å), Curie temperatures, and Co ordered moments for $SrCoO_{3-\delta}$.

Prep. (Ref.)	Poly/Xtal	a_0 (Å)	T_c (K)	Co mom. (μ_B)
HP (Ref. 22)	Xtal	3.8289	305	2.5
Electrochem (Ref. 21)	Poly	3.835(2)	280	2.1
HP (Ref. 20)	Poly	3.836	222	1.6

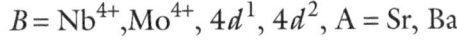

$$B = Nb^{4+}, Mo^{4+}, 4d^1, 4d^2, A = Sr, Ba$$

As might have been expected, all known $ANb(Mo)O_3$ perovskites are metallic and Pauli paramagnetic.[23] Nearly stoichiometric $SrNbO_3$ is orthorhombic *Pmna*, while $SrMoO_3$ is *Pm-3m* at 300 K but undergoes structural distortions to lower symmetry, *I4/mcm* at 266 K and orthorhombic *Imma* below 125 K.[24] These structural transitions have no apparent effect on the metallic and Pauli paramagnetic behavior. $BaMoO_3$ is *Pm-3m* at 300 K but has not been investigated at lower temperatures.[25]

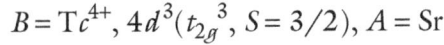

$$B = Tc^{4+}, 4d^3(t_{2g}^{\ 3}, S = 3/2), A = Sr$$

Remarkably, the structure and properties of this highly radioactive material, the *4d* analog of $SrMnO_3$, have been reported recently. $SrTcO_3$ has *Pnma* symmetry at 300 K but undergoes several phase transitions upon heating, *Imma*, *I4/mcm*, and *Pm-3m*, to 1023 K.[26] The most extraordinary feature of this material is its very high $T_N = 750$ K![27]

The magnetic structure is Type G and the ordered moment is 2.13(4) μ_B at 4 K. The large reduction from the spin only moment of 3 μ_B is attributed to strong covalency between the *4d* orbitals and the O $2p$ orbitals. Attempts to explain both the high T_N and smaller moment have been advanced, based on the idea that $SrTcO_3$ is near a metal–insulator transition, i.e. $U \sim w$.[28] There appear to be no credible reports of the existence of any $AReO_3$ perovskites, A = Ca, Sr, Ba, the *5d* analogs.

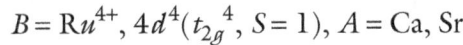

$$B = Ru^{4+}, 4d^4(t_{2g}^{\ 4}, S = 1), A = Ca, Sr$$

These are the *4d* analogs of $SrFeO_{3-\delta}$ and it is not surprising that both are metallic, although neither has the cubic *Pm-3m* structure. $CaRuO_3$ (*Pnma*) was shown quite early to have AF spin correlations in the form of a large, negative Curie–Weiss $\theta = -138$ K.

A ZFC/FC divergence occurs in the d.c. susceptibility at 87 K, but no long-range order is detected by neutron diffraction.[29] Mössbauer data indicate the onset of short-range spin correlations below 87 K, and this transition has been interpreted as spin glass like, although a.c. susceptibility data, which lack both a maximum in χ' and frequency dependence, do not support a traditional spin glass ground state.[30] As the primitive pseudo-cubic Ru sublattice is not geometrically frustrated, the lack of long-range order is still unexplained. Presumably, competing AF and F interactions are involved.

$SrRuO_3$ (*Pnma*) has been known as a ferromagnetic metal since the 1960s with $T_c = 160$ K.[31] The ordered moment is 1.6(1) μ_B from magnetization measurements on a single crystal.[32] Recent synchrotron X-ray data indicate several structural phase transitions upon heating from *Pnma–Imma–I4/mcm–Pm-3m* finally at 950 K.[33]

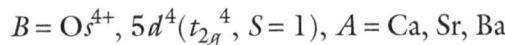

$$B = Os^{4+}, 5d^4(t_{2g}^{\ 4}, S = 1), A = Ca, Sr, Ba$$

These are all high-pressure synthesis phases and have only been known since ~ 2013.[33] The A = Ca and Sr phases are *Pnma*, while the Ba material is *Pm-3m*. All are metallic, and no sign of long-range order has been detected either by neutron diffraction or heat capacity measurements.[33]

$B = Ir^{4+}, 5d^5 (t_{2g}^5, S = 1/2)$

The ambient pressure form of $SrIrO_3$ is not a perovskite, but a *Pnma* perovskite polymorph can be quenched from high pressure.[34] This phase is metallic with no indication of magnetic order. A somewhat enhanced Pauli paramagnetism is observed.

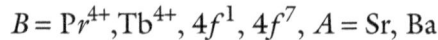

$$B = Pr^{4+}, Tb^{4+}, 4f^1, 4f^7, A = Sr, Ba$$

Two rare earth elements are stable in the +4 state, and perovskites have been reported for $SrPrO_3$, $BaPrO_3$, $SrTbO_3$, and $BaTbO_3$ at ambient pressure. All are orthorhombic (*Pnma*) although the distortion from *Pm-3m* symmetry is of course much greater for the Sr cases. $BaPrO_3$ shows AF order at 11.5 K (see ref. 35), but there is no sign of long-range order for $SrPrO_3$.[36] The magnetic structure of $BaPrO_3$ has been reported as Type G with a very small ordered moment of 0.35(5) μ_B for Pr^{4+}. This is close to the value of 0.45 μ_B expected if the crystal field ground state is $m_j = |\pm\frac{1}{2}\rangle$ arising from the $^5F_{5/2}$ free ion multiplet.[37] $SrTbO_3$ and $BaTbO_3$ show AF order at much higher temperatures due to the large expected moment of ~7 μ_B expected for a $4f^7$, $S = 7/2$, $L = 0$ configuration. Various T_N values have been reported, 37–32.5 K for $BaTbO_3$ and 33–30.5 K for $SrTbO_3$. This may be due to variable amounts of Tb^{3+} in the samples. Type G magnetic structures are found for both phases with ordered moments at the Tb^{4+} site of 6.6(1) μ_B for both.[38]

$$B = U^{4+}, 5f^2, A = Ba$$

This material shows only temperature-independent paramagnetism but is an insulator. Likely, this is due to van Vleck paramagnetism arising from a singlet crystal field ground state for this even electron system.[39]

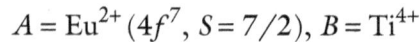

$$A = Eu^{2+} (4f^7, S = 7/2), B = Ti^{4+}$$

Eu^{2+}, $4f^7$ ($S = 7/2$) is the only A-site element which is magnetic in perovskite oxides. There was considerable excitement in the early 1960s when EuO, a rock-salt structure material, was found to be an F insulator with a remarkably high T_c (for a lanthanide-based material) of 69 K.[40] The corresponding perovskite, $EuTiO_3$, somewhat disappointingly, is AF with $T_N = 5.3$ K in spite of a positive CW $\theta = 3.8$ K.[41] Interestingly, a Eu-based perovskite hydride, $EuLiH_3$ — also an F insulator — has the second highest $T_c = 37$ K, among known Eu^{2+} insulators.[42] In fact, given that for a perovskite there only 6 nn Eu^{2+} ions while for rock salt there are 12, the derived J_{nn} for the hydride is actually larger than that for EuO, $J_{nn}/k_B = 0.83$ ($EuLiH_3$) and 0.60K (EuO).[43]

The exchange mechanism for Eu^{2+} coupling in insulators, as proposed by Goodenough and Kasuya [2,44] is unique, but somewhat related to double exchange. It involves an intra atomic excitation from the $4f^7$ ground state to a $4f^6 5d^1$ excited state, costing energy U (not to be confused with the Mott–Hubbard U). As the 5d orbitals have a large radial extent, this electron is transferred easily to an *nn* site. According to Hund's first rule, the spin of the transferred 5d electron must be parallel to those of the collective $4f^7$ spins on both sites, and thus the sign of the exchange is F. The full expression for $J_{nn} = J_{intra} b^2/U^2$, where J_{intra} is the Hund's first rule energy, b is the transfer integral — roughly proportional to the 5d–O 2p orbital overlap — and U is the aforementioned intra-atomic excitation energy 4f–5d. As the 5d orbital energy is strongly influenced by bonding with the ligands, i.e., the crystal field, J_{nn} is very sensitive to the nature of the ligands and the coordination geometry which determine U. Chien *et al.* have proposed a model, illustrated in Figure 7 to account for the remarkable difference in magnetic properties between the two perovskites, $EuTiO_3$ and $EuLiH_3$.[45] This qualitative picture has recently been confirmed by detailed DFT calculations.[46]

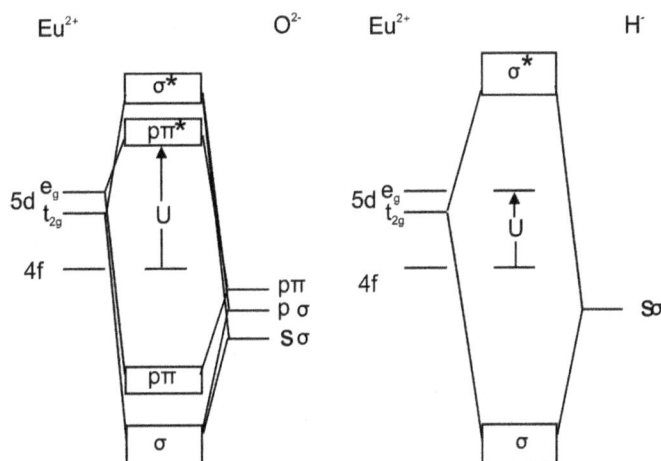

Figure 7. A proposed energy-level diagram for $EuTiO_3$ and $EuLiH_3$. The absence of *p* orbitals on the H^- ligand results in a much smaller $4f$–$5d(e_g)$ promotion energy, *U*, for the hydride, leading to a much larger J_{nn}.[45]

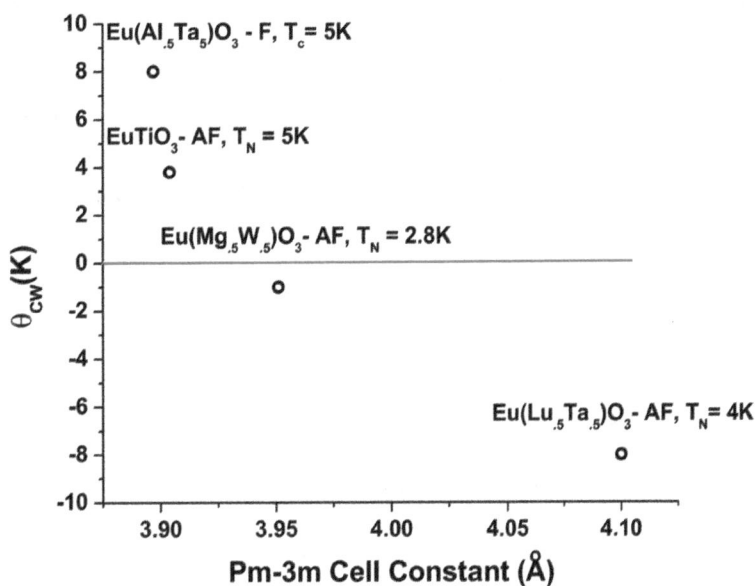

Figure 8. An apparent correlation between the CW θ and the *Pm-3m* cell constant for four cubic perovskite oxides. $Eu(Al_{0.5}Ta_{0.5})O_3$ and $Eu(Mg_{0.5}W_{0.5})O_3$ have the double perovskite structure (*Fm-3m*) with $a = 2a(Pm\text{-}3m)$. All values have been adjusted to the *Pm-3m* basis.

In addition to $EuTiO_3$, only a few perovskite oxides with Eu^{2+} exist as the Eu^{2+}/Eu^{3+} redox couple is sufficiently strong to reduce most B^{4+} ions.[47] The properties of the relatively small number of stable, cubic Eu^{2+} perovskites are shown in Figure 8 which appear to correlate with the cubic cell constant.[45]

Both $EuZrO_3$ and $EuHfO_3$ have been reported recently with *Pnma* symmetry. Both are AF with $T_N = 4.1$ K (Zr) and 3.9 K (Hf) with a Type G magnetic structure for the Zr phase.[48–50] $EuNbO_3$ has also been reported as F with $T_c = 3.9$K and $\theta = +3.9$ K.[51] As Nb^{4+} is $4d^1$, its role in the determination of the magnetic properties is unclear.

Recent studies of $Eu(Ti_{1-x}Nb_x)O_3$ indicated that for $x > 0.16$, the samples are metallic and F with $T_c = 10$ K for $x = 0.2$. The F ordering is attributed to an interaction provided by the itinerant Nb^{4+} electrons.[52] A small number of Eu^{2+} perovskites are stable with magnetic ions partially occupying the B site, such as $Eu(Mn_{0.5}{}^{2+}W_{0.5}{}^{6+})O_3$ and $Eu(Cr_{0.5}{}^{3+}Ta_{0.5}{}^{5+})O_3$. These have been characterized as ferrimagnets with the

Eu^{2+} ordering antiparallel to the Mn^{2+} and Cr^{3+} moments with relatively high T_c values of 34 and 26 K, respectively.[53]

$A^{3+}B^{3+}O_3$

This is certainly a larger family than the preceding. The +3 state is quite stable for the $3d$ elements and for a few $4d$ and $5d$ analogs as well. The rare earth elements often occupy the A-site and there are 15 of those! With a few exceptions, all ABO_3 perovskites crystallize in *Pnma* (*Pbnm*) as there are no trivalent ions of sufficient radius to fit into a 12-fold site as required by *Pm-3m*. In general the radius of the A^{3+} ion can be used to tune the B–O–B bond angle and, thus, the nn superexchange interaction. When A^{3+} is a rare earth, the lanthanide contraction permits a quite fine-grained tuning opportunity. In general, while the details differ depending on the B^{3+} ion, for the largest A^{3+}, La^{3+} the B–O–B angle is ~160° to 165°, while for Lu^{3+} angles near ~140° are found. As well, apart from La^{3+}, Y^{3+} and Lu^{3+} rare earth ions are magnetic and can interact with magnetic B^{3+} ions, which adds to the complexity of any magnetic ground state. As there are literally thousands of publications regarding the $A^{3+}B^{3+}O_3$ materials, only a very brief summary can be given here. The $REBO_3$ families which show the most conventional behavior are those for B = Cr and Fe, and these can be discussed as a group. The cases of B = Ti, V, Mn, Co, Ni, and Cu each show some unique features and will be dealt with separately. As already mentioned, for the *Pnma* space group with AF order associated with the B-site or A-sites, spin canting is permitted and usually observed. That is, the bulk d.c. susceptibility data will not show typical Nèel behavior in the sense that a relatively sharp peak near T_N will be absent, and instead one sees an F-Type enhancement of the susceptibility beginning at T_N, particularly for data taken in the field-cooled mode. Also, hysteresis in M vs H plots will arise due to the presence of the small magnetic moment, usually on the order of 10^{-2} μ_B.

Although this is somewhat outside the scope of this chapter, the $REBO_3$ perovskites have received renewed interest due to the discovery of multiferroic behavior, especially for the $RECrO_3$ and $REFeO_3$ families. There are several recent reviews.[54,55]

The presence of non-collinear magnetic order is the primary driver of the multiferroic behavior. The review by Bousquet is particularly detailed.[55]

$$B = Cr^{3+}\,(3d^3,\,t_{2g}^{\,3},\,S = 3/2)\ \text{and}\ Fe^{3+}\,(3d^5,\,t_{2g}^{\,3}e_g^{\,2},\,S = 5/2)$$

For both ions, the orbital moments are quenched, $L = 0$ to first order, and neither is JahnTeller active. The *Pnma* structure pertains for all A = RE^{3+}, although a hexagonal form can be stabilized in thin films. These will not be discussed here. Following the lanthanide contraction, the RE^{3+} ion radii decrease monotonically. Thus, the B–O–B angle also decreases monotonically. As a result, T_N for RE = La^{3+} is a maximum for the series and that for RE = Lu^{3+} is the minimum. For B = Cr^{3+}, the relevant T_N is 282 K (La), B–O–B = 161°, and 112K (Lu), B–O–B = 146° while for B = Fe^{3+}, T_N (La) is 738 K, B–O–B = 157°, and T_N (Lu) is 622 K, B–O–B = 141°. These trends are illustrated for the $RECrO_3$ series in Figure 9. In the *Pnma* structure, there are actually two distinct B–O–B angles but they rarely differ by more than ~ 1o, so an average is given here.

The basic magnetic properties of these phases have been known since been the 1960s or even earlier and are tabulated.[56] In most cases, both the RE^{3+} ion and the Fe^{3+} or Cr^{3+} ions are magnetic. There are, thus, three exchange couplings to consider, B^{3+}–O–B^{3+} superexchange which is dominant, RE^{3+}–B^{3+} which is weaker, and RE^{3+}–RE^{3+} which is by far the weakest. Thus, the B^{3+} ions order at a higher temperature and exert a 'molecular field' on the RE^{3+} sites, inducing long-range order. The sign of the inter-sublattice coupling is AF. An effect known as spin reorientation often occurs in which the spin

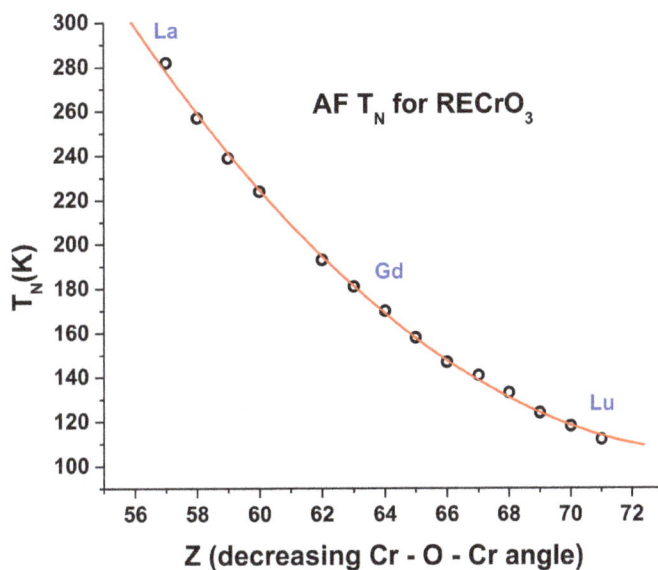

Figure 9.   The monotonic decrease in T_N with decreasing RE^{3+} radius and decreasing Cr–O–Cr angle in the $RECrO_3$ series.

direction of the B^{3+} sublattice changes, usually continuously, as the temperature is lowered. In some cases, a compensation temperature is observed as the magnetization of the RE^{3+} sublattice exceeds that of the Fe^{3+} or Cr^{3+} sublattices. At very low temperatures, $T < 4$ K usually, the RE^{3+} ions can order independently into a different configuration. These effects are manifest in the bulk magnetization as illustrated by $ErFeO_3$, Figure 10.[57] The finite magnetization from 300 to 100 K is due to the canted moment on the Fe^{3+} sublattice. The feature just below 100 K is the Fe^{3+} sublattice spin reorientation and the point, ~50 K, where the magnetization is zero is the compensation temperature, T_{comp}. Just above T_{comp}, $M_{Er} > M_{Fe}$, while just build up of M_{Er}. Not all $REFeO_3$ or $RECrO_3$ perovskites show all of these features.

Those $REFeO_3$ with non-magnetic RE, La, Lu, Y of course show no compensation point, but even some phases with magnetic RE, for example, Eu, Pr and Gd also have no T_{comp}. Most $RECrO_3$ also do not show a T_{comp}. An attempt to rationalize these systematics has recently been presented.[58]

The key to understanding the bulk magnetic behavior is of course the magnetic structures which have also been known for some time. The understanding and description of these represents one of the earliest applications of group theoretical (representation analysis) methods to this task.[59] As nearly all of the early literature and even recent publications describe the crystal structure in the non-standard *Pbnm* setting of Space Group number 62, this setting will be used here to avoid undue confusion. Just below T_N, nearly all *Pbnm* ABO_3 perovskites show a common magnetic structure which can be described in Bertaut's notation as $G_xA_yF_z$ and is illustrated in Figure 11. The spin canting gives rise to an F component, directed along the z (c) direction in the *Pbnm* setting and is the origin of the weak F behavior as well as providing the symmetry breaking needed to enable a ferroelectric response. This F component is almost never seen in the neutron diffraction data as it is extremely weak, typically ~10^{-2} to 10^{-3} μ_B. The canting angles are usually ~$1°$–$2°$. As already noted, in some cases such as $ErFeO_3$, the spins rotate such that the F_z component morphs into an F_x component with decreasing temperature and is described by the spin configuration $F_xC_yG_z$. Bertaut's approach also determines the spin configuration of the RE^{3+} ion, which is F_z for $G_xA_yF_z$ and F_xC_y for $F_xC_yG_z$. Very recent neutron diffraction studies of $ErFeO_3$ indicate that the situation at very low temperatures, where the Er^{3+} moments order independently, is even more complex.[60] This paper also provides

ErFeO$_3$

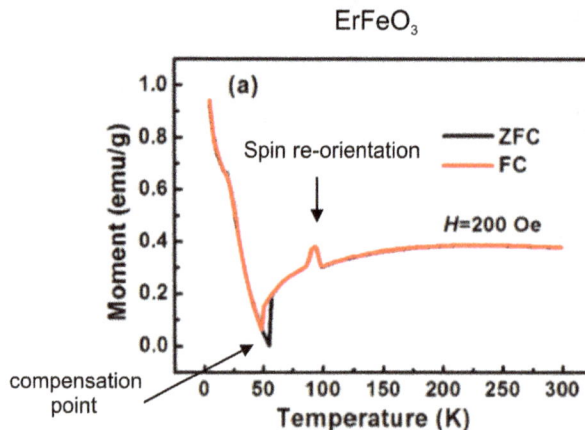

Figure 10.   Magnetization of ErFeO$_3$ showing both the spin reorientation and compensation point phenomena.[57] Reproduced with permission from AIP Publishing.

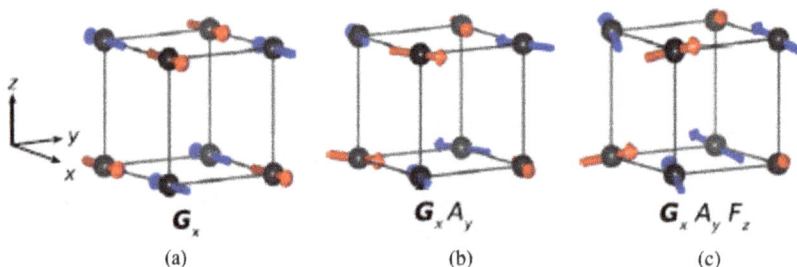

Figure 11.   The G$_x$ (*a*), G$_x$A$_y$ (*b*), and G$_x$A$_y$F$_z$ spin configurations.[55] Reproduced with permission from IOP Publishing.

an excellent discussion of both the *Pnma* and *Pbnm* descriptions of the magnetic structures along with the magnetic (Shubnikov) space groups.

$$B = (Ti^{3+}, 3d^1, t_{2g}^1, S = \tfrac{1}{2})$$

The RETiO$_3$ series is unique in that the sign of the coupling changes as a function of the RE radius and Ti–O–Ti angle.[61] For RE = La–Sm, the Ti–O–Ti coupling is AF but the sign changes to F for RE = Gd and this persists to RE = Lu. Recall that EuTiO$_3$ involves Eu^{2+} and *Pm-3m* symmetry. This is the only case in the ABO$_3$ series for which F coupling arises from a superexchange as opposed to a double exchange interaction. The details are summarized in Figure 12. Note the expected monotonic decrease of T_N in the AF region as the RE radius (and the Ti–O–Ti angle) decreases but the very scattered variation of T_C in the F region. In the latter, the properties are well described by a ferrimagnetic behavior in which the RE moments are coupled antiparallel to the F moments on the Ti^{3+} sublattice.[62] Clearly, in this region the Ti–Ti interaction and the RE–Ti interactions are similar in magnitude. All members of the RETiO$_3$ series are semiconducting, i.e. "Mott–Hubbard" insulators.

For the F semiconductor YTiO$_3$, the ordered moment on Ti^{3+} is known to high accuracy from magnetization data on a single crystal to be 0.84(1) μ_B, significantly reduced from the expected spin only value of ~1 μ_B, and the preferred moment direction is [010], i.e. along the *b*-axis in the Pbmn setting.[63]

The magnetic structures have been studied. For AF LaTiO$_3$ (T_N = 146 K in the most stoichiometric samples), the Ti sublattice orders in a G-Type configuration with an ordered Ti moment of 0.57(5) μ_B.[64,65]

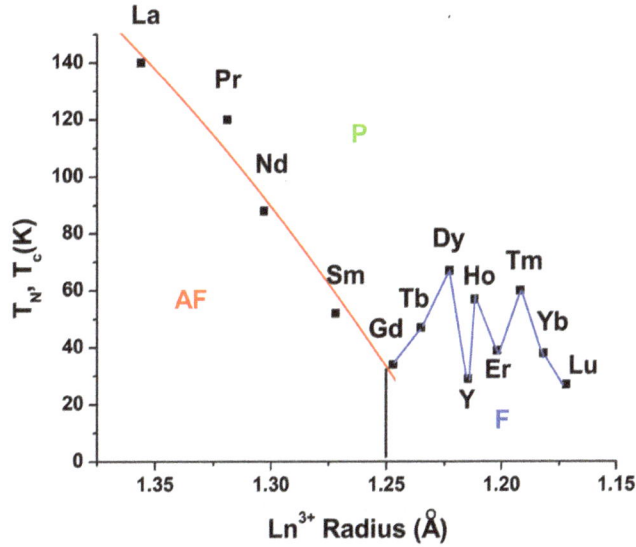

Figure 12. A magnetic phase diagram for the RETiO$_3$ series. For RE = La–Sm, the Ti sublattice is AF (G-Type) while for RE = Gd–Lu the Ti sublattice is F.

High-flux neutron diffraction data indicate that the detailed configuration is G$_x$F$_z$.[65] YTiO$_3$ shows the expected F structure but, surprisingly, with weak (~10% of the F moment) AF components, G$_x$A$_y$.[66] Magnetic structures of the ferrimagnetic materials show some systematic behavior. For RE = Er and Tm, the RE moments couple in an F$_z$ mode antiparallel to the Ti F$_z$ mode, while for RE = Tb, Dy, and Ho, one finds an F$_x$C$_y$ configuration, i.e. in the *ab*-plane.[67] These observations can be rationalized in terms of the single ion anisotropy of the RE ions.[68] In the AF part of Figure 12, NdTiO$_3$ has been studied in some detail. The availability of high-resolution neutron diffraction data has allowed the determination of both the Ti and Nd configurations to be G$_x$C$_y$ and C$_y$, respectively, with a Ti^{3+} moment of 0.45(8) μ_B and a Nd^{3+} moment of 0.82(2) μ_B at 4 K.[69]

There have been a number of efforts to explain the unprecedented magnetic behavior of the RETiO$_3$ series. So-called orbital liquid–orbital ordering models have been proposed, taking account of the Jahn–Teller nature of the $t_{2g}^1 e_g^0$ configuration.[70,71] That is, significant distortions of the Ti–O octahedra have been observed as a function of temperature, consistent with the breaking of the t_{2g} degeneracy to select a unique orbital ground state. In an attempt to explain the AF–F crossover, it has been proposed that a competition exists between the AF superexchange t_{2g}^1–O–t_{2g}^1 interaction and the F t_{2g}^1–O–e_g^0 interaction and that the latter is favored as the Ti–O–Ti angle decreases. These arguments are supported by DFT calculations.[72] However, it is fair to say that there is no generally accepted model which explains all the facts about the RETiO$_3$ perovskites, even today, more than 30 years after their discovery.

$$B = V^{3+} (3d^2, t_{2g}^2, S = 1)$$

V^{3+} is also a Jahn–Teller ion, and this is reflected in the behavior of the REVO$_3$ series. All REVO$_3$ perovskites are AF MHIs/semiconductors and crystallize in *Pnma* (*Pbnm*) at ambient temperatures. T_N varies as with the RECrO$_3$ and REFeO$_3$ series, there is no crossover to F order with decreasing RE^{3+} radius although there are competing AF and F superexchange interactions within the t_{2g} manifold. For LaVO$_3$, T_N = 145 K (V–O–V = 157°), and for LuVO$_3$, T_N = 100 K (V–O–V = 142°). The interplay between spin and orbital ordering in this series is remarkably complex, giving rise to a set of unique features in the REVO$_3$ materials such as a series of orbital-ordering phase transitions which can set in above and below T_N, one of

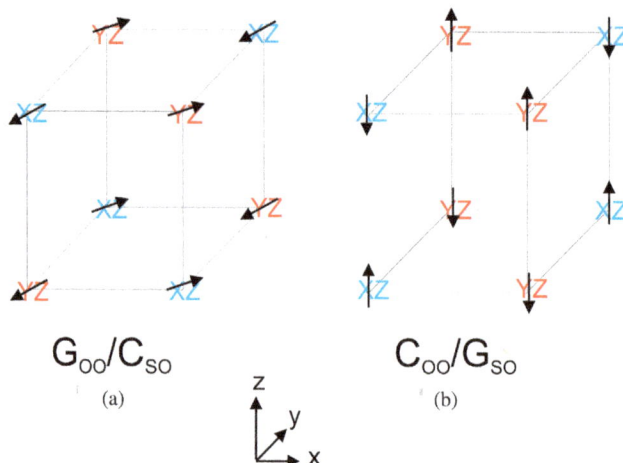

Figure 13. Two orbital ordering/spin ordering, OO/SO schemes encountered in the REVO$_3$ perovskites. The OO involves the d_{xz} and d_{yz} orbitals as shown on the primitive pseudo-cubic cell. Adapted from Ref. 77.

them being first order and involving a spin reorientation. As well, an anomalous diamagnetism is seen under some measurement conditions.

All REVO$_3$ perovskites undergo an orbital-ordering transition below T_{OO} from *Pbnm* to *P2$_1$/b*. Generally, except for LaVO$_3$, $T_{OO} > T_N$, but it varies strongly with RE^{3+}, tending to increase as the RE^{3+} radius decreases, for example, T_{OO} =154 K for CeVO$_3$ and 200 K for YVO$_3$.[73,74] The transition is continuous or second order and involves the onset of a G-Type orbital order among the previously degenerate t_{2g} orbitals. The other OO configuration seen for REVO$_3$ perovskites is C-Type. OO is observed between the d_{xz} and d_{yz} orbitals from the t_{2g} set. Two spin ordering (SO) configurations are paired with the OO configurations listed as OO/SO, GC, and CG and illustrated in Figure 13. Just below T_N, the GC picture prevails, but as the temperature decreases, a first-order transition, T_{CG}, sets in which the SO changes abruptly to G and the OO to C. A reorientation of the spins occurs.[74] For all REVO$_3$, except RE = La, the order of transitions is $T_{OO} > T_N > T_{CG}$, with T_{OO} increasing and T_N decreasing as the RE^{3+} radius decreases. In the case of LaVO$_3$, T_{OO} is difficult to detect and T_N and T_{CG} are nearly coincident with T_N = 142 K and T_{CG} = 138 K.[75] The SO schemes are rationalized according to the Goodenough–Kanamori rules.[2,8] In general, the two $t_{2g}{}^2$ electrons occupy the d_{xy} orbital and one of d_{xz} and d_{yz}. With G$_{OO}$, the superexchange interactions are xy^1–O–xy^1 in the *ab*-plane which is AF and xz^1–O–xz^0 and yz^1–O–yz^0 along *c*, which are F, consistent with C$_{SO}$, Figure 13a. When C$_{OO}$ pertains, the *c*-axis interactions now become AF, yz^1–O–yz^1, and xz^1–O–xz^1, so G$_{SO}$ results.

Magnetic structures for REVO$_3$ with RE = Dy, Ho, and Er have been reported.[76] DyVO$_3$ shows magnetically induced ferroelectricity just below T_N.[77]

As mentioned, REVO$_3$ perovskites can also show an anomalous diamagnetism below T_{CG}, which can be explained in terms of the orientation of the F moment with respect to the applied field. This effect was first reported in LaVO$_3$.[78] The arguments are quite complex and the reader is referred to Yan *et al.* for the details as applied to YVO$_3$.[74]

$$B = Mn^{3+}, 3d^4, t_{2g}{}^3 e_g{}^1, S = 2$$

With a half-filled e_g orbital, the static Jahn–Teller effect is of course much more important in this case. As well, the orthorhombic perovskite structure, *Pbnm* (*Pnma*), is found for RE = La–Dy, while for RE = Ho–Lu, Y the hexagonal *P6$_3$cm* space group is the stable form under ambient pressure. The *Pbnm*

Table 3. B–O distances and the distortion at the B–site in $LaBO_3$.

B	B–O1 (Å)	B–O2 (Å)	Ave. B–O (Å)	Δ = (Long–Ave.)/Ave.
Cr	1.974	1.967,1.970	1.970	0.004
Mn	1.968	1.906,2.180	2.018	0.162
Fe	2.009	2.002,2.007	2.006	0.003

form can be stabilized under applied pressure.[79] Alternatively, soft-chemistry techniques, such as the Pechini citrate method will yield the *Pbnm* phase, for example, $YMnO_3$ with low-temperature firing.[80]

The Jahn–Teller distortion is especially prominent for the $REMnO_3$ phases indicated, as seen in Table 3 where the Cr–O, Mn–O, and Fe–O distances are compared for $LaBO_3$. This static Jahn–Teller distortion disappears upon heating to above 800 K, but the *Pbnm* symmetry is maintained.[81]

The T_N values are also anomalously low with 282 $K(Cr^{3+})$, 140 $K(Mn^{3+})$, and 738 K (Fe^{3+}) for the $LaBO_3$ members.[82] For $REMnO_3$ with RE = Tb, Ho, Er, and Y, T_N ~40 K, the lowest values for any AF $REBO_3$ perovskite. This situation is consistent with the presence of competing AF and F superexchange, which is actually predicted by the G–K rules for 180° Mn^{3+}–O–Mn^{3+} geometry, although this angle is generally much reduced from the ideal value.[2,8] Superexchange in both $LaMnO_3$ and $CaMnO_3$ has been analyzed in detail by Millis who also argues for competing AF–F interactions, and this is supported by experimental determinations of the exchange constants via inelastic neutron scattering.[82,83]

The magnetic structures found for most $REMnO_3$ perovskites depart from the trend of either G or C bases which are obtained for most other $REBO_3$ materials. $LaMnO_3$ shows an A-Type magnetic structure, i.e. F planes coupled AF, Figure 6(b). The moment direction is along the *b*-axis (*Pbnm* setting) and its magnitude is 3.87(3) μ_B.[16,82] This magnetic structure is consistent with the experimentally determined exchange constants which show F J = 0.83 meV within the *ab*-plane and AF J = –0.58 meV along the *c*-axis.[82]

When RE^{3+} is magnetic, a more complex behavior is seen with decreasing temperature, as shown in a detailed study of $NdMnO_3$.[84] Below T_N = 78 K, the Mn^{3+} sublattice adopts the A-Type magnetic structure, but below ~65 K an F_z moment develops due to a reorientation of the Mn^{3+} moment out of the *ab*-plane (*Pbnm* setting). At 13 K, the Nd^{3+} moments order in an F_z mode. For the heavier RE^{3+} ions, such as Tb, Ho, Y, and Er, the Mn^{3+} sublattice ordering switches from collinear to a modulated sinusoidal configuration.[85] This crossover is attributed to the decreased Mn–O–Mn bond angle in the heavy $REMnO_3$ which can be as small as ~143° (Er) compared to 155° (La). It is argued that the F superexchange interaction is weakened at small Mn–O–Mn angles which destroys the A-Type structure.[85] Thus, for the $REMnO_3$ series the decrease in the Mn–O–Mn bond angle changes the magnetic structure on the Mn sublattice but the overall coupling is still net AF, in contrast to the $RETiO_3$ materials where an AF to F crossover is seen.

$$A^{2+}_{x}RE^{3+}_{1-x}MnO_3$$

Of course the greatest interest in Mn-based perovskite oxides has been for the A-site-substituted materials which show the so-called colossal magneto resistance (CMR) phenomenon. MR is defined as $(\rho(B) -\rho(0))/\rho(0)$ where $\rho(B)$ is the resistivity at applied field B and $\rho(0)$ is the zero field value. In CMR materials, MR can be as large as 10^6. Certainly, more than 10^3 papers have been published on the CMR topic, especially in the late 1990s to early 2000s, and comprehensive reviews exist.[86–89] In these systems, there is mixed Mn^{3+}/Mn^{4+} on the B-site which gives rise to strong double exchange $Mn^{3+}(t_{2g}^{3}e_g^{1})$–O–$Mn^{4+}(t_{2g}^{3}e_g^{0})$ which is always F. The lone e_g electron is regarded as itinerant and always couples F to the t_{2g}^{3} electrons on each site due to Hund's first rule. These materials are also metallic conductors for $0.2 < x < 0.4$. The competition between AF superexchange and F double exchange gives rise to a very complex magnetic behavior.

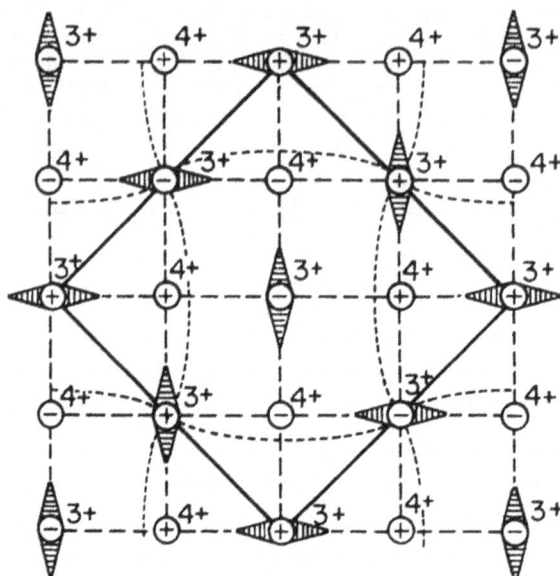

Figure 14.   Charge ordering in the *ab*-plane of a A$^{2+}_x$A$^{3+}_{1-x}$MnO$_3$ perovskite.[2]

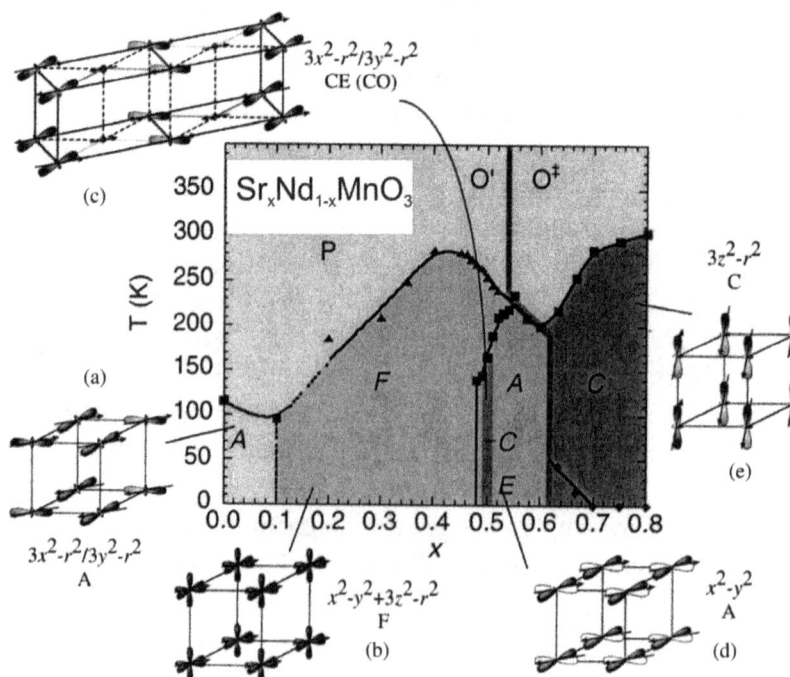

Figure 15.   The remarkably complex magnetic phase diagram for the system Sr$_x$Nd$_{1-x}$MnO$_3$. P = paramagnetic, A, F, C, and E are the magnetic structures, and the various SO/OO schemes are shown at the edges of the diagram.[90]

Another important phenomenon contributing to this complexity is the tendency for charge ordering (CO) between Mn^{4+} and Mn^{3+} ions which is illustrated in Figure 14 and is important when Mn^{3+}/Mn^{4+} ~0.5. In addition, OO is in the mix, and the interplay of all of these influences gives rise to remarkably complex behavior which is illustrated for the case of Sr$_x$Nd$_{1-x}$MnO$_3$ in Figure 15.[90] It is important to note that the sequence of magnetic structures shown in Figure 15 had been discovered much earlier by Wollan and Koehler in the 1960s in their seminal study of the Ca$_x$La$_{1-x}$MnO$_3$ system.[16]

Note that the very rare CE magnetic structure, Figure 14, is found in this system. The mechanism for CMR has been discussed in many reviews and is beyond the scope of this chapter.[86–88]

$$B = Co^{3+}, 3d^6, t_{2g}^6 e_g^0, S = 0 \text{ or } t_{2g}^4 e_g^2, S = 2 \text{ or } t_{2g}^5 e_g^1, S = 1.$$

The situation for B = Co^{3+} is more complex than might have been expected. If the low-spin (LS) state, $S = 0$ is obtained at all temperatures, then the magnetic properties are somewhat trivial. However, the other two possibilities, high spin (HS), $S = 2$ and intermediate spin (IS) $S = 1$ present different situations.

$LaCoO_3$ has been studied since the 1950s. Alone among $RECoO_3$ phases, it has *R-3c* symmetry while for the others the familiar *Pnma* symmetry is found. Interest was revived by Senaris and Goodenough in 1995, and the earlier work is summarized.[91] The bulk susceptibility data for $LaCoO_3$ show a broad maximum near 100 K for most samples, Figure 16, although there are differences among various samples. Many samples contain Co_3O_4 as a minor impurity phase which complicates the analysis by showing an F behavior at applied fields of order 0.1 T which is attributed to interfaces between $LaCoO_3$ particles and Co_3O_4.[92] There is no evidence that the F order is long range. There seems to be general agreement that at low temperature, the ground state is LS $S = 0$.[91] The rise in the susceptibility with increasing temperature has been attributed to the population of either the HS or IS states.[91] Note that the IS state would involve a JT distortion, either static or dynamic. There is no evidence for a static distortion at any temperature, but a dynamic JT effect has never been ruled out.[93] Curie–Weiss fits to the data above 100 K suggest AF correlations as $\theta_C \sim -145$ K to -182 K, but again no long-range AF order is detected.[93]

As mentioned, all other $RECoO_3$ perovskites crystallize in *Pmna* (*Pbnm*), but the spin-state transition persists. In $PrCoO_3$ and $NdCoO_3$, the spin-state transition appears to shift to higher temperatures.[94] As the Co^{3+} sublattice is not ordered, the RE^{3+} ions display mainly paramagnetic behavior over the accessible temperature range.

Another important feature of these materials is the occurrence of a gradual metal–insulator transition at relatively high temperatures, > 500 K in most cases. Heat capacity studies show that for the $RECoO_3$ series, this transition moves to higher temperatures as the RE^{3+} radius decreases.[95]

$$Ni^{3+}, 3d^7, t_{2g}^6 e_g^1, S = 1/2$$

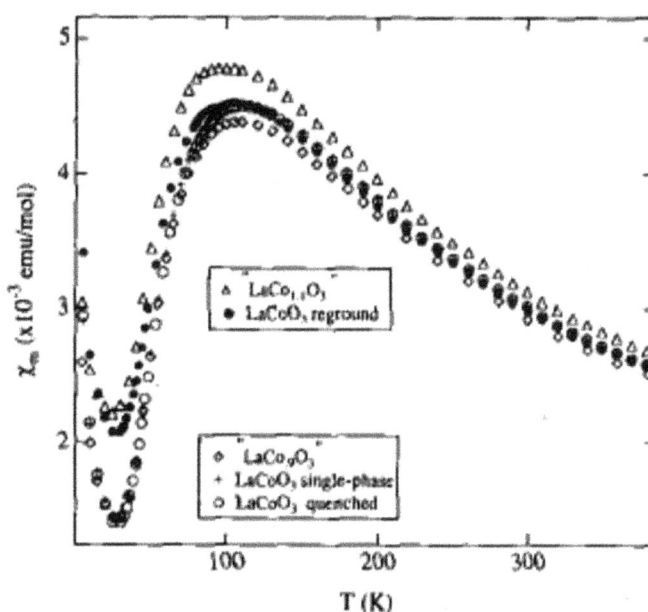

Figure 16. $LaCoO_3$. Susceptibility data for various $LaCoO_3$ samples.[91] Reproduced from Ref. 91 with permission from Elsevier.

The $RENiO_3$ perovskites require high-pressure synthesis using either very high-pressure techniques such as a belt or anvil apparatus or much milder approaches such as high oxygen over pressure.[96,97] The need for very high-pressure methods increases as the RE^{3+} radius decreases.[97] The main interest in these materials is two-fold, (1) that for the large radius $RENiO_3$ phases, metallic behavior is observed at ambient temperatures, a quite rare situation in undoped perovskites and (2) in some cases an abrupt metal–insulator transition occurs upon cooling.[98] In terms of the ZSA model, the $RENiO_3$ perovskites must lie near the boundary between low Δ metals, $W > \Delta$, and charge-transfer insulators, $W < \Delta < U$ (see figure MH-CT). As for $LaCoO_3$, $LaNiO_3$ has $R\text{-}3c$ symmetry, while for RE = Pr–Gd the usual $Pnma$ symmetry is found, but for RE = Dy–Ho,Y a subtle distortion to $P2_1/n$ is reported.[97] Note that in $P2_1/n$ there are two Ni sites. This has been interpreted as a charge disproportionation, $2Ni^{3+} \rightarrow Ni^{2+} = Ni^{4+}$.

AF magnetic order at the Ni and RE sites occurs in the insulating phase for all $RENiO_3$. Figure 17 shows the evolution of both $T_{M/I}$ and T_N for the series. These temperatures are coincident for RE = PR and Nd but quite different for the other RE. Note that $T_{M/I}$ increases monotonically with decreasing RE radius but that T_N shows a maximum at RE = Sm. For RE = Pr and Nd, T_N appears to be coincident with $T_{M/I}$.[98]

Magnetic structures are known for all members of the series. Although G-Type order might have been expected in analogy with the $3d^1$ $RETiO_3$ series, a much more complex order is found on the Ni sublattice described by $k = (1/2\ 0\ 1/2)$ based on the $Pbnm$ cell. In terms of the pseudo-cubic subcell, this k requires a $4x$ expansion along all three axes. It is argued that OO also occurs between the d_z^2 and $d_{x^2-y^2}$ orbitals in the e_g set.[98] The magnetic structure of $NdNiO_3$, for example, is remarkably complex and is shown in Figure 18. For any direction in the crystal, the Ni–O–Ni–O–Ni chain of interactions varies as F–AF–F. Below ~16 K half of the Nd^{3+} moments also order in layers which alternate between ordered Nd^{3+} moments and paramagnetic moments. The refined Ni^{3+} moment is ~0.9 μ_B determined using polarized neutrons. This is truly one of the most complex magnetic structures found for any $REBO_3$ perovskite.

The magnetic structure of $HoNiO_3$ has also been solved.[99] The same $k = (1/2\ 0\ 1/2)$ is found, but here there are two Ni sites with different <Ni–O> distances. Permitting each Ni site to have different moments, Ni(1) = 1.4 μ_B and Ni(2) = 0.6 μ_B gave the best refinement results, which supports only a partial charge disproportionation model. For full $2Ni^{3+}(t_{2g}^6 e_g^1) \rightarrow Ni^{2+}(t_{2g}^6 e_g^2) + Ni^{4+}(t_{2g}^6)$ disproportionation, the Ni^{4+}

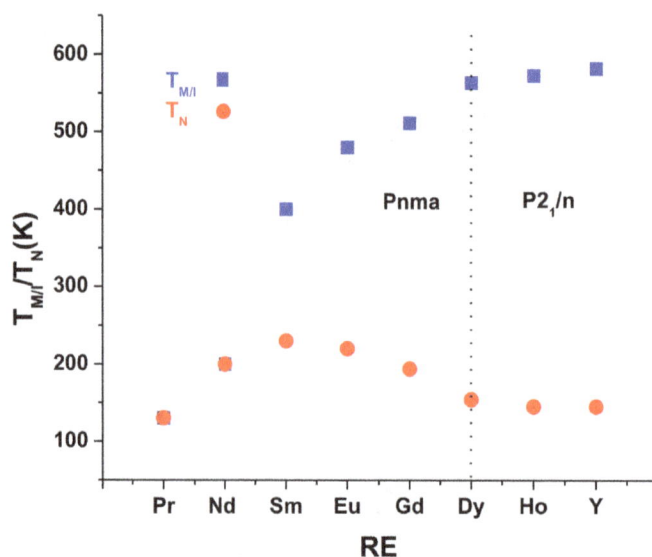

Figure 17.   The variation of $T_{M/I}$ and T_N for the $RENiO_3$ perovskites as a function of the RE ion. Note that these temperatures are coincident for RE = Pr and Nd. The boundary between $Pnma$ and $P2_1/n$ is indicated.

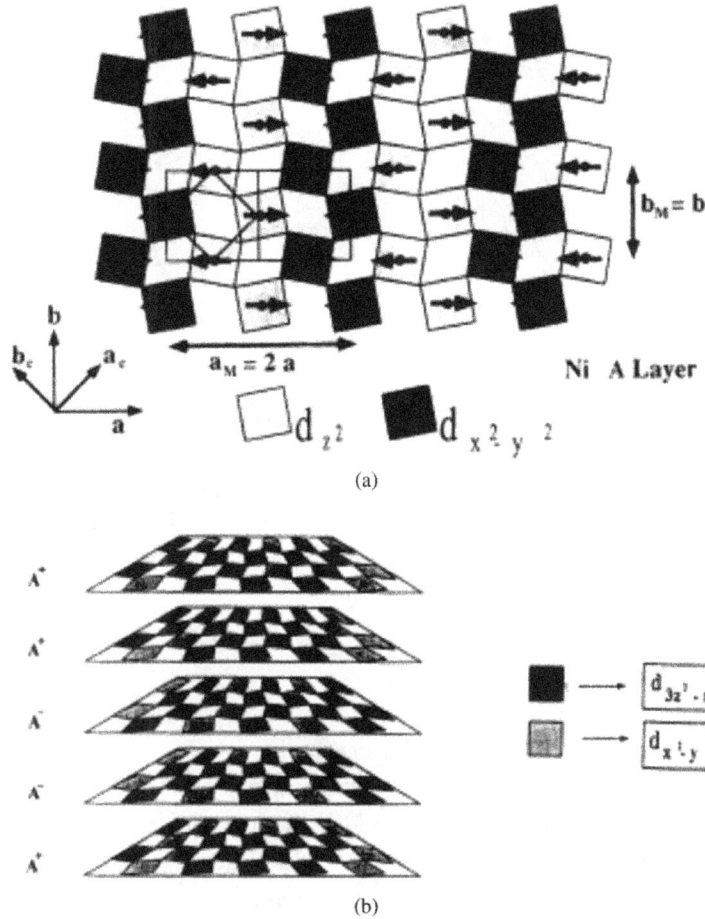

Figure 18. The magnetic structure of the Ni^{3+} sublattice of $NdNiO_3$. (a) Ordering of the Ni^{3+} moments in one layer of the structure, noted as A^+. The octahedra are shaded to indicate the orbital order between d_z^2 and $d_{x^2-y^2}$ orbitals. The dimensions of the magnetic and crystallographic cells are indicated. (b) The stacking of A^+ and A^- layers in the full magnetic structure. In the A^- layers, the Ni^{3+} spins are inverted relative to the A^+.[98] Reproduced from Ref. 98 with permission from the American Physical Society.

site should have $S = 0$. The Ho^{3+} moments order below 3 K with a different $k = (0\ 0\ 0)$ pattern. The full magnetic structure at 1.5 K is shown in Figure 19.

Finally, the magnetic structure of $DyNiO_3$ has also been solved.[100] The crystal structure is also described in $P2_1/n$ and for the magnetic structure $k = (1/2\ 0\ 1/2)$ for both the Ni and Dy sublattices, unlike the situation for $HoNiO_3$, resulting in the proposed magnetic structure shown in Figure 19.[99] It should be noted that in both cases, there exists ambiguity regarding the ordering on the Ni sublattices between a collinear and non-collinear model.

$$Cu^{3+}, 3d^8, t_{2g}^6 e_g^2, S = 1.$$

$LaCuO_3$ can only be synthesized under very high-pressure conditions and was first reported by Demazeau.[101] Non-stoichiometric versions, $LaCuO_{3-x}$ are common. The synthesis conditions for the preparation of these phases have been discussed.[102] $LaCuO_3$ has $R\text{-}3c$ symmetry while the oxygen-deficient phases show a variety of space groups from tetragonal to monoclinic. Stoichiometric $LaCuO_3$ and most of the $LaCuO_{3-x}$ materials are metallic and a Pauli paramagnet.[102] Among the other $RECuO_3$ perovskites, only $NdCuO_3$ has been reported, which is also metallic.[103]

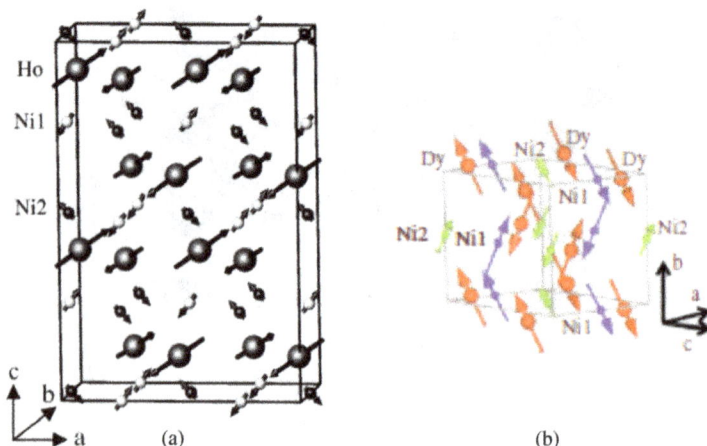

Figure 19.   (a) The Magnetic structure of $HoNiO_3$ at 1.5 K showing the Ho moments, large grey spheres and the two Ni moments in black and white.[99] (b) The magnetic structure of $DyNiO_3$ at 1.5 K.[100] (a) Reproduced from reference 99 with permission from the American Physical Society. (b) Reproduced from reference 100 with permission from Elsevier.

$$Ru^{3+}, 4d^5, t_{2g}^5 e_g^0, S = 1/2.$$

$RERuO_3$ perovskites, apart from $LaRuO_3$, can only be prepared using very high-pressure methods. The entire series, RE = La–Lu, has been reported.[104] All crystallize in *Pnma* but are oddly Ru deficient, a very rare situation for ABO_3 perovskites or for any perovskite family in fact, and the average Ru oxidation state is ~3.3–3.5. All are paramagnetic down to 2 K, and show only RE^{3+} magnetism. For $LaRuO_3$, a large temperature-independent susceptibility (TIP) is found which could suggest metallic/Pauli behavior which is supported by a 1970s study that found metallic transport behavior.[105]

$$Rh^{3+}, 4d^6, t_{2g}^6 e_g^0, S = 0.$$

The entire series $RERhO_3$ can be prepared under ambient conditions.[106] All have *Pbnm* symmetry and are insulating or semiconducting, indicating a filled t_{2g} subband. Magnetic susceptibility data are dominated by RE^{3+} magnetism. No long-range order is detected to 5 K. $LaRhO_3$ shows largely TIP behavior with a small Curie tail at low temperatures, suggestive of magnetic impurities.

A remarkable magnetic phenomenon is found in the series $LaCo_{1-x}Rh_xO_3$ between the two $S = 0$ end members. For x~0.1–0.4, an F-like signal is induced as shown in Figure 20.[107] T_C is estimated as ~15 K for $x = 0.2$. M vs H data indicate hysteresis and a magnetization of > 0.2 μ_B/mole at 2 K, so canted AF behavior can be ruled out.

It is noted that Vegard's law is not obeyed, the observed cell volume is always greater than the expected linear behavior. Data above 200 K were fitted to the Curie–Weiss law and μ_{eff} values were extracted. A model for the electronic state of Co^{3+}, which fits the μ_{eff} vs x trend, is that the substitution of Rh^{3+} forces half of the Co^{3+} ions into the HS state, $S = 2$, while the others remain as LS, $S = 0$. A model involving the IS configuration is ruled out, but one in which charge transfer occurs, i.e. $Co^{3+} + Rh^{3+} \rightarrow Co^{2+} + Rh^{4+}$, is consistent with the data. Oddly, the θ_C values from the CW fit are < 0, suggesting AF spin correlations. PND data show no sign of long-range order, either AF or F, and evidence is presented for a spin glass ground state for $x > 0.4$.[108] Optical data indicate that HS Co^{3+} is indeed present in the solid solution.[109] It is argued that the substitution of the larger Rh^{3+} ion for LS Co^{3+} induces the transformation to the larger HS Co^{3+} state via elastic relaxation.[110]

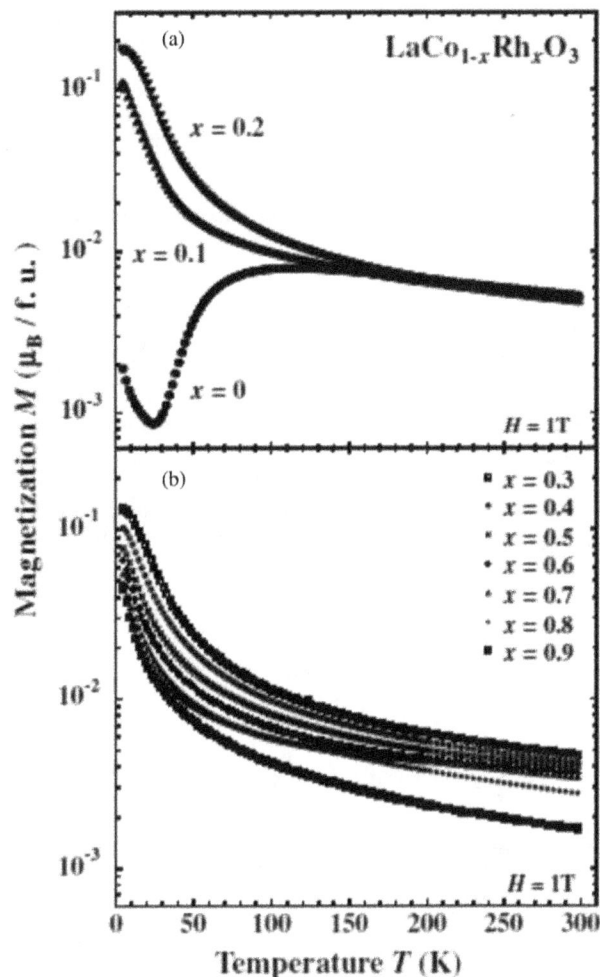

Figure 20. Induction of ferromagnetic response in the solid solution $LaCo_{1-x}Rh_xO_3$.[107] Reproduced from reference 107 with permission from the Physical Society of Japan.

$A_2BB'O_6$ — B-site ordered double perovskites

A large number of perovskites, more than 1000 have been reported, exist in which the B-site is occupied equally by two ions, B and B'.[111] If B and B' differ sufficiently in ionic radius (IR) and formal charge (FC), the ions will order crystallographically to form an ordered double perovskite (DP) as illustrated in Figure 21. In this class of perovskites, the A-site, which may be occupied by more than one ion also, is not ordered. The B and B' sites form interpenetrating f.c.c. lattices. The f.c.c. lattice can be described as an array of edge-sharing tetrahedra. The criteria for B-site ordering have been discussed.[112]

In general, if $\Delta(IR) > 0.26$ Å and $\Delta(FC) > 2$, site order will exist. For smaller $\Delta(IR)$, one often finds partial site order or antisite disorder (ASD). Due to the tendency of $4d$ and $5d$ ions to form stable oxidation states of +4 and greater with relatively small IR, the DP structure is an ideal host for these ions such as Mo^{5+}, Ru^{5+}, Os^{5+}, Re^{5+}, Re^{6+}, Os^{6+}, Os^{7+}, etc.

A group theory family tree has been derived for $A_2BB'O_6$ perovskites as shown in Figure 22. The aristotype has, of course, *Fm-3m* symmetry. In fact, of the large number of possible symmetries, only three are commonly found, *Fm-3m, I4/m,* and *P2$_1$/n (P2$_1$/c)*.

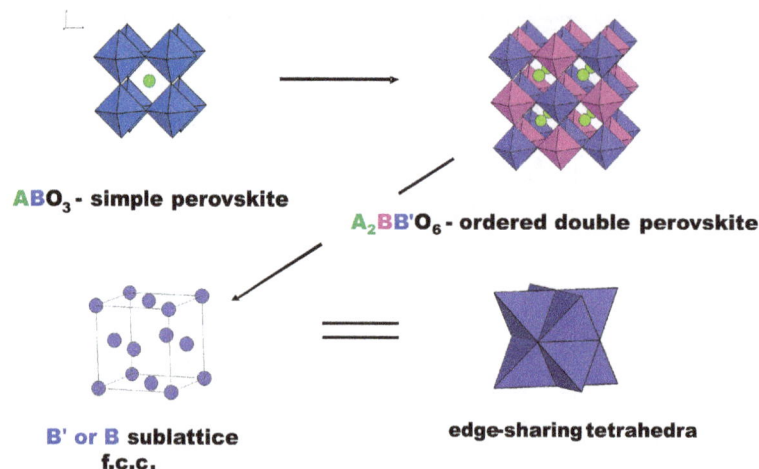

Figure 21. Formation of the B-site ordered double perovskite structure by substitution of two ions, B and B′, which order crystallographically to form two interpenetrating f.c.c. lattices. The f.c.c. lattice can be described as an array of edge-sharing tetrahedra.

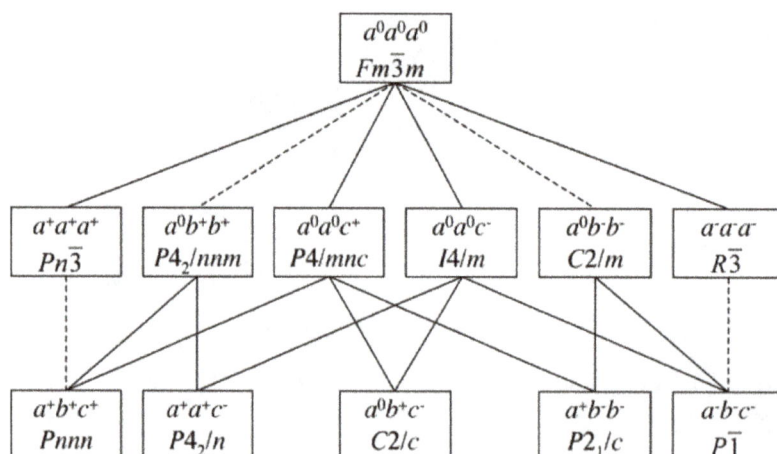

Figure 22. The space group 'family tree' for B-site ordered double perovskites.[113]

Two classes of magnetic DP oxides are of interest, (1) both B and B′ are magnetic and (2) only B′ is magnetic. The first group is the potential 'half-metallic ferromagnets' which often show large MR with potential for spintronics applications and the second group can display geometric magnetic frustration (GMF).

B and B′ magnetic

Sr_2FeMoO_6

This DP has been known since 1963.[114] Interest was revived when a giant MR effect was reported.[115] Sr_2FeMoO_6 is a ferromagnetic or ferrimagnetic with $T_C > 400$ K and is metallic.[116] This effect can be understood in terms of the band structure, Figure 23. The Fermi level, E_F, lies within a gap with respect to the

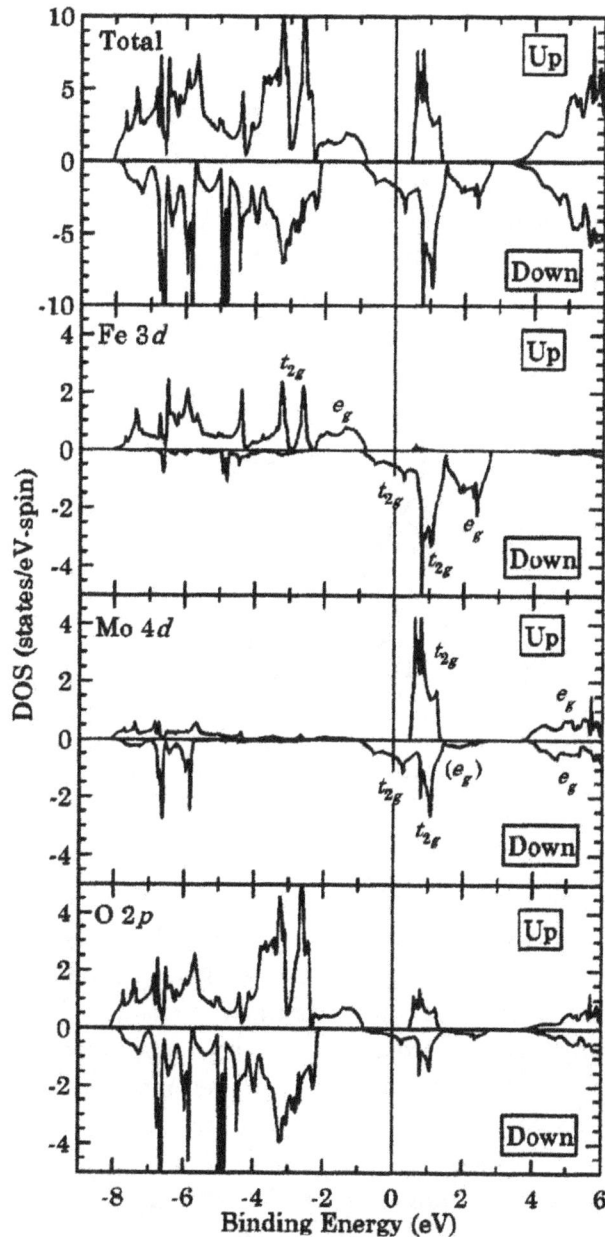

Figure 23. Spin polarized band structure of Sr_2FeMoO_6, showing the origin of the half-metallic magnetic behavior.[117] Reproduced from reference 117 with permission from the American Physical Society.

spin up DOS, but within the t_{2g} DOS from both the Fe and Mo states. Thus, the material is insulating with respect to the spin up states but metallic with respect to the spin down states. Both the extent of B-site order and the oxidation states of Fe and Mo have been debated. B-site order varies widely from sample to sample. In single crystals, it can be as high as ~95% but samples with only 20% are also known.[118] The consensus is that a model with Fe^{3+}/Mo^{5+} is a good approximation to the formal charges.[119,120] The observed μ_{sat} in the best samples is 3.8 μ_B/mole, which agrees well with a ferrimagnetic model, $Fe^{3+}\uparrow(5\ \mu_B)Mo^{5+}\downarrow(1\ \mu_B)$.[117] Note that the G–K rules predict F coupling for a t_{2g}^1–O–$t_{2g}^3 e_g^2$ 180° pathway.[2,8] Other examples of such violations will be noted in later sections.

$A_2Fe(Cr)ReO_6$

Table 4. Symmetry, T_C, site order and saturation moments for selected DP.

DP	SG	T_c (K)	B-site order (%)	μ_{sat} (μ_B)/F.U.	Refs.
Ca_2FeReO_6	$P2_1/n$	540	—	2.4	121,122
Sr_2FeReO_6	$I4/m$	419	83–85	1.9–2.7	121,123
Ba_2FeReO_6	Fm-$3m$	340	90–97	2.1–2.4	124
Sr_2CrReO_6	$I4/m$	635	87–90	0.82–1.29	125

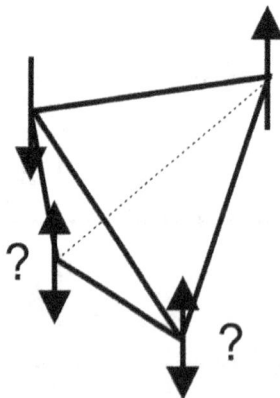

Figure 24.   Geometric frustration on the tetrahedron. With nn AF exchange constraints, two of the four spins will be frustrated.

Several other related materials with $T_C > 300$ K have been studied, as displayed in Table 4. In these materials, Re^{5+} and Fe^{3+}/Cr^{3+} are assumed. Reasonable B-site ordering is usually found, in spite of the fact that Δ(IR) values are quite small, ~0.05 or so, and there is a strong dependence on preparation technique. Low-temperature, soft-chemistry methods seem to give the best results.[126] The crystal symmetry follows the expected trend of decreasing symmetry with decreasing tolerance factor. Ca_2FeReO_6 shows a complex behavior below T_C with phase separation into at least two other monoclinic phases.[122]

The μ_{sat} values are roughly consistent with ferrimagnetic coupling between Re^{5+} and Fe^{3+}/Cr^{3+}, but the Re^{5+} moment is not well determined. Note that the value for Sr_2CrReO_6 is lower than for the $Fe^{3+}(S=5/2)$ DP, consistent with the smaller ordered moment on $Cr^{3+}(S=3/2)$. Of this group, Ca_2FeReO_6 appears to be non-metallic, while the other Fe-based DP are metals. Reports vary for Sr_2CrReO_6 but high-quality thin films are semiconducting.[127] Due to the very high T_C, the Cr-based phase has received much attention for potential application in spintronics.

Only B′ magnetic

As illustrated in Figure 21, the B′ sublattice is f.c.c., which is one of the canonical geometrically frustrated lattices, based on edge-sharing tetrahedra.[128] It is well known that, under the constraint of nearest neighbor AF exchange, two of the four spins on the corners of a tetrahedron will be frustrated, Figure 24. A useful criterion to judge the presence of frustration, but one which should be used with caution, is the frustration index, $f = |\theta_C|/T_N$, first introduced by Ramirez.[128] Here, θ_C is obtained from fitting susceptibility data to the Curie–Weiss law, $\chi = C/(T-\theta_C)$. According to the mean field theory, $\theta_C = 2S(S+1)/3k_B \Sigma_m z_m J_m$.[129] Here, k_B is Boltzmann's constant, S is the spin, $z_m J_m$ indicate the number and the exchange constant of the mth neighbors respectively. Thus, θ_C is the weighted, algebraic sum of all of the exchange

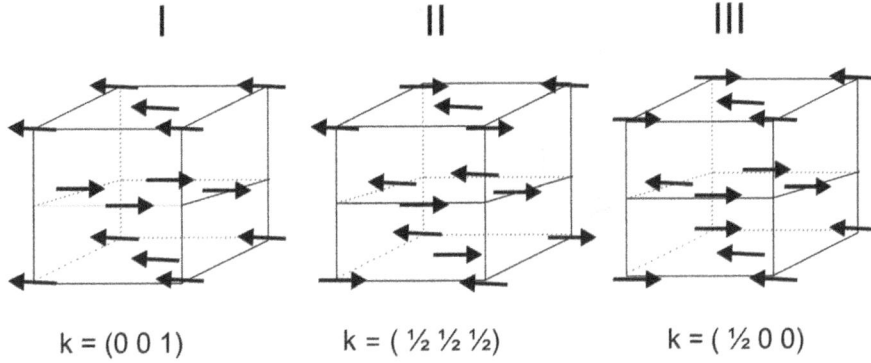

Figure 25. The three f.c.c. AF magnetic structures, Types I, II, and III, predicted by mean field theory. Only one f.c.c. cell is shown and the ordering wave vectors are given below. The spin directions are arbitrary.

Table 5. Spin–spin correlations for nn and nnn spins in the three f.c.c. AF magnetic structures. The ordering wave vectors, k, are also given along with the magnetic cell dimensions.

Type	nn	nnn	k	Mag. Cell
I	4F, 8AF	6F	(001)	$a_M = a$, $b_M = b$, $c_M = c$
II	6F, 6AF	6AF	(½ ½ ½)	$a_M = 2a$, $b_M = 2b$, $c_M = 2c$
III	4F, 8AF	4F, 2AF	(½ 0 0)	$a_M = 2a$, $b_M = b$, $c_M = c$

interactions. The product $\theta_C k_B$ sets the energy scale for the magnetic exchange interactions in any material. Thus, one expects to find magnetic order at a temperature near θ_C or at some large fraction of θ_C. For the f.c.c. lattice, mean field theory predicts four possible magnetic structures, one F and three AF.[129,130] The three AF structures are shown in Figure 25.

In Table 5, some properties of the three structures are presented. A remarkable fact, gleaned from perusal of Table 5, is that in none of the predicted magnetic structures are the nn spin correlations purely AF. In spite of the fact that all of the nn exchange pathways are equivalent as shown in Figure 26. This expresses the frustrated nature of the f.c.c. magnetic lattice. It is common to describe Type I as consisting of F planes stacked in an AF manner along a (001) direction and Type II as F sheets stacked AF along (111) direction. As a historical note, the first magnetic structure solved from neutron diffraction data was that of MnO with Mn^{2+} ions on a f.c.c. lattice.[131] MnO shows the Type II magnetic structure.

Returning to the frustration index, f, a phase diagram for the f.c.c. magnetic lattice which relates the observable $|\theta_C|/T_N$ ratio to the J_{nn}/J_{nnn} ratio was determined in the 1960s.[129] The maximum $|\theta_C|/T_N$ found is 5, and more typical values are in the range of 1–3. Thus, it is common to attribute $f \gg 5$ to the effects of geometric frustration.

As mentioned, great caution should be applied as there can be other contributions to θ_C apart from exchange. This is particularly true when crystal field (CF) levels lie within or near to the temperature range over which the magnetic data are collected. To see the origin of this problem, consider the van Vleck equation given as follows for the susceptibility:[132]

$$\chi = N_0 \left[\Sigma_i \left(E_i^{(1)2}/k_B T - 2 E_i^{(2)} \right) \exp(-E_i^0/k_B T) \right] / \Sigma_i \exp(-E_i^0/k_B T)$$

where E_i^0 is the energy of the ith state. $E_i^{(1)} = \langle n|\mu|n \rangle$ where μ is the magnetic moment operator and $|n \rangle$ is the wave function of the ith state. These are the 'first order Zeeman' terms and involve only diagonal

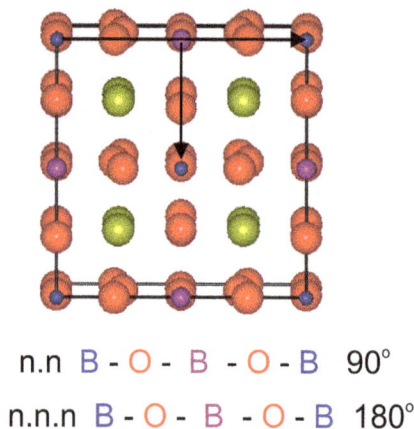

n.n B - O - B - O - B 90°

n.n.n B - O - B - O - B 180°

Figure 26. Super–superexchange pathways in $A_2BB'O_6$ DP (black arrows) The 12 nn interactions involve a 90° B′–O–B–O–B′ pathway and the 6 nnn a 180° pathway. Both include the diamagnetic B ion. The gold spheres are the A-site ions.

matrix elements. Note that these terms make a temperature-dependent contribution. $E_i^{(2)} = \Sigma_m (<n|\mu|m> <m|\mu|n>)/(E_m-E_n)$ where Σ_m includes contributions from excited states, m, and off-diagonal elements are involved. These are known as the second-order Zeeman terms. Note that these will be temperature independent and the magnitude will depend strongly on the energy difference, E_m-E_n, between the ground and excited states. Each CF level will contribute a temperature dependent magnetic moment through the first-order Zeeman terms and there will be TIP contributions from the second order terms. As the temperature is lowered, the occupation of the various levels changes as some are depopulated and, of course, the susceptibility changes giving a departure from the pure Curie law. While the above-mentioned equation does not resemble the C–W law in any obvious way, the data at high temperatures can often be analyzed to fit the C–W law and the derived θ_C can be strongly < 0, due only to CF effects. An excellent case history is afforded by the non-perovskite material $Nd_3Ga_5SiO_4$ in which a $\theta_C = -52$ K was attributed to CF effects.[133]

Relevant magnetic properties of selected DP in this category are listed in Table 6. Clearly, a wide variety of f values, from ~1 to >295 and ground states are observed. AFLRO = long range AF, FLRO = long range F, SF = spin frozen, GCSING = gapped collective singlet.
Some trends are clear. All $S = 5/2$ systems order with moderate f values. In the $S = 3/2$ case, f values are much larger, indicating a stronger frustration effect. In general, ordering temperatures are greater for $Fm\text{-}3m$ symmetry DP than for $P2_1/n$, as for the latter the exchange pathway angles deviate from the ideal 90° and 180° values. As well, T_N is greater for $5d$ elements, Os^{5+}, than for their $4d$ counterparts, Ru^{5+}, for the same symmetry, consistent with the greater radial extent of the $5d$ vs $4d$ orbitals. For the more quantum spin DP, $S = 1$ and $1/2$, states which lack long-range order, such as spin freezing, appear with greater frequency. The most remarkable case is Ba_2YMoO_6 for which a gapped, collective singlet ground state occurs.[150–152] The $P2_1/n$ version, La_2LiMoO_6, does not form a singlet state.[150] Currently, there is no theoretical consensus regarding Ba_2YMoO_6.[153]

A singular feature of the $S = 1$ and $1/2$ DP is the occurrence of what can be called doppelgängers. These are materials which are isostructural, isoelectronic, and have essentially the same lattice dimensions, yet their magnetic properties are remarkably and unaccountably different. Selected examples are shown in Table 7. With cell constants within 0.12%, it is truly remarkable that Ba_2LiOsO_6, Os^{7+}, $5d^1$, shows AFLRO and Ba_2MgReO_6, Re^{6+}, $5d^1$, FLRO. For Ba_2MgOsO_6 and Ba_2ZnOsO_6, both Os^{6+}, $5d^2$, the cell constant difference is only slightly larger, 0.27%, but the ground state of the former is AFLRO with a high T_N while the latter forms an SF state below 28 K. Perhaps even more astonishingly, Ba_2CaOsO_6 shows AFLRO below 48 K but Ba_2CdOsO_6 does not order to 0.4 K. In all of these cases, it seems clear that the non-magnetic B ion must play a larger than

Table 6. Magnetic properties of selected DP with B′ only magnetic.

S	DP	S.G.	θ_C(K)	T_N, T_C, T_f (K)	f	Grd. state	Ref.
5/2	Ba_2MnWO_6	*Fm-3m*	−64	7.5	9	AFLRO	134,135
5/2	Sr_2MnWO_6	*P2₁/n*	−71	14	5	AFLRO	136
3/2	Ba_2YRuO_6	*Fm-3m*	−571	36	16	AFLRO	137,138
3/2	Ba_2YOsO_6	*Fm-3m*	−700	68	10	AFLRO	139
3/2	La_2LiRuO_6	*P2₁/n*	−170	24	7	AFLRO	137,140
3/2	La_2LiOsO_6	*P2₁/n*	−154	30	5	AFLRO	141
1	Sr_2NiWO_6	*I4/m*	−175	54	3	AFLRO	142
1	Ba_2YReO_6	*Fm-3m*	−616	35	18	SF	143
1	Ba_2CaOsO_6	*Fm-3m*	−156	49	3	AFLRO	144
1	Ba_2CdOsO_6	*Fm-3m*	−118	0.4<	>295	?	145
1/2	Sr_2CaReO_6	*P2₁/n*	−443	14	32	SF	146
1/2	Ba_2MgReO_6	*Fm-3m*	−373	21	9	FLRO	147,148
1/2	Ba_2LiOsO_6	*Fm-3m*	−40	8	5	AFLRO	149
1/2	Ba_2NaOsO_6	*Fm-3m*	−10	7	~1	FLRO	149
1/2	Ba_2YMoO_6	*Fm-3m*	−219	—	—	GCSING	150–152

Table 7. Doppelgängers among selected *Fm-3m* Double Perovskites.

DP	a (Å)	% diff.	Grd. state	$T_{N,C}$(K)
Ba_2LiOsO_6	8.1046(2)	0.12	AFLRO	8
Ba_2MgReO_6	8.09498(2)	0.12	FLRO	18
Ba_2MgOsO_6	8.0757(1)	0.27	AFLRO	52
Ba_2ZnOsO_6	8.0975(1)	0.27	SF	~28
Ba_2CaOsO_6	8.3190(1)	0.53	AFLRO	48
Ba_2CdOsO_6	8.3633(2)	0.53	no order	0.4<

expected role in determining the sign and strength of the super–superexchange interactions. Note that both Zn^{2+} and Cd^{2+} have filled *d*-levels, while for Ca^{2+} empty *d*-levels, are available. This issue has also been noticed and discussed by Feng *et al.* in the context of DP such as $Ba_2BO_sO_6$ where B = Sc, Y, and In.[154] While some theory has emerged focused on DP with 4*d* and 5*d* ions, and thus fairly strong spin orbit coupling, there remains much scope for both experiment and theory in this class of magnetic perovskites.[153,155]

$AA′BB′O_6$, $AA′B_2O_{5+x}$, $AA_2′BB_2′O_{9-x}$: A-site (and B-site) ordered DP

A-site ordering is much less common than B-site ordering in DP, as explained by King and Woodward.[156] The most common ordering is layering of A and A′ ions, and most involve Mn or Fe on the B-site with a with a diamagnetic B ion or occupying both B and B′ sites. Ordered O^{2-} vacancies are also often present. Typical compositions are $NaLnMn(Co)(Fe)WO_6$, $LnBaFe_2O_{5+x}$, or $LnBaMnO_{5+x}$, with x ~0.5. These structures are described in detail by King and Woodward and involve square pyramidal coordination of the transition elements as illustrated in Figure 27.[156] Also covered in this section are perovskites with both A-site order, B-site order, and O^{2-} vacancies such as $LnBa_2Fe(Cu)_3O_{9-x}$, where x = 1–3, and the high-pressure phases such as Mn_2FeReO_6 and $CaCu_3Fe_2Re_2O_{12}$.

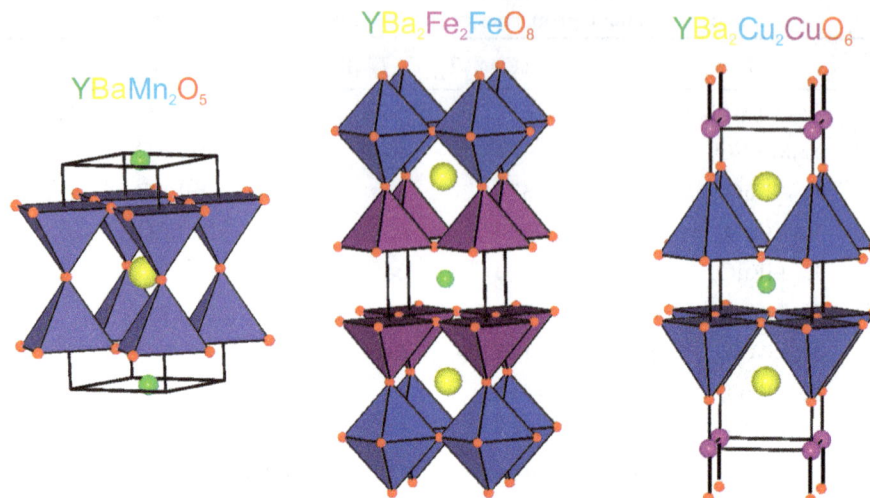

Figure 27.   The crystal structures of selected perovskites with A-site, B-site, and O^{2-} vacancy ordering. All crystallize in *P4/mmm*. Note the presence of square pyramidal coordination of the magnetic ions and for the cuprate, linear coordination for Cu^+.

Table 8.   Selected magnetic properties for $NaREMnWO_6$ perovskites.[157]

RE	θ_C (K)	μ_{eff} (μ_B) Obs.	μ_{eff} (μ_B) theory	T_N (K)
La	−42	5.91	5.92	10
Ce	−40	6.02	6.44	13
Pr	−34	6.63	6.92	11,8,5
Nd	−35	6.89	6.94	11
Sm	−43	5.77	5.98	12,3
Gd	−27	8.28	9.90	13
Tb	−27	11.49	11.38	15,9
Dy	−19	11.7	12.17	13,4
Ho	−20	11.8	12.14	11

AA′BB′O$_6$

The series $NaREMnWO_6$ is exemplary of this class and has been studied for a range of RE.[157] Here, the B-site ions are Mn^{2+} and W^{6+}. The crystal structures are monoclinic, $P2_1$.[158] Selected data are shown in Table 8. All show $\theta_C < 0$ and AFLRO in a narrow range from 10 K to 15 K. Several phases undergo more than one magnetic transition; indeed, there are three for the RE = Pr material. Magnetic structures are reported for RE = La, Nd, and Tb.[158] Apart from $NaLaMnWO_6$ for which the magnetic structure is described by the commensurate wave vector, $k = (1/2\ 0\ 1/2)$, the others have complex, incommensurate magnetic structures with $k = (0\ 0.48\ 1/2)$ for RE = Nd at all temperatures below T_N. In this case, the Nd^{3+} and Mn^{2+} moments order at the same temperature, 11 K. In $NaTbMnWO_6$, two k-vectors are found near $T_N = 15$ K, $(1/2\ 0\ 1/2)$ and $(0\ 0.47\ 1/2)$, but below 6 K the incommensurate k vanishes.

REBaB$_2$O$_{5+x}$

These phases were first reported by Chapman for B = Mn and by Karen and Woodward for B = Fe.[159,160] For $x = 0$, a mixed valence B^{2+}/B^{3+} exists, while for $x = 0.5$, B^{3+} only is present. The crystal symmetry

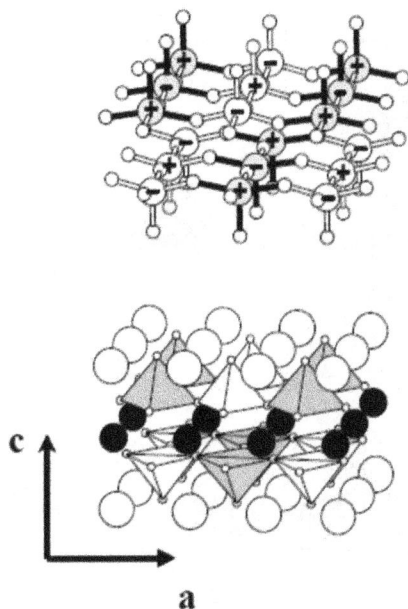

Figure 28. TOP: The magnetic structure of $YBaCo_2O_5$. The grey spheres are Co^{2+} and the white spheres Co^{3+} while +/− show the spin correlations. BOTTOM: The Co^{2+}/Co^{3+} charge ordering pattern (same color correlations). The black and large white spheres show the Y and Ba ions, respectively.[161] Reproduced from reference 161 with permission from the American Physical Society.

without charge ordering is generally *P4/mmm*, which persists up to $x \sim 0.5$.[160] B^{2+}/B^{3+} charge ordering is often observed in perovskites of this composition as in, for example, $YBaCo_2O_5$,[161] $YBaMn_2O_5$,[162] and $TbBaFe_2O_5$.[163] The $YBaCo_2O_5$ situation is particularly complex as an HS–LS transition on the Co^{2+} site also occurs along with the charge and magnetic order. The G-Type magnetic structure and charge ordering for $YBaCo_2O_5$ is shown in Figure 28.

The magnetism of $YBaMn_2O_5$ is also surprising. The space group here is *P4/mmm* in the paramagnetic regime but changes to *P4/nmm* upon magnetic ordering below 167 K which allows for the Mn^{2+}/Mn^{3+} charge ordering.[162] A G-Type structure is also found but this leads to a ferrimagnetic ground state with the Mn^{2+} moments ($S = 5/2$) antiparallel to the Mn^{3+} ($S = 2$) along the *c*-axis resulting in a net moment at 5 K of the expected 1 μ_B.[159] For $TbBaFe_2O_5$, T_N occurs at 450 K, while T_{CO} sets in below ~300 K. $TbBaMn_2O_5$ has the same charge ordered and magnetic structure as $YBaCo_2O_5$ and is not ferrimagnetic.[163]

Among the $x = 0.5$ phases, $YBaMn_2O_{5.5}$ has been studied in detail.[164] This perovskite crystallizes in a large supercell, $2a \times 2a \times 4a$, with respect to the primitive perovskite cell dimension, a. The space group is *Icma* (*Ibam*) which allows for two Mn^{3+} sites, one square pyramidal and one octahedral, Figure 29. $YBaMn_2O_{5.5}$ undergoes AFLRO below $T_N = 140$ K with a canted moment. The magnetic structure is again G-Type for both of the Mn sites with moments along the *a*-axis. Ordered moments are nearly the same for the two Mn sites, 3.5(1) μ_B for the square pyramidal site and 3/7(1) μ_B at the octahedral site.

$$REBa_2B_3O_{9-x}$$

$REBa_2Fe_3O_8$ and $REBa_2Cu_3O_6$ are the most investigated perovskites of this class. In both systems, oxygen ions can be intercalated with relative ease, as with the preceding $REBaB_2O_5$ perovskites, and indeed they are of greatest interest as oxygen storage/source materials.

The $REBaFe_3O_8$ series has been investigated in detail.[165] For RE = La, Pr, and Nd, cubic *Pm-3m* symmetry is reported, indicating that the O^{2-} vacancies are not ordered. All others crystallize in *P4/mmm* as

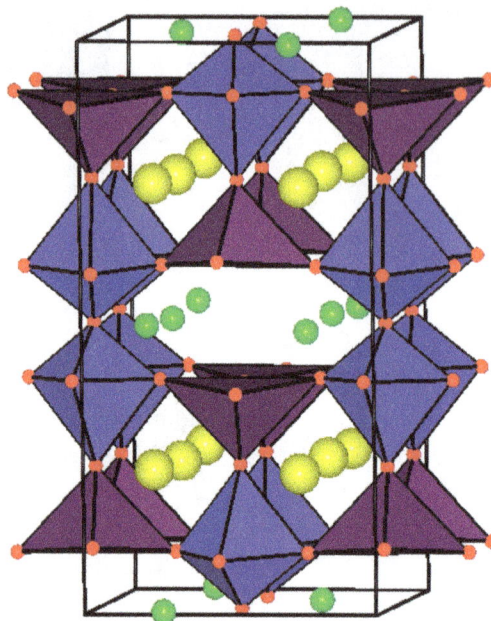

Figure 29. The crystal structure of $YBaMn_2O_{5.5}$. The green and gold spheres are Y and Ba, respectively. The two Mn sites, one square pyramidal(magenta) and one distorted octahedral(blue), are shown. The *c*-axis is vertical.[164]

illustrated in Figure 27. Note that the average Fe valence is +3.67 for $x = 0$, which is often partitioned as $2Fe^{3+}$ and Fe^{5+} although some presence of Fe^{4+} cannot be ruled out. True charge ordering is not reported.

For the *P4/mmm* phases, AFLRO occurs with $T_N \sim 650$ K, relatively independent of the RE. In the cubic, disordered versions such as $LaBa_2Fe_eO_8$, $T_N = 188$ K, the much lower value being ascribed to Fe valence disorder on the magnetic sites as well as O^{2-} vacancy disorder. G-Type magnetic structures are observed for RE = Y, Dy, and Er, with ordered moments on the Fe sites of 3.3–3.4 μ_B at ambient temperature which increases to 3.8 μ_B at 10 K. The RE ions do not order above 10 K. The ordered moment on the Fe sites at 10 K decreases strongly as *x* decreases from unity, reaching values of 1.3 μ_B for $x = 0.2$.

$YBa_2Cu_3O_{6+y}$ can be regarded as the 'parent' AF form of the famous superconductor $YBa_2Cu_3O_7$. In the former, formal Cu^{2+}/Cu^+ ordering clearly occurs as illustrated in Figure 27, although, as the cuprates are charge-transfer insulators, holes represented by O^- are active as well. The crystal symmetry is *P4/mmm*.[166] For small values of *y*, AFLRO is observed described by $k = (1/2\ 1/2\ 0)$, i.e. a magnetic unit cell 2a × 2a × c with respect to the chemical cell.[167] Only the Cu^{2+} site has an ordered moment of 0.60(5) μ_B. For y = 0.1, $T_N = 410$ K which decreases to 368 K for $y = 0.18$.[168] The moments lie in the *ab*-plane, Figure 30. The Cu^{2+} moment also decreases with increasing *y*, for $y = 0.18$ it is 0.44(3)μ_B. These values are much lower than the $\sim 1.1\ \mu_B$ usually found in Cu^{2+} insulators and has been ascribed to a combination of quantum fluctuations and covalency with the O^{2-} ligands.

A and B sites both magnetic

A few perovskites exist in which both the A and B sites contain magnetic *d*-group transition metal ions. These are generally high-pressure phases such as $CaCu_3Fe_2Re_2O_{14}$, a so-called 'quadruple perovskite' and Mn_2FeReO_6.[169,170] The remarkable magnetic structure of $CaCu_3Fe_2Re_2O_{12}$ is shown in Figure 31. The chemical space group is *P-3n*, not the expected *Fm-3m* due to Ca/Cu ordering on the A-site and significant octahedral tilting. Cu^{2+} has square planar coordination, while Fe and Re ions have the usual octahedral

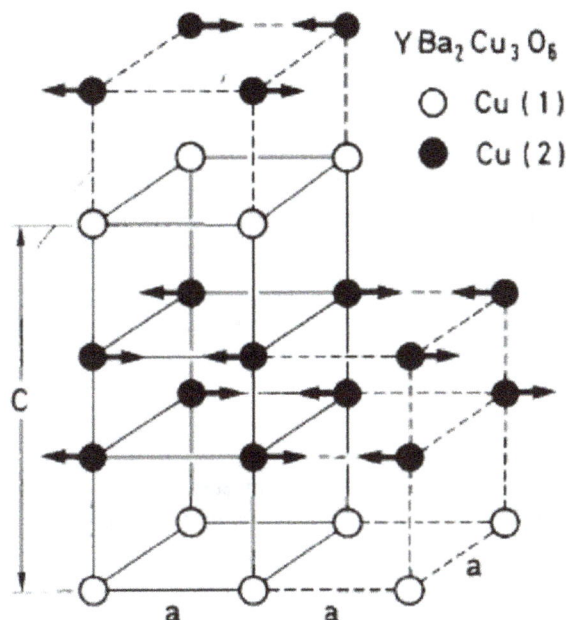

Figure 30. The magnetic structure of $YBa_2Cu_3O_6$. The magnetic unit cell is 2a × 2a × c as indicated and the moments, 0.60(5) μ_B, only on the Cu^{2+} site, lie in the *ab*-plane.[167] Reproduced from reference 167 with permission from Elsevier.

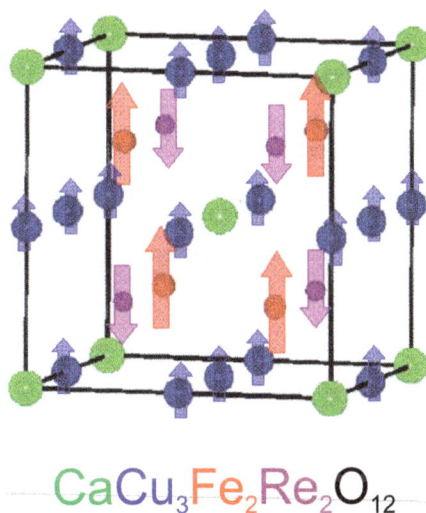

Figure 31. The magnetic structure of $CaCu_3Fe_2Re_2O_{12}$.[169] The oxide ions have been omitted for clarity.

coordination. The bulk susceptibility and magnetization indicate ferrimagnetic behavior, $T_C = 560$ K and the saturation moment is 8.7 μ_B / formula unit. Note that the Cu^{2+} spins couple ferromagnetically with the Fe^{3+} spins and antiparallel with the small Re^{5+} spins giving such a large saturation magnetization. It appears that the individual site moments were not refined from ND data.

The magnetic properties of Mn_2FeReO_6 are more complex.[170] The chemical symmetry is $P2_1/n$. It is also a ferrimagnet with a high $T_C = 520$ K, but the Mn^{2+} spins are decoupled from the $Fe^{3+}\downarrow Re^{5+}\uparrow$ pattern until 75 K. This is ascribed to a frustration effect. The spin ordering sequence is shown in Figure 32.

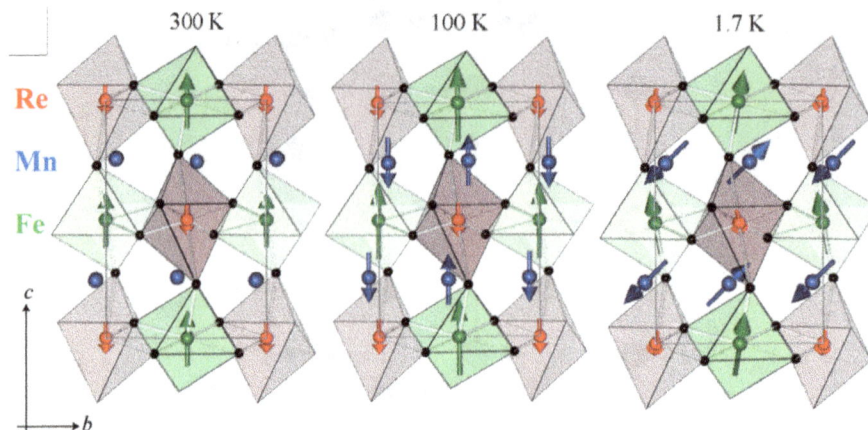

Figure 32. Spin coupling patterns for Mn_2FeReO_6 at selected temperatures. Note that the coupling of the Mn^{2+} spins to the $Fe^{3+}\downarrow Re^{5+}\uparrow$ order is frustrated.[170] Reproduced from reference 170 with permission from John Wiley and Sons.

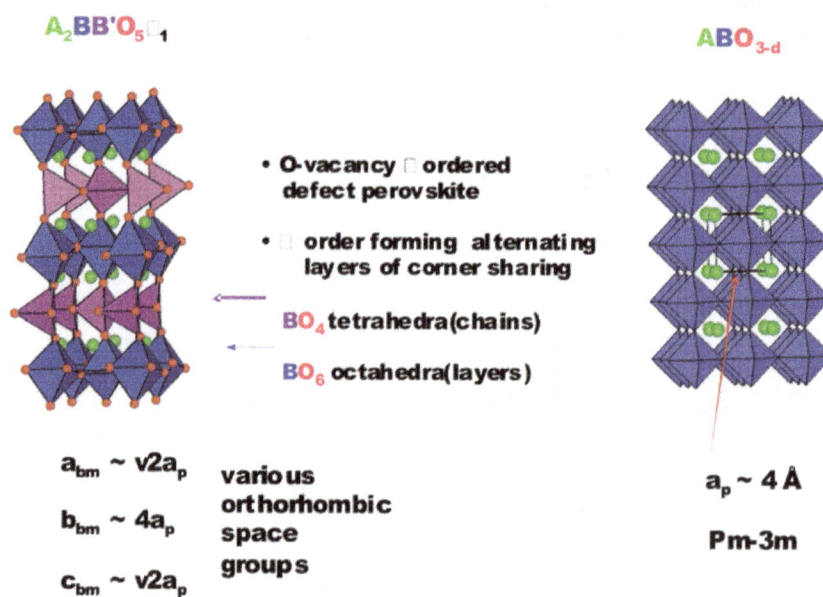

Figure 33. Comparison of the O^{2-} vacancy ordered brownmillerite and the primitive perovskite structures. Note the large supercell for brownmillerite.

$AA'BB'O_{6-x}$, x = 1. Brownmillerites

Many perovskites exist with large concentrations of vacancies on the O^{2-} site. Most of these have the composition $AA'BB'O_5$ and crystallize in some form of the brownmillerite structure. A-site ordering is not known, but BB' site ordering does occur. A comparison of the brownmillerite structure with a disordered perovskite structure is shown in Figure 33.

Note the occurrence of a very large supercell with dimensions $\sqrt{2}a \times \sqrt{2}a \times 4a$ with respect to a primitive perovskite cell. There are now two B sites, one octahedral and one tetrahedral. The octahedra share four corners forming a layer, while the tetrahedra share two corners with adjacent tetrahedra, forming chains, and two corners with the octahedral layers. Brownmillerites crystallize in several orthorhombic space groups and one monoclinic. These details have only a minor influence on the magnetic properties, and the reader is referred to the literature for more information.[171] The brownmillerite 'periodic table' is much less extensive than that for perovskites.

The only A-site ions are Ca^{2+}, Sr^{2+}, Ba^{2+}, and La^{3+}, while the B-site ions are restricted to stable trivalent ions, Sc^{3+}, Cr^{3+}, Mn^{3+}, Fe^{3+}, Co^{3+}, Al^{3+}, Ga^{3+}, and In^{3+} and can include the divalent ions Mn^{2+}, Ni^{2+}, and Cu^{2+} when La^{3+} is also present. By far, most brownmillerites have divalent ions on the A-site and trivalent ions on the B-sites. Cation ordering on the B-sites occurs according to site preference energies or radius matching, with Sc^{3+}, Mn^{3+}, and Cr^{3+} showing strong preference for the octahedral site. Co^{3+} is generally HS but shows little site preference and Al^{3+}, Ga^{3+}, and In^{3+} prefer the tetrahedral site.

A = Ca, Sr

Many magnetic brownmillerites involve phases with Ca on the A-site with a few examples of A = Sr. Some relevant properties are listed in Table 9. All show AFLRO in spite of the presence of two distinct B sites which might suggest the possibility of ferrimagnetism. The space group symmetry usually allows for spin canting.

Of course, T_N values are high for the mostly Fe^{3+}-based materials. Note that only G or C-Type magnetic structures are found with moments along either *b* or *a*. Figure 34 shows the relationship between these structures demonstrating that the G-Type can be understood as two interpenetrating C structures on the O_h and T_d sites, which explains the absence of ferrimagnetism. Magnetic anisotropy is generally controlled by the O_h site ion. For Mn^{3+}, the spins are along *b* giving C_y or G_y, while for Fe^{3+} one finds G_x. A spin re-orientation occurs in Ca_2FeCoO_6 between 100 K and 200 K from G_y to G_x and in $Ca_2Fe_{1.5}Mn_{0.5}O_5$ between 400 K and 450 K.[183]

Substitutions on either the O_h or T_d sites have a significant effect on the magnetic properties of $Ca_2Fe_2O_5$ as shown in Figure 35. Note that Co substitution has the least influence, diamagnetic substitutions on either the T_d or O_h sites have a greater effect, and the largest decrease occurs for the O_h site ions, Mn^{3+} and Cr^{3+}. An attempt has been made to rationalize these results in terms of a layered model for the magnetism in brownmillerites.[182] The following expression has been proposed for layered magnets:

$$k_B T_C \sim \xi^2 J_{inter}\, S^2$$

where ξ is the intralayer correlation length, J_{inter} is the interlayer exchange constant and S is the spin.[183] First, Co^{3+} ($t_{2g}^4 e_g^2$), $S = 2$ will have only AF SE with Fe^{3+} ($t_{2g}^3 e_g^2$), $S = 5/2$ according to the G – K rules, so ξ^2 will be little change and the main effect is a small reduction of $<S^2>$. Diamagnetic Al^{3+} and Ga^{3+} with T_d site substitution will have little effect on ξ^2 but $<S^2>$ is more strongly diminished than for Co^{3+}. Note that Sc^{3+} substitution on the O_h site is more effective in reducing T_c than the Al/Ga substitutions on the T_d site as some decrease in ξ^2 via a dilution effect is expected. Cr^{3+} (t_{2g}^3), $S = 3/2$ and Mn^{3+} ($t_{2g}^3 e_g^1$), $S = 2$ will both have F SE interactions with Fe^{3+} (G–K rules), and thus $\xi^2 J_{inter}$, and $<S^2>$ will all be diminished.

La-based brownmillerites

Brownmillerite phases are reported in La-substituted materials such as $La_{2-x}A_x Mn_2O_5$ where A = Ca, Sr, and Ba.[171] In the Parsons study, *x* does not reach 1. However, in another work, $LaSrMn_2O_5$ is claimed with *I2mb* symmetry.[184] An F signal is found below ~300 K. The magnetization at 5 K is $0.25\mu_B$/F.U., much too small to be true ferromagnetism but too large to attribute to spin freezing, normally. In principle, this

Table 9. Some magnetic properties of selected brownmillerite structure oxides, $A_2BB'O_5$.

A	B,B'	S.G.*	T_N (K)	Mag. Strc.	Ref.
Ca	Fe,Fe	*Pnma*	725	G_x	172
Ca	Fe(T_d),Mn(O_h)	*Pnma*	407	G_y	173
Ca	Fe,Co	*Pbcm*	580	G_y,G_x	174
Ca	Al(T_d),Mn(O_h)	*I2mb*	152	C_y	175
Ca	Ga(T_d),Mn(O_h)	*Pnma*	160	C_y	175
Ca	1.5Fe,0.5Cr(O_h)	*Pnma*	455	G_x	176,177
Sr	Fe,Fe	*Imma***	693	G_x	178,179
Sr	Fe,Co	*Imma*	> 300	G_x	180

Notes. *These are for the settings where $b>c>a$. Many other settings are used in the literature. **This is the average structure symmetry. Electron diffraction finds domains with *Pbcm* and *C2/c* symmetry.[181]

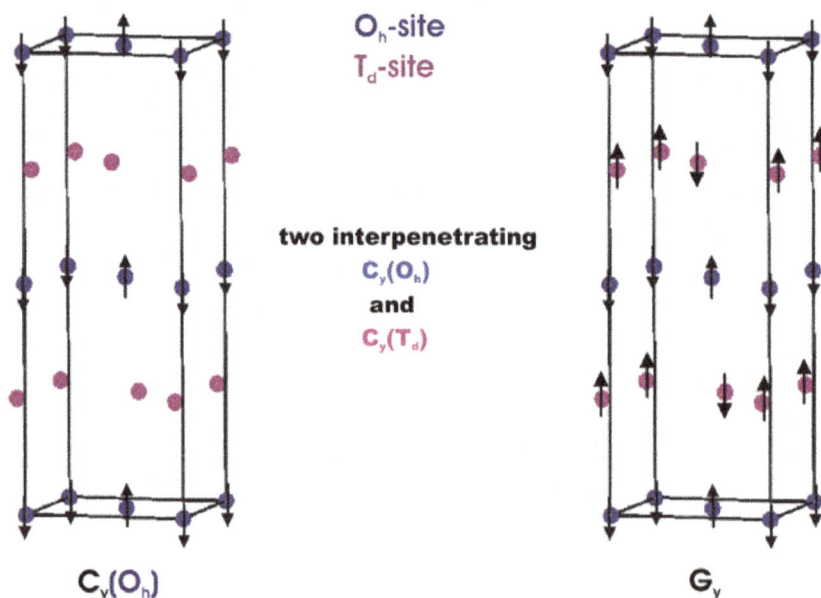

Figure 34. The relationship between the C and G-Type magnetic structures for brownmillerites. G is an interpenetration of C structures on both the O_h and T_d sites.

material should contain Mn^{2+} ($S = 5/2$) on the T_d site and Mn^{3+} ($S = 2$) on the O_h site. No magnetic structure seems to have been reported.

$Sr_4Fe_4O_{11}(Sr_2Fe_2O_{5.5})$

This material is compositionally related to brownmillerite and crystallizes in *Cmmm*.[185] The unit cell is roughly $2\sqrt{2}a_P \times 2a_P \times \sqrt{2}a_P$. In principle, it contains both Fe^{3+} and Fe^{4+}, formulated as d^5L^{-1}, i.e. involving a ligand hole. There are two Fe sites, one square pyramidal (SP) and one O_h, and charge ordering is proposed with the Fe^{4+} ions on the SP site. It shows a $T_N = 240$ K, and a magnetic structure has been proposed in which only half of the Fe spins are ordered. This is attributed to frustration as each Fe spin on the SP site

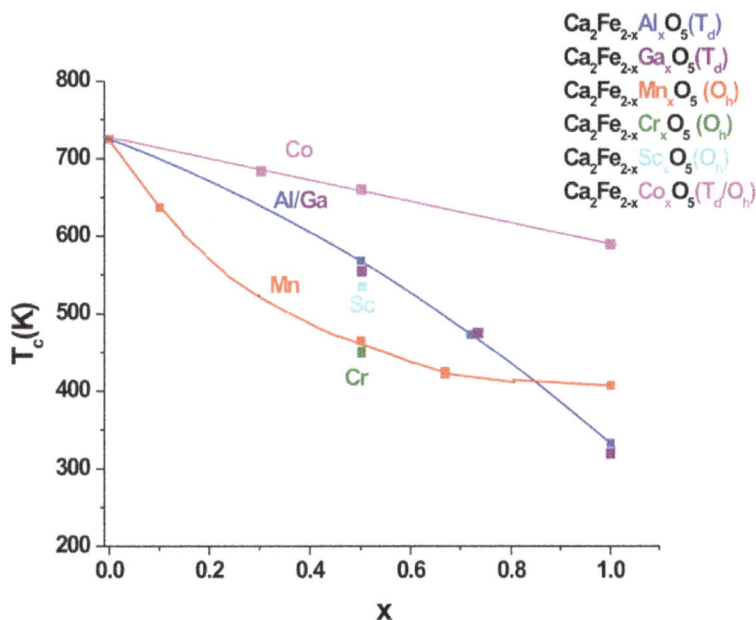

Figure 35. The effect on T_c by various substitutions on O_h and T_d sites in $Ca_2Fe_{2-x}B_xO_5$. The site preference for the B-ions is indicated.[182] Reproduced from reference 182 with permission of the Royal Society from Chemistry.

has four nn spins on the O_h site, two ↑ and two ↓ and vice versa.[186] It is not possible to distinguish between models in which the order is restricted to either the SP or O_h sites from existing powder neutron diffraction data.[185,186]

Sr_2FeMnO_5 and $Sr_2Fe_{1.5}Cr_{0.5}O_5$

A few materials exist with a brownmillerite stoichiometry but in which the O^{2-} vacancies are disordered resulting in *Pm-3m* symmetry. Two of these are Sr_2FeMnO_5 and $Sr_2Fe_{1.5}Cr_{0.5}O_5$.[187,188] While the cell constants, $a_0 = 3.8933$ Å for Fe/Mn and 3.9449 Å for Fe/Cr, differ by only 1.3%, the magnetic properties are remarkably divergent. Sr_2FeMnO_5 is a rare example of a superparamagnet based on underlying AF spin correlations with a blocking temperature, $T_B \sim 50$ K, while $Sr_2Fe_{1.5}Cr_{0.5}O_5$ shows AFLRO, G-type, with $T_N = 565$ K, as illustrated in Figure 36.

In both cases, the local structure as determined by neutron pair distribution function data (NPDF) differs strongly from the average *Pm-3m* symmetry. An attempt has been made to rationalize these stark differences in terms of the local structure.[189]

Pillared Perovskites. $RE_5B_3B'O_{16}$

Perovskites can form intergrowths with other structural units. One such intergrowth involves a dimeric unit, $B_2O_{10}^{10-}$ or $B_2O_{10}^{-12}$. The overall structure is shown in Figure 37.

The B'B'' ions in the perovskite layers are ordered as in DP. The B_2O_{10} units link the perovskite layers only through the B'' sites with overall symmetry *C2/m* or *C-1*. The *a* and *b* axes are $\sim 2a_P$ and the perovskite layers are separated by $c \sim 10$ Å.[190,191] Table 10 shows the elements involved which are from the 3*d*, 4*d*, and 5*d* series along with RE = La. The most extensive series involves $Re^{5+}(5d^2)$ on the B and B' sites with various divalent 3*d* ions on the B'' site.[191–194]

Typical magnetic properties are illustrated by $La_5Re_3MnO_{16}$, Figure 38.

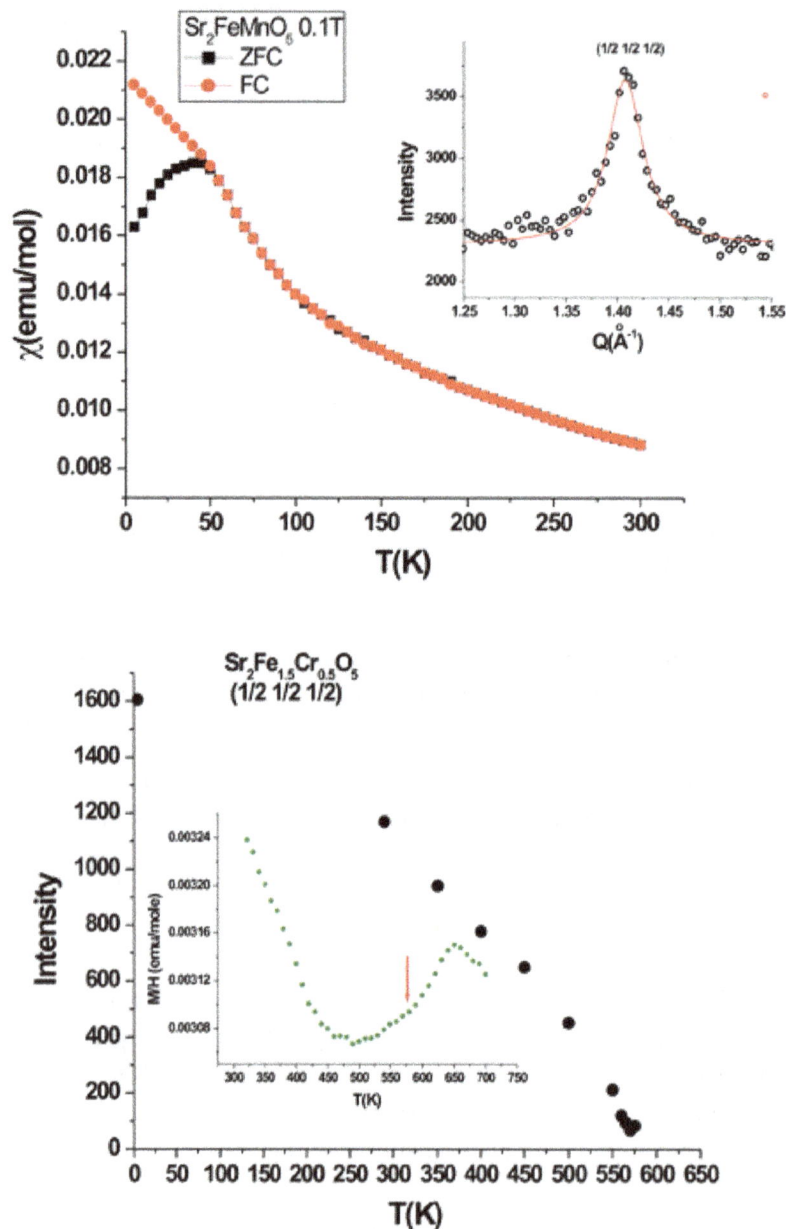

Figure 36. TOP: Susceptibility data for Sr_2FeMnO_5 showing T_B ~50 K. The inset shows the (1/2 1/2 1/2) magnetic peak and a fit to a Lorentzian line shape giving ξ~50 Å. BOTTOM: Temperature dependence of the (1/2 1/2 1/2) magnetic peak of $Sr_2Fe_{1.5}Cr_{0.5}O_5$ showing T_N= 565 K. The inset shows the susceptibility and T_N is indicated by the red arrow.[187,188]

Plotting the data as χT vs T shows the build up of short-range ferrimagnetic intraplanar correlations just above T_C, the sharp spike, followed by an even sharper decrease when the planes couple AF below T_C. Metamagnetic transitions are usually seen at relatively low applied fields, indicating the weak interplanar coupling.

Magnetic structures are known from ND measurements and all show k = (0 0 1/2) indicating A-Type ordering as illustrated in Figure 39. For the Re–Mn case, the Re(1.0 μ_B) spins appear to be parallel to c while the Mn(3.9 μ_B) spins make an angle of ~ 18°. In all other cases, the Re and B″ moments are strictly antiparallel. The properties of all of the members of this series are summarized in Table 11.

$$RE_5B_2B'B''O_{16}$$

Figure 37.   The structure of $RE_5B_2B'B''O_{16}$. Perovskite layers, $B'B''O_6^{5-}$ are 'pillared' by edge-sharing dimeric $B_2O_{10}^{10-}$ units. The crystal symmetry is *C-1* and the long *c*-axis is vertical and the view is nearly along the *a*-axis.

Table 10.   Ions which occupy the various sites in $La_5\ B_2B'B''O_{16}$ pillared perovskites.

B	B'	B''	Ref.
Mo^{4+}	Mo^{4+}, Mo^{5+}, V^{4+}	Mo^{5+}, V^{4+}	190, 195
Re^{5+}	Re^{5+}	$Mg^{2+}, Mn^{2+}, Fe^{2+}, Co^{2+}, Ni^{2+}$	191–194
Os^{5+}	Os^{5+}	Mn^{2+}	196

The $B'' = Ni^{2+}$ phase appears to have two transitions. As well, the intraplanar spin correlations seem to be F rather than AF as for the others. Note also from the last column that all of the intraplanar SE interactions, Re–O–B'', are predicted by the G–K rules to be at least weakly F, an expectation which is realized perhaps only for the Ni phase. This situation also arises with certain DP materials such as Ca_2CoOsO_6 and Ca_2NiOsO_6 for which the G–K rules predict F SE between Os^{6+} (t_{2g}^2) and Co^{2+} and Ni^{2+} but for which DFT calculations indicate AF coupling.[197]

Perhaps the most remarkable material in this series is $La_5Re_3MgO_{16}$, which shows a very large, negative θ_C but no LRO to 2 K giving $f > 1100$! The data with the low *T* Curie tail subtracted can be fitted to a 2D, $S = 1$ model for short-range spin correlations, Figure 39.[192] The decoupling of the layers may reflect the very convoluted SE pathway: Re^{5+}–O–Mg–O–Re=Re–O–Mg–O–Re^{5+}. Further studies are surely warranted here.

The Ruddlesden–Popper phases. $(AO)(ABO_3)_n$, $n = 1$–3

The Ruddlesden–Popper (RP) phases are the most common examples of the intergrowth mechanism involving perovskites. Here, a layer with AO stoichiometry and NaCl symmetry intergrows with one or more

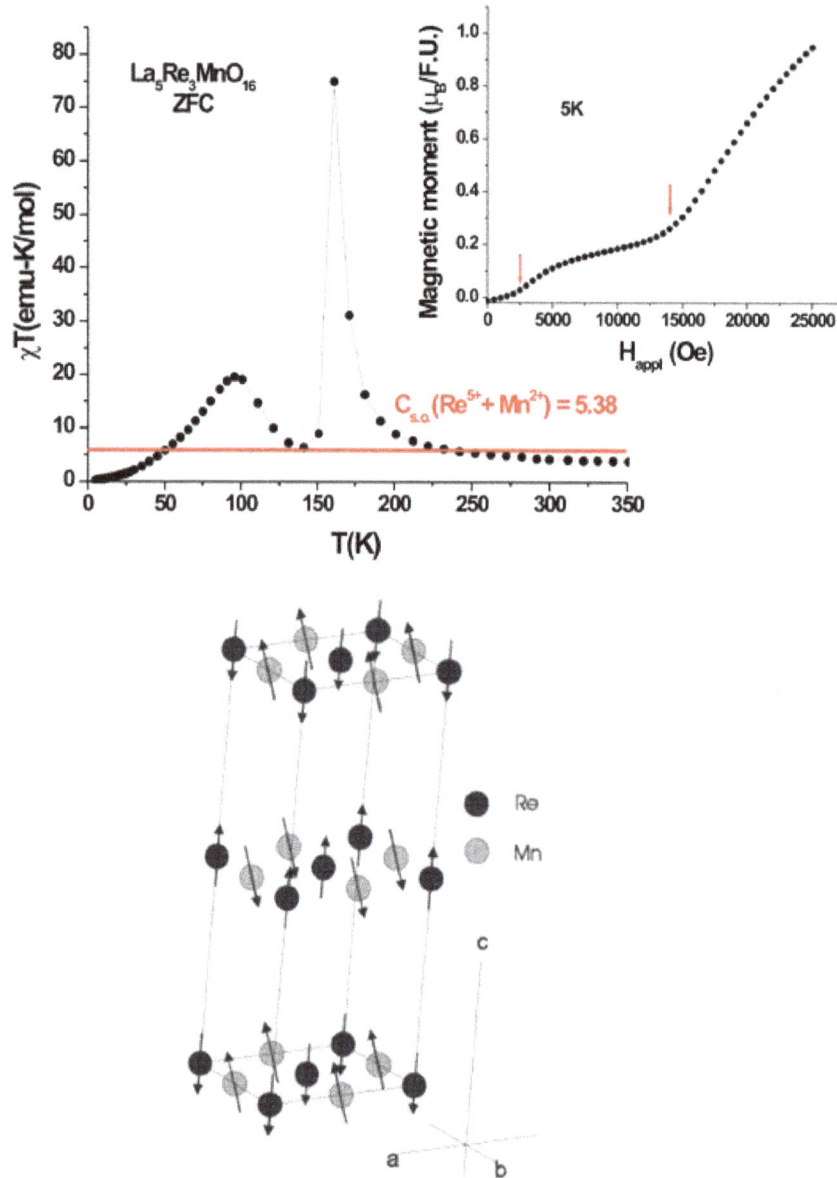

Figure 38. TOP: Plot of χT vs T for $La_5Re_3MnO_{16}$ showing the build up of ferrimagnetic short-range spin correlations within the *ab*-plane followed by AF coupling of the planes below $T_C = 161$ K. The inset shows metamagnetic transitions indicated by red arrows in *M* vs *H* plots. BOTTOM: The magnetic structure indicating ferrimagnetic intraplanar Re–Mn coupling and overall AF coupling of the planes described by $k = (0\ 0\ 1/2)$.

layers of perovskite (ABO_3) symmetry and stoichiometry, see Figure 4. In general, RP series members with $n = 1$ or 2 are quite stable with $n = 3$ phases being more unusual. In this chapter, the discussion will be confined to the $n = 1$ phases.

A_2BO_4, $n = 1$

The $n = 1$ structure is sometimes named after K_2NiF_4, an isostructural fluoride with symmetry *I4/ mmm*. As the magnetic sublattice has body-centered tetragonal symmetry, interlayer magnetic frustration is a feature of this structure type, given nn AF interactions, as illustrated in Figure 40(a). 2D spin

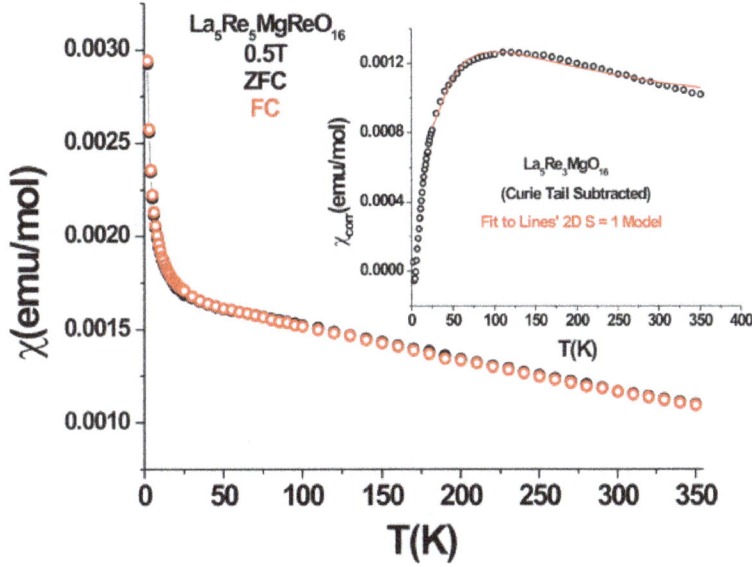

Figure 39. Magnetic susceptibility for $La_5Re_3MgO_{16}$. Note that ZFC and FC data are superimposed at all temperatures. The upturn at low T is attributed to paramagnetic impurities. There is no sign of LRO to 2 K. The inset shows a fit to the data with the Curie tail subtracted for a 2D, $S = 1$ model.[192]

Table 11. Summary of the magnetic properties of the $La_5Re_3B''O_{16}$ series.

B''	θ_C (K)	T_C (K)	J(Re–O–B'')	Configs. G–K rules
Mn	−48	161	AF	$t_{2g}^2 - t_{2g}^3 e_g^2$ (F)
Fe	−84	155	AF	$t_{2g}^2 - t_{2g}^4 e_g^2$ (F)
Co	−71	33	AF	$t_{2g}^2 - t_{2g}^5 e_g^2$ (F)
Ni	−217	14, 36	F?	$t_{2g}^2 - t_{2g}^6 e_g^2$ (F)
Mg	−575	—	—	—

correlations build up within the planes giving rise to a χ_{max} which is related to the 2D exchange constant as in Figure 40(b). This is not T_N, but the true 3D ordering temperature can be located from susceptibility data by application of the Fisher heat capacity analysis, Figure 40(c). Values for J_{2D} can be extracted from data such as Figure 40(b) by fitting the curves to models obtained from high-temperature series expansions.[198]

Three A/B valence combinations are known, $A^{2+}_2B^{4+}O_4$, $A^{2+}A^{3+}B^{3+}O_4$, and $A^{3+}_2B^{2+}O_4$.

$A^{4+}_2B^{4+}O_4$

$$B = Ti^{4+}, A = Eu^{2+}$$

Recall that the perovskite $EuTiO_3$ is an AF insulator, $T_N = 5.3$ K. It is, thus, surprising to find that Eu_2TiO_4 is an F insulator with $\theta_C = 10$ K, $T_C = 9$ K, and $\mu_{SAT} = 6.0$ μ_B.[201] This result was explained in terms of the crystal fields acting on the Eu^{2+} ion which reduces the $4f$–$5d$ energy gap for Eu_2TiO_4 relative to $EuTiO_3$, thereby enhancing the F Goodenough–Kasuya exchange in a manner similar to that depicted in Figure 7 which compares $EuTiO_3$ and $EuLiH_3$.[202]

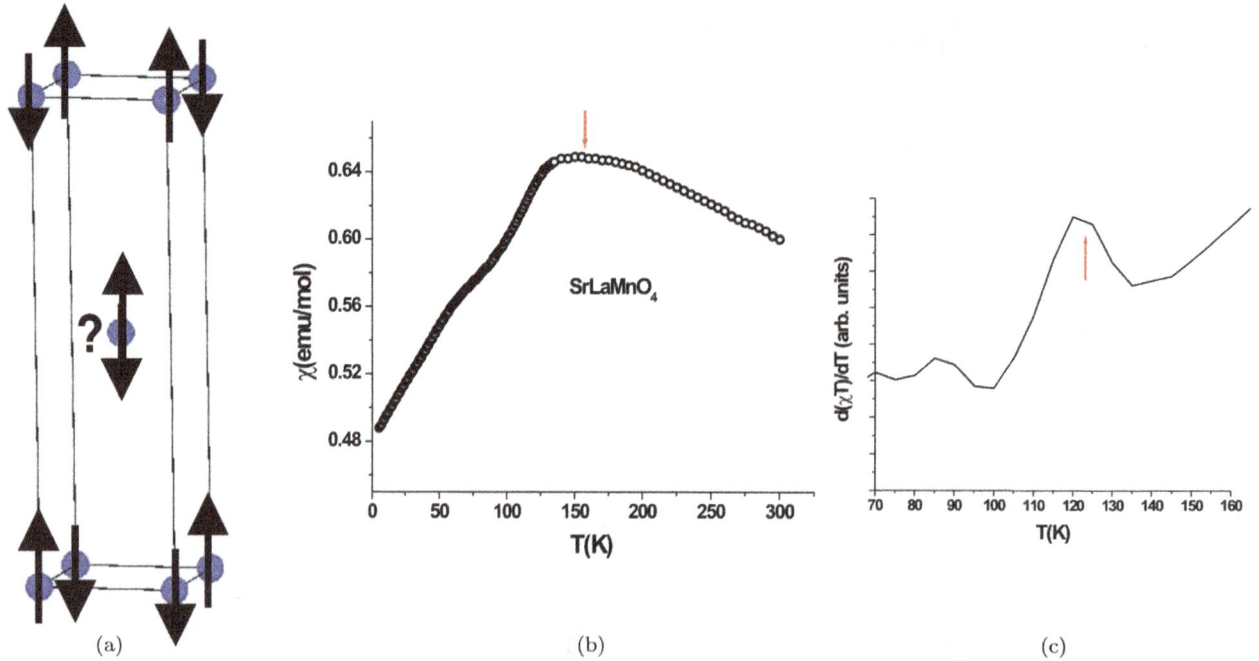

Figure 40.   (a)Interlayer magnetic frustration resulting from the body-centered tetragonal magnetic sublattice of K_2NiF_4 structure, $n =1$ RP phases. (b) Typical susceptibility data exemplified by $SrLaMnO_4$ showing a broad maximum, $T_{max} \sim 160$ K, due to 2D spin correlations which is not T_N. (c) Plot of the Fisher heat capacity, $d(\chi T)/dT$ vs T, to locate the true $T_N = 128$ K.[199,200]

$$B = V^{4+}, 3d^1, t_{2g}^{1}, S = 1/2$$

Sr_2VO_4 is a Mott insulator but shows no sign of magnetic LRO in spite of several anomalies in the susceptibility. These effects are attributed to orbital ordering brought about by the interaction of a tetragonal CF component and SOC resulting in a ground-state Kramers doublet comprised of |xz> and |yz> orbitals with unquenched orbital angular momentum.[203] Structural phase transitions result upon cooling from *I4/mmm* to *Immm* back to *I4/mmm*.[204] It has been proposed that a 'hidden' order exists in this material and perhaps in other d^1 systems.[205]

$$B = Cr^{4+}, 3d^2, t_{2g}^{2}, S = 1$$

Ba_2CrO_4 and Sr_2CrO_4 do not normally crystallize in the $n = 1$ RP structure.[206] However, Sr_2CrO_4 with the *I4/mmm* structure has been reported under ambient conditions.[207] From spectroscopic studies, it appears that the valences Cr^{3+}/Cr^{5+} are present rather than Cr^{4+}. The susceptibility appears to be roughly TIP, suggesting either Pauli or Van Vleck paramagnetism.

$$B = Mn^{4+}, 3d^3, t_{2g}^{3}, S = 3/2.$$

Sr_2MnO_4 shows classic 2D behavior with a $\chi_{max} \sim 250$ K and $T_N = 170(5)$ K obtained from ND.[208] A value of $J_{2D}/k_B = -80$ K is obtained by fitting to a series expansion for $S = 3/2$ on a square lattice. The magnetic structure is also typical for many K_2NiF_4 oxides, described by $k = (1/2 \ 1/2 \ 0)$. The magnetic unit cell has dimensions $a_{mag} = \sqrt{a_{cryst}}$ and $c_{mag} = c_{cryst}$ as shown in Figure 41.The Mn^{4+} moments are along c, and the refined moment is $2.42(5) \ \mu_B$ slightly reduced from the expected $3 \ \mu_B$.

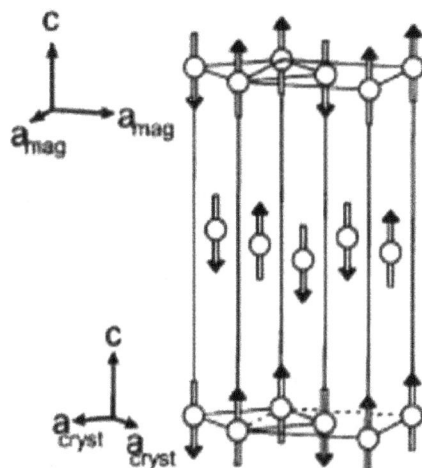

Figure 41.   Typical magnetic structure for K_2NiF_4 structure oxides. The magnetic cell dimensions are $a_{mag} = \sqrt{2}a_{cryst}$, $c_{mag} = c_{cryst}$. The moments can also lie in the *ab*-plane.[199] Reproduced from Ref. 199 with permission from the Royal Society of Chemistry.

$$B = Fe^{4+}, 3d^4$$

Sr_2FeO_4 has been reported, and a CW fit to the susceptibility data support a HS, $t_{2g}^3e_g^1$ configuration.[209] A broad $\chi_{max} \sim 60$ K is observed, but no transition to AFLRO was detected in ND down to 4 K. However, Mössbauer spectra show a very complex hyperfine splitting at 4.2 K, indicative of four magnetic sites and suggest that at least magnetic SRO is present.

$$B = Co^{4+}, 3d^5$$

High-pressure synthetic methods are needed to stabilize the Co^{4+} state, even in a perovskite-related structure. Sr_2CoO_4 is known and Co^{4+} appears to be in the IS state, $t_{2g}^4e_g^1$, $S = 3/2$, as the observed $\mu_{eff} = 3.72$ μ_B is close to the spin only value of 3.87 μ_B.[210] Polycrystalline samples have a very low resistivity, $\rho \sim 10^{-2}$ ohm-cm at 300 K and might be metallic or at least semimetallic. Remarkably, Sr_2CoO_4 is ferromagnetic with $T_C = 250$ K but with a saturation moment of only ~ 1 μ_B instead of 3 μ_B which might be expected. A recent report indicates a very complex magnetization behavior for this material at low temperatures.[211]

$$B = Mo^{4+}, 4d^2$$

Sr_2MoO_4 is metallic and Pauli paramagnetic.[23]

$$B = Ru^{4+}, 4d^4$$

Sr_2RuO_4 is metallic, as is ferromagnetic $SrRuO_3$, but the RP phase is a superconductor with $T_c = 0.93$ K.[212] Nonetheless, this discovery has sparked great interest in the condensed matter physics community in part because of a very large anisotropy in the critical fields.[213]

$$B = Rh^{4+}, 4d^5$$

The space group of Sr_2RhO_4 is not *I4/mmm* but *I4$_1$/acd* due to octahedral tilting.[214] Recent studies on single crystals shows that the material is a highly anisotropic metal.[215,216] The magnetic susceptibility is also somewhat anisotropic and not strictly TIP as might have been expected. Some of these anomalies are attributed to strong correlation effects.

$$B = Ir^{4+}, 5d^5$$

Counterintuitively, Sr_2IrO_4 is insulating rather than metallic as are the $4d$ analogs just discussed. This situation has been attributed to strong SOC effects.[217] The application of strong CF and SOC results in a $J_{eff} = 1/2$ ground electronic state in this one electron system and a gap opens between upper and lower $J_{eff} = 1/2$ Hubbard bands. See ref. 215 for a detailed explanation. The space group is also $I4_1/acd$ as for the Rh analog. AFLRO is reported with a very high $T_N = 220–250$ K from both powder and single-crystal studies.[218,219] There are some discrepancies in reported ordered moments which range from $0.25–0.35$ $\mu_B/$ Ir^{4+}. Very recent work points out significant sample to sample variability, which is likely attributable to oxygen non-stoichiometry.[220]

$A^{2+}A'^{3+}B^{3+}O_4$

$$B = V^{3+}, 3d^2, t_{2g}^2, S = 1$$

The series $SrREVO_4$ has been reported.[221] The susceptibility for $SrLaVO_4$ is approximately TIP over a wide range, but $SrLaVO_4$ is clearly insulating.[222] No long-range order of any type has been detected to 4 K for any A = RE.[221] A spin-orbital liquid state has been proposed for this material.[223] These results point to some sort of spin singlet ground state and are difficult to understand, indicating that further work is needed.

$$B = Cr^{3+}, 3d^3, t_{2g}^3, S = 3/2$$

$SrLaCrO_4$ has been known for some time, but $BaLaCrO_4$ is not reported .[224,227] It shows classic 2D spin correlations with $T(\chi_{max})$ ~470 K and $T_N = 242$ K. A value of $J_{2D}/k_B = -62$ K was obtained by fitting the data to an appropriate model. The magnetic structure is that of Figure 41, with the exception that the Cr^{3+} spins are in the *ab*-plane with a refined Cr^{3+} ordered moment of 2 μ_B, although an error was not given.[225]

$CaYCrO_4$ and $SrYCrO_4$ have been studied recently.[226] Unlike $SrLaCrO_4$ which crystallizes in *I4/mmm*, these materials show orthorhombic distortions to *Bmab* for $CaYCrO_4$ and *Pccn* for $SrYCrO_4$. Due to the lower symmetry, the D–M interaction is in play for these materials and AFLRO is accompanied by a canted AF signal, manifested by a ZFC/FC divergence as shown in Figure 42. $T_N = 170$ K for SrY and 140 K for CaY.

$$B = Mn^{3+}, 3d^4, t_{2g}^3 e_g^1, S = 2$$

Some susceptibility data are already shown for $SrLaMnO_4$ in Figure 40(b), indicating a broad χ_{max} near 160 K and AFLRO with $T_N = 128$ K with the $k = (1/2\ 1/2\ 0)$ magnetic structure, Figure 41 and an ordered Mn^{3+} moment of 3.3 μ_B.[199] $BaLaMnO_4$ does not show signs of long range order to 2 K.[199] Solid solutions $Sr_{1-x}Ba_xLaMnO_4$ show a remarkable decrease in the ordered Mn moment with increasing x, and T_N appears to vanish near $x = 0.35$, as illustrated in Figure 43, while the paramagnetic moment is unaffected. There is currently no detailed explanation for this behavior.

$$B = Fe^{3+}, 3d^5, t_{2g}^3 e_g^2, S = 5/2$$

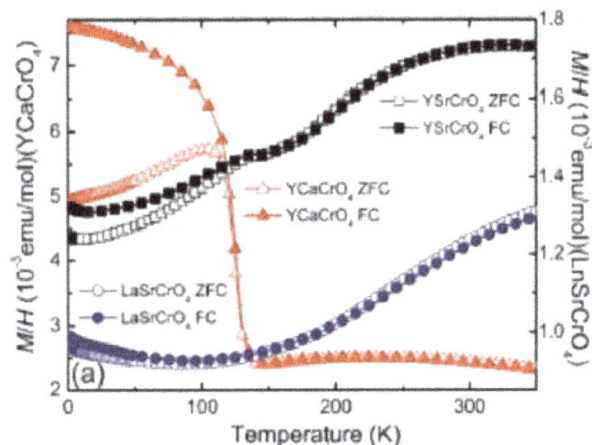

Figure 42.   Susceptibility data for $SrLaCrO_4$, $YCaCrO_4$, and $YSrCrO_4$. Note the ZFC/FC divergences for the latter two materials, consistent with a canted AF ordering below T_N.[226] Reproduced from reference 226 with permission from the Royal Society of Chemistry.

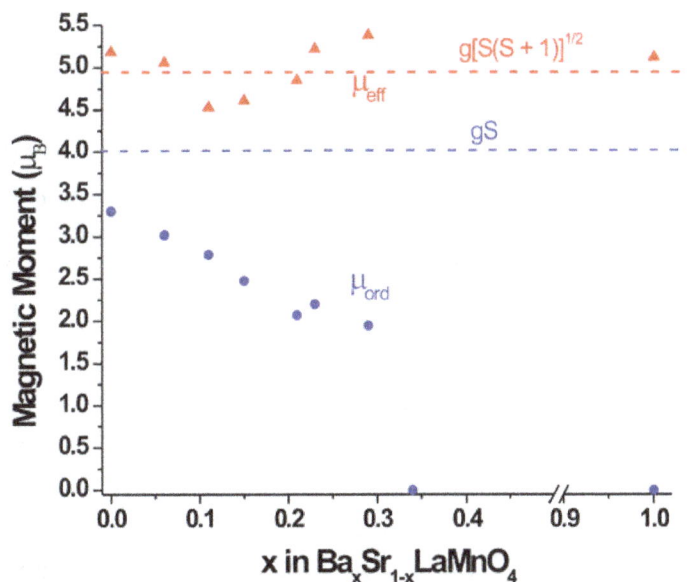

Figure 43.   Quenching of the Mn^{3+} ordered moment as a function of Ba^{2+} substitution in $Sr_{1-x}Ba_xLaMnO_4$ which reaches ~0 μ_B near $x = 0.35$. The paramagnetic moment is unaffected. T_N also vanishes near $x = 0.35$.[199]

A K_2NiF_4 structure phase is known for A = Ca^{2+}, Sr^{2+}, and Ba^{2+} where A' = La^{3+}. Both A = Sr^{2+} and Ba^{2+} phases show *I4/mmm* symmetry, while the situation for A = Ca^{2+} is ambiguous. An X-ray powder study from 1980 reports *Cmca*, but the distortion from *I4/mm* is apparently small.[227] All three show 2D magnetism with $T(\chi_{max})$ values near or above 600 K and $T_N = 370$ K, 380 K, and 341 K for A = Ca, Sr, and Ba, respectively. Magnetic structures are the expected $k = (1/2\ 1/2\ 0)$ type with the Fe^{3+} moments, 4.96(1) μ_B as determined on a single crystal of $SrLaFeO_4$ essentially equal to the spin only value, which lie in the *ab*-plane.[228] J_{2D} was measured from analysis of spin waves in $SrLaFeO_4$.

At this stage, it is of interest to compare results for $SrLaBO_4$ where B = Cr, Mn, and Fe as in Table 12 . Note that values of $T(\chi_{max})$, T_N, and J_{2D} are anomalously small for B = Mn.

Table 12. Comparison of selected magnetic properties of $SrLaBO_4$ phases.

B	S	$T(\chi_{max})$ (K)	T_N (K)	J_{2D}/k_B (K)*
Cr^{3+}	3/2	470	242	60
Mn^{3+}	2	160	128	13
Fe^{3+}	5/2	650	380	36

Note: *J_{2D} is estimated from the expression $J_{2D} = k_B T(\chi_{max})/\tau S(S+1)$ where $\tau = 2.07$.[198]

This is traceable to competing intraplanar AF and F SE interactions as predicted by the G–K rules for Mn^{3+}.[2]

$$B = Co^{3+},$$

Of the several reports on the magnetic properties of $SrLaCoO_4$, a clear consensus does not appear to exist. One area of agreement is that the LS, $S = 0$ is not the ground state. There are reports of F behavior below 250 K and spin glass transitions below 10 K.[229] A recent study of an apparently high-quality single crystal supports a spin glass transition at 7 K and a high population of HS $Co^{3+}(t_{2g}^4 e_g^2)$[230]

$$B = Ni^{3+}, 3d^7$$

$SrLaNiO_4$ appears to be highly conducting, probably metallic, and roughly Pauli paramagnetic.[231,232] $BaLaNiO_4$ on the other hand is reported to be insulating.[232] There are no anomalies in the susceptibility or specific heat indicating no long-range magnetic order. This strange state of affairs has been dubbed a 'Fermi Glass'.[232]

$A^{3+}_2B^{2+}O_4$

$$B = Co^{2+}, 3d^7, A = La^{3+}$$

Oxygen stoichiometry is a major issue with this and other materials in this class. They are often formulated $A_2BO_4 + \delta$ where δ can approach ~0.2 or higher. In fact, $La_2CoO_{4.0}$ can take up excess oxygen even at ambient temperature, and very complex modulated structures have been unraveled from combined ND and synchrotron X-ray studies.[233] An attempt will be made to confine these remarks to cases where $\delta \sim 0$. As with other La_2BO_4 RP phases, this material undergoes three crystallographic phase transitions with decreasing temperature at temperatures T_1 and T_2 with $T_1 > T_2$. Above T_1, the ideal $I4/mmm$ structure is found. Below T_1, a continuous transition to an orthorhombic phase, often described by $Cmca$ or a variant, occurs. Below T_2, a first-order transition to another tetragonal phase (LT T) is reported described in $P4_2/ncm$. The cell constants of the orthorhombic phase, related to the $I4/mmm$ phase, a, c, is $a_o \sim \sqrt{a}$, $b_o \sim \sqrt{a}$, $c_o \sim c$ and for the LTT, $a_{LTT} \sim \sqrt{a}$, $c_{LLT} \sim c$. For La_2CoO_4, $T_1 = 900$ K and $T_2 = 135$ K.

AFLRO is found in the orthorhombic phase with $T_N = 275$ K.[234,235] The 135 K structural transition induces a spin flip from the higher temperature AF structure. The ordered Co^{2+} moment is 2.9(1) μ_B, consistent with HS Co^{2+}, $t_{2g}^5 e_g^2$, $S = 3/2$. Notably, 2D spin correlations are observed above T_N. A discussion of the possible magnetic structures is given and the moments lie in the ab-plane.[236]

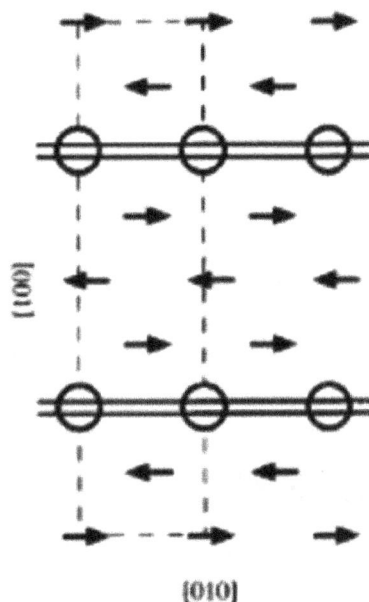

Figure 44. Stripe order between charge (holes — **O**) and magnetic order in a NiO_2 plane of $La_2NiO_{4+\Delta}$ with $\Delta = 0.133$.[239] Reproduced from reference 239 with permission from the American Physical Society.

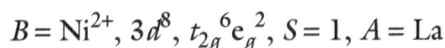

$$B = Ni^{2+}, 3d^8, t_{2g}^6 e_g^2, S = 1, A = La$$

$La_2NiO_{4+\delta}$ where $0 < \delta < 0.25$ has been studied quite intensively. Single crystals can be grown by the optical float zone method and have been employed in neutron-scattering experiments. The magnetic properties are of course a strong function of δ. For $\delta \sim 0$ classic 2D AF behavior is found, including 'rods' of scattering above T_N in ND data.[236] $T_N = 327$ K for $\delta \sim 0$, although there is sample dependence. As with La_2CoO_4, a series of crystallographic transitions occurs as follows: $T > 770$ K ($I4/mmm$), 770–80 K ($Bmab$), $T < 80$ K ($P4_2/ncm$).[237,238] The 80 K transition is first order, while a continuous phase transition occurs at the higher temperature. The magnetic structure is essentially that of Figure 42 but with the Ni^{2+} moments in the *ab*-plane. As for La_2CoO_4, the spins rotate 45° as a result of the 80 K transition.[238] The ordered moment for $Ni^{2+} = 1.68(6)$ μ_B, much smaller than the ~ 2 μ_B expected for $S = 1$.[238]

For $\delta > 0$, i.e. 'hole–doping,' both charge and magnetic order are entwined in the so-called 'stripe' phases, Figure 44. T_N decreases sharply upon hole doping reaching 110 K for $\delta = 0.133$.[239]

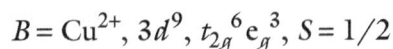

$$B = Cu^{2+}, 3d^9, t_{2g}^6 e_g^3, S = 1/2$$

La_2CuO_4 is of course the parent compound for the first high-temperature cuprate superconductors. As is now very well known (a database search of 'La_2CuO_4' will yield $\sim 4 \times 10^3$ hits), small levels of substitution on the A-site with Sr^{2+} or Ba^{2+} or oxygen excess as in $La_2CuO_4+\delta$ results in hole-induced superconductivity at temperatures $T_c > 30$ K. In this section, attention will be confined to $\delta \sim 0$ systems.

As with the other La_2BO_4 materials, the B = Cu phase has $I4/mmm$ symmetry for $T_1 > 500$ K but transforms to $Cmca$ ($Bmab$) at lower temperatures.[240,241] $T_N = 250(5)$ K with an ordered Cu^{2+} moment

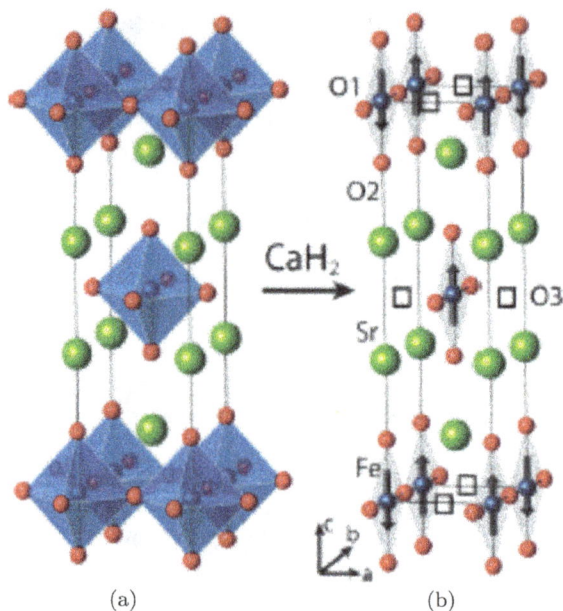

Figure 45.   The reduction of the RP phase Sr_2FeO_4 with CaH_2 resulting in Sr_2FeO_3 with reduced dimensionality.[244] Sr ions are green, Fe ions blue, and O^{2-} ions red. The arrows on the Fe^{2+} ions show the G-Type magnetic structure. Reproduced from reference 244 with permission from the American Chemical Society.

of 0.40(4) μ_B, again much reduced from the ~ 1 μ_B expected for $S = 1/2$.[238] The Cu^{2+} moments lie in the ab-plane. There is evidence that AFLRO persists in $\delta > 0$ materials.[242] In this case, phase separation occurs between a superconducting phase with $T_c = 30$ K and an AF phase with $T_N = 210$ K. As mentioned above, there exists a vast literature of 'doped' cuprates which lies outside the scope of this chapter.

Hydride reduced perovskites

The final section in this already long chapter will concern, briefly, some new perovskite-related materials prepared by hydride reduction of perovskite oxides belonging to some of the families already discussed above. This technique was pioneered by Rosseinsky and Hayward, and there is by now a fairly extensive literature on these phases; thus, only a few examples will be given here.[243] The basic synthetic strategy is to react a perovskite or RP phase with, for example, CaH_2 at moderate temperatures. Often, a new phase with reduced spatial dimensionality results. There are two types of products, one in which O^{2-} ions are removed without H^- substitution and the other in which hydride ions are part of the new structure. An example of the first type is the reduction of Sr_2FeO_4 with CaH_2 to form Sr_2FeO_3, as shown in Figure 45.[244]

Note the square planar coordination of the Fe^{2+} ions in the reduced material. $T_N = 179$ K and the ordered Fe^{2+} moment is 3.1(1) μ_B. Similarly, $SrFeO_3$ can be reduced by CaH_2 to $SrFeO_2$ with infinite layers of FeO_2.[245] Here, $T_N = 473$ K, and again the magnetic structure is G-Type.

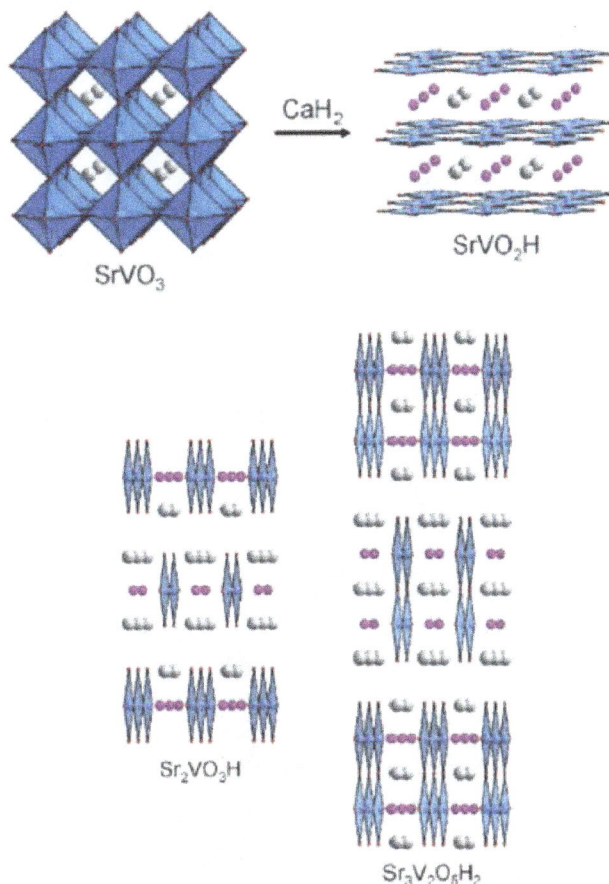

Figure 46. Reaction of RP phases $Sr_{n+1}V_nO_{3n+1}$, $n = \infty$, 1, 2 with CaH_2 resulting in a variety of lower dimensional oxy-hydride phases.[246] Sr ions are grey, V blue, O^{2-} red, and H^- pink. Note that O^{2-}/H^- site ordering occurs. Reproduced from reference 246 with permission from John Wiley and Sons.

An example of the second type is the reduction of members of the RP series $Sr_{n+1}V_nO_{3n+1}$, $n = \infty$, 1, 2, shown in Figure 46.[246]

In most cases, O^{2-}/H^- site ordering occurs, which has important implications for the SE pathways. The symmetry is generally reduced from *Pm-3m* to *P4/mmm* for $SrVO_2H$ and from *I4/mmm* to *Immm* for the RP phases.[244,245]

While the parent phases are mostly metallic or non-magnetic, these new oxy-hydrides usually show some form of magnetic order. $SrVO_2H$, now based on V^{3+}, is reported to show AFLRO with T_N well above 300 K.[246] It has been noted that infinite VO_2 sheets are present in this phase analogous to infinite CuO_2 sheets in $SrCuO_2$. O^{2-}/H^- site ordering leads to some unexpected low dimensional magnetic properties for Sr_2VO_3H, as shown in Figure 47.[247] The absence of *p*-orbitals on H^- favors σ–Type SE pathways and weakens π–Type pathways as shown. The high-temperature susceptibility data above $T_N = 160$ K can be well fitted with a 1D $S = 1$ chain model.[247]

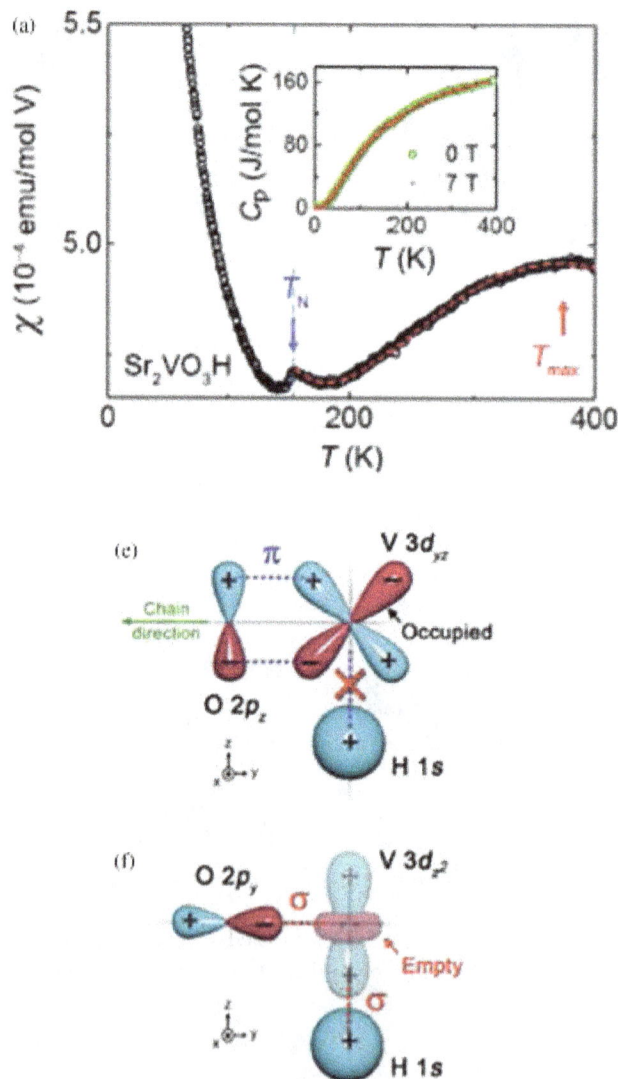

Figure 47.   TOP: Magnetic susceptibility for Sr_2VO_3H showing a broad χ_{max} just below 400 K and $T_N = 160$ K. The dashed red line is a fit to an $S = 1$, 1D chain model. BOTTOM: The role of site ordered H^- ions on SE pathways leading to the 1D magnetism.[247] Reproduced from reference 247 with permission from the American Physical Society.

References

1. Greedan, J. E., Magnetic Oxides. In *Encyclopedia of Inorganic and Bioinorganic Chemistry*, Scott, R. A. Ed. Wiley and Sons: 2017.

2. Goodenough, J. B., *Magnetism and the Chemical Bond.* Interscience Publishers: New York–London, 1963.

3. Goldschmidt, V. M., *Naturwissenschaften* 1926, 21, 477.

4. Glazer, A. M., *Acta. Cryst.* B 1972, 28, 3384; P. M. Woodward, *Acta. Cryst. B* 1997, 53, 32.; Howard, C. J.; Stokes, H. T., *Acta. Cryst. B* 1998, 54, 782.

5. Lufaso M.W.; Woodward, P. M., *Acta. Cryst. B* 2001, 57, 725.

6. Zaanen, J.; Sawatzky, G. A.; Allen, J. W., *Phys. Rev.* Lett. 1985, 55, 418.

7. Koehler, W.C.; wollan, E. O.; Wilkinson, M. K., *Phys. Rev.* 1960, 118, 58.

8. Kanamori, J.; *J. Phys. Chem. Solids* 1959, 10, 87.

9. Inoue, I. H.; Makino, H.; Ishikawa, M.; Husseyn, E.; Rozenberg, M. G., *Physica B* 1997, 237–238, 61.

10. Onoda, M.; Ohta, H.; Nagasawa, H., *Solid State Comm.* 1991, 79, 281.
11. Nishimura, K.; Yamada, I., Oka, K.; Shimakawa, Y.; Azuma, M., *J. Phys. Chem. Solids* 2014, 75, 710.
12. Komarek, A. C.; Moller, T.; Isobe, M.; Drees, Y.; Ulbrich, H.; Azuma, M.; Fernandez-Dıaz, M. T.; Senyshyn, A.; Hoelzel, M.; Andr´e, G.; Ueda, Y.; Gruninger, M.; Braden, M., *Phys. Rev. B* 2011, 84, 125114.
13. MacChesney, J. B.; Williams, H. J.; Potter, J. F.; Sherwood, R. C., *Phys. Rev.* 1967, 64, 779.
14. Bozin, E. S.; Sartbaeva, A.; Zheng, H.; Wells, S. A.; Mitchell, J. F.; Proffen, Th.; Thorpe, M. F.; Billinge, S. J. L., *J. Phys. Chem. Solids* 2008, 69, 2146.
15. Takeda, T., Ohara, S., *J. Phys. Soc.* Jpn. 1974, 37, 275.
16. Wollan, E. O.; Koehler, W. C., *Phys. Rev.* 1955, 100, 545.
17. Ravi, S.; Manoranjan Kar.; Borah S. M.; Krishna, P. S. R., *Cryst. Res. Technol* 2008, 43, 1318.
18. Kawasaki, S.; Takano, M.; Takeda, Y., *J. Solid State Chem.*1996, 121, 174.
19. Takeda, T.; Yamaguchi, Y.; Watanabe, H., *J. Phys. Soc. Jpn.* 1972, 33, 967.
20. Taguchi, H.; Shimada, M.; Koizumi, M.; *J. Solid State Chem.*1979, 29, 221.
21. Bezdicka, P.; Wattiaux, A.; Grenier, J. C.; Pouchard, M.; Hagenmuller, P.; *Z. anorg. allg. Chem.* 1993, 619, 7.
22. Long, Y.; Kaneko, Y.; Ishiwata, S.; Taguchi, Y.; Tokura, Y., *J. Phys. Cond. Matter* 2011, 23, 245601.
23. Ikeda, S. I.; Shirakawa, N., *Physica C* 2000, 341–348, 785.
24. Macquart, R. B.; Kennedy, B. J.; Avdeev, M., *J. Solid State Chem.* 2010, 183, 249.
25. Nassif, V.; Carbonio, R. E., *J. Solid State Chem.* 1999, 146, 266.
26. Thorogood, G. J.; Avdeev, M.; Carter, M. L.; Kennedy, B. J.; Ting, J.; Wallwork, K. S., *Dalton Trans.* 2011, 40, 7228.
27. Rodriguez, E. E.; Poineau, F.; Llobet, A.; Kennedy, B. J.; Avdeev, M.; Thorogood, G. J.; Carter, M. L.; Seshadri, R.; Singh, D. J.; Cheetham, A. K.; *Phys Rev. Lett.* 2011, 106, 067201.
28. Mravlje, J.; Aichhorn, M.; Georges, A., *Phys Rev. Lett.* 2012, 108, 197202.
29. Felner, I.; Asaf, U., *Physica B* 2003, 337, 310.
30. Yoshii, K.; Abe, H., *Physica B* 2002, 312–313, 791.
31. Callaghan, A.; Moeller, C. W.; Ward, R., *Inorg. Chem.* 1966, 9, 1572.
32. Kanbayashi, A.; *J. Phys. Soc. Jpn.* 1976, 41, 1876.
33. Kennedy, B. J.; Hunter B. A.; Hester, J. R., *Phys. Rev. B* 2002, 65, 224103.
34. Blanchard, P. E. R.; Reynolds, E.; Kennedy, B. J.; Kimpton, J. A.; Avdeev, M.; Belik, A. A., *Phys. Rev B.* 2014, 89, 214106.
35. Bickel, M.; Goodman, G. L.; Soderholm L.; Kanellakopulos, B., *J. Solid State Chem.* 1988, 76, 178.
36. Hinatsu, Y.; Itoh, M.; Edelstein, N., *J. Solid State Chem.* 1996, 132, 337.
37. Rosov, N.; Lynn, J. W.; Lin, Q.; Cao, G.; O'Reilly, J.W.; Pernambuco-Wise, P.; Crow, J. E., *Phys. Rev. B* 1992, 45, 982.
38. Tezuka, K.; Hinatsu, Y.; Shimojo, Y.; Morii, Y., *J. Phys. Cond. Matter* 1998, 10, 11703.
39. Hinatsu, Y., *J. Alloys Compd.* 1993, 193, 113.
40. Matthais, B. T.; Bozorth, R.M.; van Vleck, J. J., *Phys. Rev. Lett.* 1961, 7, 160.
41. McGuire, T. R.; Shafer, M. W.; Joenk, R. J.; Alperin, H. A., Pickart, S. J., *J. Appl. Phys.* 1966, 37, 981.
42. Greedan, J. E., *J. Phys. Chem. Solids* 1971, 32, 819.
43. Chien, C.L.; Greedan, J. E., *Phys. Lett. A* 1971, 36, 197.
44. Kasuya, T., *IBM J. Res. Dev.* 1970, 14, 214.
45. Greedan, J. E.; Chien, C. L.; Johnston, R. G., *J. Solid State Chem.* 1976, 19, 155.
46. Ranjan, R.; Nabi, H. S.; Pentcheva, R., *J. Appl. Phys.* 2009, 105, 053905.
47. McCarthy, G. J.; Greedan, J. E., *Inorg. Chem.* 1975, 4, 772.
48. Zong, Y.; Fujita, K.; Akamatsu, H.; Murai, S.; Tanaka, K., *J. Solid State Chem.* 2010, 183, 168.
49. Avdeev, M.; Kennedy, B. J.; Kolodiazhnyi, T., *J. Phys. Cond. Matter* 2014, 26, 095401.
50. Akamatsu, H.; Fujita, K.; Hayashi, H.; Kawamoto, T.; Kumagai, Y.; Zong, Y.; Iwata, K.; Oba, F.; Tanaka, I.; Tanaka, K., *Inorg. Chem.* 2012, 51, 4560.
51. Zubov, V. G.; Tyutyunnik, A. P.; Perelaiev, V. A.; Shveikin, G. P.; Kohler, J.; Kremer, R. K.; Simon, A.; Svensson, G.; *J. Alloys Compd.* 1995, 226, 24.
52. Li, L.; Zhou, H.; Yan, J.; Mandrus, D.; Keppens, V., *APL Mater.* 2014, 2, 110701.
53. Greedan, J. E., *AIP Conf. Proc.* 1974, 18, 749.
54. Sana, R.; Sundaresan, A.; Rao, C. N. R., *Mater. Horiz.* 2014, 1, 20.
55. Bousquet, E.; Cano, A., *J. Phys. Cond. Matter* 2016, 28, 123001.
56. Burzo, E., *Magnetic Properties of Non-Metallic Inorganic Compounds Based on Transition Elements: Perovskites 1 (Parts α and β).* Landolt-Bornstein New Series. Vol. 111/27F1α and β, Ed. Wijn, H. P. J.; Springer: 1996.
57. Shen, H.; Chen, Z.; Hong, F.; Xu, J.; Yuan, S., Cao, S., Wang, X., *Appl. Phys. Lett.* 2013, 103, 192404.

58. Zhao, H. J.; Iniguez, J.; Chen, X.M.; Bellaiche, L., *Phys Rev. B.* 2016, 93, 014417.

59. Bertaut, E. F., *Spin Configuration of Ionic Structures: Theory and Practice. Magnetism* Vol 111. Eds. Rado G. T.; Suhl, H., Academic Press: New York and London, 1963, 150–205.

60. Deng, G.; Guo, P.; Ren, W.; Cao, S.; Maynard-Casely, H. E.; Avdeev, M.; MacIntyre, G. J., *J. Appl. Phys.* 2015, 117, 164105.

61. Greedan, J. E., *J. Less-Comm. Met.* 1985, 111, 335, and references therein.

62. Turner C. W.; Greedan, J. E., *J. Solid. State. Chem.* 1980, 34, 203.

63. Garrett, J. D.; Greedan J. E.; MacLean, D. A., *Mater. Res. Bull.* 1981, 16, 145.

64. Goral J. P.; Greedan, J. E., *J. Magn. Magn. Mater.* 1983, 37, 315.

65. Cwik, M.; Lorenz, T.; Baier, J.; Muller, R.; Andre, G.; Bouree, F.; Lichtenberg, F.; Freimuth, A.; Schmitz, R.; Muller-Hartmann, E.; Braden, M., *Phys. Rev. B* 2003, 68, 060401R.

66. Ulrich, C.; Khaiullin, G.; Okamoto, S.; Reehuis, M.; Ivanov, A.; He, H.; Taguchi, Y.; Tokura, Y.; Keimer, B., *Phys. Rev. Lett.* 2002, 89, 167202.

67. Turner, C. W.; Greedan, J. E.; Collins, M. F., *J. Magn. Magn. Mater.* 1980, 20, 165.

68. Turner, C. W.; Greedan, J. E., *J. Magn. Magn. Mater.* 1983, 36, 242.

69. Sefat, A. S.; Greedan J. E.; Cranswick, L., *Phys. Rev. B* 2006, 74, 104418.

70. Cheng, J.-G.; Sui, Y.; Zhou, J.-S.; Goodenough, J. B.; Su, W. H., *Phys. Rev. Lett.* 2008, 101, 087205.

71. Takubo, K.; Shimuta, M.; Kim, J. E.; Kato, K.; Takata, M.; Katsufuji, T.; *Phys. Rev. B* 2010, 82, 020401.

72. Mochizuki M.; Imada, M., *J. Phys. Soc. Jpn.* 2004, 73, 1833.

73 Ren, Y.; Nugroho, A. A.; Menovsky, A. A.; Strempfer, J.; Rutt, U.; Iga, F.; Takabatake, T.; Kimball, C. W., *Phys. Rev. B* 2003, 67, 014017.

74. Yan, J.-Q.; Zhou, J.-S.; Goodenough, J. B., *Phys. Rev. B* 2005, 72, 094412.

75. Nguyen, H. C.; Goodenough, J. B.; *J. Solid State Chem.* 1995, 119, 24.

76. Reehuis, M.; Ulrich, C.; Prokes, K.; Mat'as, S.; Fujioka, J.; Miyasaka, S.; Tokura, Y.; Keimer, B., *Phys. Rev. B* 2011, 83, 064404.

77. Zhang, Q.; Singh, K.; Simon, C.; Tung, L. D.; Balakrishnan, G.; Hardy, V., *Phys. Rev. B* 2014, 90, 024418.

78. Mahajan, A. V.; Johnston, D. C.; Torgeson, D. R.; Borsa, F., *Phys. Rev. B* 1992, 46, 10966.

79. Ussi-Esko, K.; Malm, J.; Imamura, N.; Yamauchi, H.; Karppinen, M., *Mater. Chem. Phys.* 2008, 112, 1029.

80. Quezel, S.; Rossat-Mignod, J.; Bertaut, E. F., *Sol. State Comm.* 1974, 14, 941.

81. Rodriguez-Carvajal, J.; Hennion, M.; Moussa, F.; Moudden, A. H., *Phys. Rev. B* 1998, 57, R3189.

82. Moussa, F.; Hennion, M.; Rodriguez-Carvajal, J.; Moudden, H.; Pinsard, L.; Revcolevschi, A., *Phys. Rev. B* 1996, 54, 15149.

83. Millis, A. J., *Phys. Rev. B* 1997, 55, 6405.

84. Munoz, A.; Alonso, J. A.; Martinez-Lope, M. J.; Garcia-Munoz, J. K.; Fernandez-Diaz, M. T., *J. Phys. Cond. Matter* 2000, 12, 1361.

85. Munoz, A.; Alonso, J. A.; Casais, M. T.; Martinez-Lope, M. J.; Martinez, J. L.; Fernandez-Diaz, M. T., *J. Phys. Cond. Matter* 2002, 14, 3285.

86. Ramirez, A. P., *J. Phys. Cond. Matter* 1997, 9, 8171.

87. Tokura, Y.; Tomioka, Y., *J. Magn. Magn. Matter* 1999, 200, 1.

88. Salamon, M. B.; Jaime, M., *Rev. Mod. Phys.* 2001, 73, 83.

89. Rao, C. N. R.; Arulraj, A.; Cheetham, A.; Raveau, B., *J. Phys. Cond. Matter* 2000, 12, 83.

90. Okuda, T.; Kimura, T.; Kuwahara, H.; Tomioka, Y.; Asamitsu, A.; Okimoto, Y.; Saitoh E.; Tokura, Y., *Mater. Sci. Eng.* 1999, B63, 163.

91. Senaris-Rodriguez, M. A.; Goodenough, J. B., *J. Solid State Chem.* 1995, 116, 224.

92. Durand, A. M.; Belanger, D. P.; Booth, C. H.; Ye, F.; Chi, S.; Fernandez-Baca, J.A.; Bhat, M., *J. Phys. Cond. Matter* 2013, 25, 382203.

93. Belanger, D. P.; Keiber, T.; Bridges, F.; Durand, A. M.; Mehta, A.; Zeng, H.; Mitchell, J. F.; Borzenets, V., *J. Phys. Cond. Matter* 2016, 28, 025602.

94. Yan, J.-Q.; Zhou J.-S.; Goodenough, J. B., *Phys. Rev. B* 2004, 69, 134409.

95. Tachibana, M.; Yoshida, T.; Kawaji. H.; Atake, T., Takayama-Muromachi, E., *Phys. Rev. B* 2008, 77, 094402.

96. Demazeau, G.; Marbeuf, A.; Pouchard, M.; Hagenmuller, P., *J. Solid State Chem.* 1971, 3, 582.

97. Alonso, J. J.; Martinez-Lope, M. J.; Casais, M. T.; Aranda, M. A. G.; Fernandez-Diaz, M.T., *J. Am. Chem. Soc.* 1999, 121, 4754.

98. Garcia-Munoz, J. L.; Rodriguez-Carvajal, J.; LaCorre, P., *Phys. Rev. B* 1994, 50, 978.

99. Fernandez-Diaz, M. T.; Alonso, J. A., Martinez-Lope, M. J.; Casais, M. T.; Garcia-Munoz, J. L., *Phys. Rev. B* 2001, 64, 144417.

100. Munoz, A.; Alonso, J. A.; Martinez-Lope, M. J.; Fernandez-Diaz, M. T., *J. Solid State Chem.* 2009, 182, 1982.

101. Demazeau, G.; Parent, C.; Pouchard, M.; Hagenmuller, P., *Matter Res. Bull.* 1972, 7, 913.

102. Karppinen, M.; Yamauchi, H.; Suematsu, H.; Isawa, I.; Nagano, M.; Itti, R.; Fukunaga, O., *J. Solid State Chem.* 1997, 130, 213.

103. Chen, B.-H.; Walker, D.; Suard, E.; Scott, B. A.; Mersey, B.; Hervieu, M.; Raveau, B., *Inorg. Chem.* 1995, 34, 2077.

104. Sinclair, A.; Rodgers, J. A.; Topping, C. V.; Misek, M.; Stewart, R. D.; Kockelmann, W.; Bos J.-W.; Attfield, J. P., *Angew. Chem. Int. Ed.* 2014, 53, 8343.

105. Bouchard, R. J.; Weiher, J. F., *J. Solid State Chem.* 1972, 4, 80.

106. Taniguchi, T.; Iizuka, W.; Nagata, Y.; Uchida, T.; Samata, H., *J. Alloys Compd.* 2003, 350, 24.

107. Asai, S.; Furuta, N.; Yasui, Y.; Terasaki, I., *J. Phys. Soc. Jpn.* 2011, 80, 104705.

108. Guo, H.; Manna, K.; Luetkens, H.; Hoetzel, M.; Komarek, AS. C., *Phys. Rev. B* 2016, 94, 205128.

109. Terasaki, I.; Asai, S.; Taniguchi, H.; Okazaki, R.; Yasui, Y.; Ikemoto Y.; Moriwaki, T., *J. Phys. Cond. Matter* 2017, 29, 235802.

110. Knizek, K.; Hejtmanek, J.; Marysko, M.; Jirak, Z.; Bursik, J., *Phys. Rev. B* 2012, 85, 134401.

111. Vassala, S.; Karppinen, M., *Prog. Solid State Chem.* 2015, 43, 1.

112. Anderson, M. T.; Greenwood, K. B.; Taylor, G. A.; Poeppelmeier, K. R., *Prog. Solid State Chem.* 1993, 22, 197.

113. Howard, C. J.; Kennedy, B. J.; Woodward, P. M., *Acta.Cryst. B* 2003, 59, 463.

114. Patterson, F. K.; Moeller, C. W.; Ward, R., *Inorg. Chem.* 1963, 2, 196.

115. Kobayashi, K.-I.; Kimura, T.; Sawada, H.; Terakura, K., *Nature* 1998, 677.

116. Borges, R. P.; Thomas, R. M.; Coey, J.; M. D.; Suryanarayanan, R.; Ben-Dor, L.; Pinsault-Gaudart, L.; Revcolevschi, A., *J. Phys. Cond. Matter* 1999, 11, L445.

117. Saitoh, T.; Nakatake, M.; Kakizaki, A.; Nakajima, H.; Morimoto, O.; Xu, S.; Morimoto, Y.; Hamada, N.; Ajura, Y., *Phys. Rev. B,* 2002, 66, 035112.

118. Sanchez, D.; Alonso, J. A.; Garcia-Hernandez, M.; Martinez-Lope. M. J.; Martinez, J. L.; Mellergård, A., *J. Magn. Magn. Mater.* 2002, 242–245, 729.

119. Moreno, M. S.; Gayone, J. E.; Abbate, M.; Caneiro, A.; Niebieskikwiat, D.; Sanchez, R. D.; de Siervo, A.; Landers, R.; Zampieri, G., *Solid State Commun.* 2001, 120, 161.

120. Karppinen, M.; Yamauchi, H.; Yasukawa, Y.; Linden, J.; Chan, T. S.; Liu, R. S.; Chen, J. M., *Chem. Mater.* 2003, 15, 4118.

121. Alamelu, T.; Varadaraju, U. V.; Venkatesan, M.; Douvalis, A. P.; Coey, J. M. D., *J. Appl. Phys.* 2002, 91, 8909.

122. Westerberg, W.; Lang, O.; Ritter, C.; Felser, C.; Tremel, W.; Jakob, G., *Sol. State Comm.* 2002, 122, 201.

123. Abe, M.; Nakagawa, T.; Nomura, S., *J. Phys. Soc. Jpn.* 1973, 35, 1360.

124. Rammeh, N.; Ehrenberg, H.; Fuess, H.; Cheikkh-Rouhou, A., *Phys. Stat. Sol.* 2006, 9, 3225.

125. Kato, H.; Okuda, T.; Okimoto, Y.; Tomioka, Y.; Takenoya, Y.; Ohkubo, A.; Kawasaki M.; Tokura, Y., *Apl. Phys. Lett.* 2002, 81, 328.

126. Fuoco, L.; Rodriguez, D.; Peppel, T.; Maggard, P. A., *Chem. Mater.* 2011, 23, 5409.

127. Hauser, A. J.; Solitz, J. R.; Dixit, M.; Williams, R. E. A.; Susner, M. A.; Peters, B.; Mier, L. M.; Gustafson, T. L.; Sumpton, M. D.; Fraser, H. L.; Woodward, P. M.; Yang, F. Y., *Phys. Rev. B* 2012, 85, 161201(R).

128. Ramirez, A. P., *Ann. Rev. Mater. Sci.* 1994, 24, 453.

129. Smart, J. S., *Effective Field Theories of Magnetism.* W.B. Saunders Co.: Philadelphia and London, 1966, 64, 77.

130. Lefmann, K.; Rischel, C., *Eur. Phys. J. B* 2001, 21, 313.

131. Shull, C. G.; Strausser.; Wollan, E. O., *Phys. Rev.* 1951, 83, 333.

132. van Vleck, J. H., *The Theory of Electric and Magnetic Susceptibilities.* Oxford University Press: 1965, 181ff.

133. Simonet, V.; Ballou, R.; Robert, J.; Canals, B.; Hippert, F.; Bordet, P.; Lejay, P.; Fouquet, P.; Ollivier, J.; Braithwaite, D.; *Phy. Rev. Lett.* 2008, 100, 237204.

134. Azad, A. K.; Ivanov, S. A.; Eriksson, S.-G.; Eriksen, J.; Rundlof, H.; Mathieu, R.; Svedlindh, P., *Matter Res. Bull.* 2001, 36, 2215.

135. Khattah, C. P.; Cox, D. E.; Wang, F. Y. Y., *J. Solid State Chem.* 1976, 17, 323.

136. Azad, A. K.; Ivanov, S. A.; Eriksson, S.-G.; Eriksen, J.; Rundlof, H.; Mathieu, R.; Svedlindh, P., *J. Magn. Magn. Mater.* 2001, 237, 124.

137. Aharen, T.; Greedan, J. E.; Ning, F.; Imai, T.; Michaelis, V.; Kroeker, S.; Zhou, H.; Wiebe, C. R.; Cranswick, L. M. D., *Phys. Rev. B* 2009, 80(13), 134423.

138. Battle, P. D.; Jones, C.; W., *J. Solid State Chem.* 1989, 78,108.

139. Kermarrec, E.; Marjerrison, C. A.; Thompson, C. M.; Maharaj, D.; Levin, K.; Kroeker, S.; Granroth, G. E.; Flacau, R.; Yamani, Z.; Greedan, J. E.; Gaulin, B. D., *Phys Rev. B* 2015, 91, 075133.

140. Battle, P. D.; Grey, C. P.; Hervieu, M.; Martin, C.; Moore, C.A.; Paik, Y., *J. Solid State Chem.* 2003 ,175, 20.

141. Thompson, C. M.; Marjerrison, C. A.; Sharma, A. Z.; Wiebe, C. R.; Maharaj, D.; Sala, G.; Flacau, R.; Hallas, A.; Cai, Y.; Gaulin, B. D.; Luke, G. M.; Greedan, J. E., *Phys. Rev. B* 2016, 93, 014431.

142. Iwanaga, D.; Inaguma, Y.; Itoh, M., *Mat. Res. Bull.* 2000, 35, 449.

143. Aharen, T.; Greedan, J. E.; Bridges, C. A.; Aczel, A. A.; Rodriguez, J.; MacDougall, G.; Luke, G. M.; Michaelis, V. K.; Kroeker, S.; Wiebe, C. R.; Zhou, H.; Cranswick, L. M. D., *Phys. Rev. B* 2010, 81, 064436.

144. Thompson, C. M.; Carlo, J. P.; Flacau, R.; Aharen, T.; Leahy, I. A.; Pollichemi, J. R.; Munsie, T. J. S.; Munevar, J.; Medina, T.; Cheung, S.; Goko, T.; Uemura , Y. J.; Luke, G. M.; Greedan. J. E., *J. Phys. Cond. Matter* 2014, 26, 306003.

145. Marjerrison, C. A.; Thompson, C. M.; Sharma, A. Z.; Hallas, A. M.; Wilson, M. N.; Munsie, T. J. S.; Flacau, R.; Wiebe, C. R.; Gaulin, B. D.; Luke, G. M.; Greedan, J. E., *Phys. Rev. B* 2016, 94, 134429.

146 Wiebe, C. R.; Greedan, J. E.; , Luke, G. M.; Gardner, J. S., *Phys. Rev B* 2002, 65, 144413.

147. Marjerrison, C. A.; Thompson, C. M.; Sala, G.; Maharaj, D. D.; Kermarrec, E.; Cai, Y.; Hallas, A.; Munsie, T. J. S.; Granroth, G. E.; Flacau, R.; Greedan, J. E.; Gaulin, B. D.; Luke, G. M., *Inorg.Chem.* 2016, 55, 10701.

148. Brannik, K. G.; Ehrenerg, H.; Dehn, J. K.; Fuess, H., *Sol. State Sci.* 2003, 5, 235.

149. Stitser, K. E.; Smith, M. D.; zur Loye, H.-C., *Sol. State Sci.* 2002, 4, 311.

150. Aharen, T.; Greedan, J. E.; Imai, T.; Bridges, C. A.; Aczel, A. A.; Rodriguez, J.; MacDougall, G.; Luke, G. M.; Michaelis, V. K.; Kroeker, S.; Wiebe, C. R.; Zhou, H.; Cranswick, L. M. D., *Phys. Rev. B* 2010, 81, 224409.

151. de Vries, M. A.; Mclaughlin, A. C.; Bos, J.-W., *Phys. Rev. Lett.* 2010, 104, 177202.

152. Carlo, J. P.; Clancy, J. P.; Aharen, T.; Yamani, Z.; Ruff, J. P. C.; Van Gastel, G. J.; Granoth, G. E.; Greedan, J. E.; Dabkowska, H. A.; Gaulin, B. D., *Phys. Rev. B* 2011, 84 , 100404.

153. Chen, G.; Pereira, R.; Balents, L., *Phys. Rev. B* 2010, 82, 174440.

154. Feng, H. L.; Yamaura, K.; Tjeng, L. H.; Jansen, M., *J. Solid State Chem.* 2016, 243, 119.

155. Dodds, T.; Choy, T.-P.; Kim, Y. B., *Phys. Rev. B* 2011, 84, 104439.

156 King, G., Woodward, P. M., *J. Mater. Chem.* 2010, 20, 5785.

157. King, G.; Wayman, L. M.; Woodward, P. M., *J. Solid State Chem.* 2009, 182, 1319.

158. King, G.; Wills, A. S.; Woodward Patrick, M., *Phys. Rev. B* 2009, 79, 224428.

159. Chapman, J. P.; Attfield, J. P.; Molgg, M.; Friend, C. M.; Beales, T. P., *Angew. Chem. Int. Ed. Engl.* 1996, 35, 2482.

160. Karen, P.; Woodward, P. M., *J. Mater. Chem.* 1999, 9, 789.

161. Vogt, T.; Woodward, P. M.; Karen, P.; Hunter, B. A.; Henning, P.; Moodenbaugh, A. R., *Phys. Rev. Lett.* 2000, 84, 2969.

162. Millange, F.; Suard, E.; Caignaert, V.; Raveaul, B., *Mater. Res. Bull.* 1999, 34, 1.

163. Karen, P.; Woodward, P. M.; Linden, J.; Vogt, T.; Studer, A.; Fischer, P., *Phys. Rev. B* 2001, 64, 214405.

164. Perca, C.; Pinsard-Gaudart L.; Daoud-Aladine, A.; Fernandez-Dıaz, M, T.; Rodrıguez-Carvajal, J., *Chem. Mater.* 2005, 17, 1835–1843.

165. Karen, P.; Kjekshus, A.; Huang, Q.; Lynn, J. W.; Rosov, N.; Sora, N.; Karen, V. L.; Mighell, A. D.; Santoro, A., *J. Solid State Chem.* 1998, 136, 21.

166. Swinnea, J. S.; Steinfink, H., *J. Matter. Res.* 1987, 2, 424.

167. Petitgrand, D.; Collin, G., *Physica C.* 1988, 153–155, 192.

168. Casalta, H.; Schleger, P.; Montfrooij, W.; Andersen, N. H.; Lebech, B.; Liang, R.; Hardy, W. N., *J. Magn. Magn. Mater.* 1995, 140–144, 1297.

169. Chen, W.-T.; Mizumaki, M.; Seki, H.; Senn, M. S.; Saito, T.; Kan, D.; Attfield, J. P.; Shimakawa, Y., *Nature Comm.* 2014, DOI: 10.1038/ncomms4909.

170. Arevalo-Lopez, A, M.; McNally, G. M.; Attfield, J. P., *Angew. Chem. Int. Ed.* 2015, 54, 12074.

171. Parsons, T. G.; D'Hondt, H.; Hadermann, J.; Hayward, M. A., *Chem. Mater.* 2009, 21, 5527.

172. Takeda, T.; Yamaguchi, Y.; Tomiyoshi, S.; Fukase, M.; Sugitomo, M.; Watanabe, H., *J. Phys. Soc. Jpn.* 1968, 24, 446.

173. Ramezanipour, F.; Cowley, B.; Derakhshan, S.; Greedan, J. E.; Cranswick, L. M. D., *J. Solid State Chem.* 2009,182, 153.

174. Ramezanipour, F.; Greedan, J. E.; Grosvenor, A. P.; Britten, J. F.; Cranswick, L. M. D.; Garlea, V. O., *Chem. Mater.* 2010, 22, 6008.

175. Wright, A. J.; Palmer, H. M.; Anderson, P. A.; Greaves, C., *J. Mater. Chem.* 2002, 12, 978.

176. Battle, P. D.; Bollen, S. K.; Gibb, T. C.; Matsuo, M., *J. Solid State Chem.* 1991, 90, 42.

177. Gibb, T. C.; Matsuo, M., *J. Solid State Chem.* 1990, 88, 485.

178. Takeda, T.; Yamaguchi, Y.; Tomiyoshi, S.; Watanabe, H.; Yamamoto, H., *J. Phys. Soc. Jpn.* 1969, 26, 1320.

179. Schmidt, M.; Campbell, S. J., *J. Solid State Chem.* 2001, 156, 292.

180. Battle, P. D.; Gibb, T. C.; Lightfoot, P., *J. Solid State Chem.* 1988, 76, 334.

181. D'Hondt, H.; Abakumov, A. M.; Hadermann, J.; Kalyuzhnaya, A. S.; Rozova, M. G.; Antipov, E. V.; Van Tendeloo, G., *Chem. Mater.* 2008, 20, 7188.

182. Ramezanipour, F.; Greedan, J. E.; Cranswick, L. M. D.; Garlea, V. O.; Donaberger, R. L.; Siewenie, J., *J. Am. Chem. Soc.* 2012, 134, 3215.

183. de Jongh L. J., *Magnetic Properties of Layered Transition Metal Compounds*. Ed. de Jongh, L.J., Kluwer Academic Publishers: Dordrecht, Boston,London, 1990, 19.

184. Cortes-Gil, R.; Ruiz-Gonzalez, M. L.; Alonso, J. M.; Vallet-Regi, M.; Hernando, A.; Gonzalez-Calbet, J. M.; *Chem. Eur. J.* 2007, 13, 4246.

185. Hodges, J. P.; Short, S.; Jorgensen, J. D.; Xiong, X.; Dabrowski, B.; Mini, S. M.; Kimball, C.W., *J. Solid State Chem.* 2000, 151, 190.

186. Schmidt, M.; Hofmann. M.; Campbell, S. J., *J. Phys. Cond. Matter* 2003,15, 869.

187. Ramezanipour, F.; Greedan, J. E.; Siewenie, J.; Proffen, Th.; Ryan, D. H.; Grosvenor, A. P.; Donaberger, R. L., *Inorg. Chem.* 2011, 50, 7779.

188. Ramezanipour, F.; Greedan, J. E.; Siewenie, J.; Donaberger, R. L.; Turner, S.; Botton, G. A., *Inorg. Chem.* 2012, 51, 2638.

189. King, G.; Ramezanipour, F.; Llobet A.; Greedan, J. E., *J. Solid State Chem.* 2013, 198, 407.

190. Ledesert, M.; Labbe, Ph.; McCarroll, W. H.; Leligny H.; Raveau, B., *J. Solid State Chem.* 1993, 105, 143.

191. Wiebe, C. R.; Gourrier, A.; Langet, T.; Britten, J. F.; Greedan, J. E., *J. Solid State Chem.* 2000, 151, 31.

192. Chi, L.; Green, A. E. C.; Hammond, R.; Wiebe, C. R.; Greedan, J. E., *J. Solid State Chem.* 2003, 170, 165.

193. Green, A. E. C.; Wiebe, C. R.; Greedan, J. E., *Solid State Sci.* 2002, 4, 305.

194. Cuthbert, H. L.; Greedan, J. E.; Lachlan Cranswick., *J. Solid State Chem.* 2006, 179, 1938.

195. Ramezanipour, F.; Derahkshan, S.; Greedan, J. E.; Cranswick, L. M. D., *J. Solid State Chem.* 2008, 181, 3366.

196. Chi, L.; Swainson, I. P.; Greedan, J. E., *J. Solid State Chem.* 2004, 177, 3086.

197. Morrow, R.; Samanta, K.; Saha Dasgupta, T.; Xiong, J.; Freeland, J. W.; Haskel, D.; Woodward, P. M., *Chem. Mater.* 2016, 28, 3666.

198. Navarro, R., Application of High- and Low-Temperature Series Expansions to Two- Dimensional Magnetic Systems. In *Magnetic Properties of Layered Transition Metal Compounds*. Ed. de Jongh, L. J., Kluwer Academic Publishers: Dordrecht, Boston, London, 1990, 105–190.

199. Bieringer, M.; Greedan, J. E., *J. Mater. Chem.* 2002, 12, 279.

200. Fisher, M. E., *Phil Mag.* 1962, 7, 1731.

201. Greedan, J. E.; McCarthy, G. J., *Mater. Res. Bull.* 1972, 7, 531.

202. Chien, C.-L., De Benedetti S.; De, F.; Barros, S., *Phys. Rev. B* 1974, 10, 3913.

203. Zhou, H. D.; Jo, Y. J.; Fiore Carpino, J.; Munoz, G. J.; Wiebe, C. R.; Cheng, J. G.; Rivadulla, F.; Adroja, D. T., *Phys. Rev. B* 2010, 81, 212401.

204. Yamauchi, I.; Nawa, K.; Hiraishi, M.; Miyazaki, M.; Koda, A.; Kojima, K. M.; Kadono, R.; Nakao, H.; Kumai, R.; Murakami, Y.; Ueda, H.; Yoshimura, K.; Takigawa, M., *Phys. Rev. B* 2015, 92, 064408.

205. Jackeli, G.; Khaliullin, G., *Phys. Rev. Lett.* 2009, 103, 067205.

206. Liu, G.; Greedan, J. E.; Gong, W.-H., *J. Solid State Chem.* 1993, 105, 78.

207. Baikie, T.; Ahmad, Z.; Srinivasan, M.; Maignan, A.;Pramana, S. S.; White, T. J., *J. Solid State Chem.* 2007, 180, 1538.

208. Bouloux, J.-C.; Soubeyroux, J.-L.; LeFlem G.; Hagenmuller, P., *J. Solid State Chem.* 1981, 38, 34.

209. Dann, S. E.; Weller, M. T.; Currie, D.; Thomas, M. F.; Al-Rawwas, Ahmed D., *J. Mater. Chem.* 1993, 3, 1231.

210. Wang, X. L.; Takayama-Muromachi1, E., *Phys. Rev. B* 2005, 72, 064401.

211. Li, Q.; Yuan, X.; Xing, L.; Xu, M., *Nat. Sci. Rep.* 2016, 6, 27712. DOI: 10.1038/srep27712.

212. Maeno, Y.; Hashimoto, H.; Yoshida, K.; Nishizaki, S.; Fujita, T.; Bednorz, J. G.; Lichtenberg, F., *Nature* 1994, 372, 532.

213. Yoshida, K.; Maeno, Y.; Nishizaki S.; Fujita, T., *Physica C* 1996, 263, 519.

214. Subramanian, M. A.; Crawford, M. K.; Harlow, R.; Ami, T.; Fernandez-Baca, J. A.; Wang, Z. R.; Johnston, D. C., *Physica C*, 1994, 235–240, 743.

215. Perry, R. S.; Baumberger, F.; Balicas, L.; Kikugawa, N.; Ingle, N. J. C.; Rost, A.; Mercure, J. F.; Maeno, Y.; Shen, Z. X.; Mackensie, A. P., *New J. Phys.* 2006, 8, 175.

216. Nagai, I.; Shirakawa, N.; Umeyama, N.; Ikeda, S.-I., *J. Phys. Soc. Jpn.* 2010, 79, 114719.

217. Kim, B. J.; Jin, H.; Moon, 1S. J.; Kim, J.-Y.; Park, B.-G.; Leem, C. S.; Jaejun Yu, Noh, T. W.; Kim, C.; Oh, S.-J.; Park, J.-H.; Durairaj, V.; Cao, G.; Rotenberg, E., *Phys. Rev. Lett.* 2008, 101, 076402.

218. Ye, F.; Chi, S.; Chakoumakos, B. C.; Fernandez-Baca, J. A.; Qi, T.; Cao, G., *Phys. Rev. B* 2013, 87, 144405.

219. Dhital, C.; Hogan, T.; Yamani, Z.; de la Cruz, C.; Chen, X.; Khadka, S.; Ren, Z.; Wilson, S. D., *Phys. Rev. B* 2013, 87, 144405.

220. Sung, N. H.; Gretarsson, H.; Proepper, D.; Porras, J.; Le Tacon, M.; Boris, A. V.; Keimer, B.; Kim, B. J., *Phil. Mag.* 2016, 96, 413.

221. Greedan, J. E.; Gong, W. H., *J. Alloys Compd.* 1992, 180, 281.

222. Suzuki, N.; Noritake, T.; Hioki, T., *J. Alloys Compd.* 2014, 612, 114.

223. Dun, Z. L.; Garlea, V. O.; Yu, C.; Ren, Y.; Choi, E. S.; Zhang, H. M.; Dong, S.; Zhou, H. D., *Phys. Rev. B* 2014, 89, 235131.

224. Aso, K., *J. Phys. Soc. Jpn.* 1978, 44, 1083.

225. Collomb, A.; Samaras, D.; Joubert, J. C., *Phys. Stat. Sol.* 1978, 50, 635.

226. Kao, T.-H.; Sakurai, H.; Kolodiazhnyi, T.; Suzuki, Y.; Okabe, M.; Asaka, T.; Fukuda, K.; Okubo, S.; Ikeda, S.; Hara, S.; Sakurai, T.; Ohta H.; Yang, H.-D., *J. Mater. Chem. C* 2015, 3, 3452.

227. Soubeyroux, J. L.; Courbin, P.; Fournes, L.; Fruchart, D.; Le Flem, G., *J. Solid State Chem.* 1980, 31, 313.

228. Qureshi, N.; Ulbrich, H.; Sidis, Y.; Cousson, A.; Braden, M., *Phys. Rev. B* 2013, 87, 054433.

229. Liu, Y. Y.; Chen, X. M.; Liu, X. Q., *Sol. State Comm.* 2005, 136, 576.

230. Guo, H.; Hu, Z.; Pi, T.-W.; Tjeng, L. H.; Komarek, A. C., *Crystals* 2016, 6, 98.

231. Cava, R. J.; Batlogg, B.; Palstra, T. T.; Krajewski, J. J.; Peck Jr., W.F.; Ramirez, A. P.; Rupp, L. W., *Phys. Rev. B* 1999, 43, 1229.

232. Schilling, A.; Dell'Amore1, R.; Karpinski, J.; Bukowski, Z.; Medarde, M.; Pomjakushina, E.; Muller, K. A., *J. Phys. Condens. Matter* 2009, 21, 015701.

233. Le Dreau, L.; Prestipino, C.; Hernandez, O.; Schefer, J.; Vaughan, G.; Paofai, S.; Perez-Mato, J. M.; Hosoya, S.; Paulus, W., *Inorg. Chem.* 2012, 51, 9789.

234. Yamada, K.; Matsuda, M.; Endoh, Y.; Keimer, B.; Birgeneau, R. J.; Onodera, S.; Mizusaki, J.; Matsuura, T.; Shirane, G., *Phys. Rev. B* 1989, 39, 2336.

235. Babkevich, P.; Prabhakaran, D.; Frost, C. D.; Boothroyd, A. T., *Phys. Rev. B* 2010, 82, 184425.

236. Nakajima, K.; Yamada, K.; Hosoya, S.; Endoh, Y.; Birgeneau, R. J., *Z. Phys. B* 1995, 96, 479.

237. Lander, G. H.; Brown, P. J.; Spalek, J.; Honig, J. M., *Phys. Rev. B* 1989, 40, 4463.

238. Rodriguez-Carvajal, J.; Fernandez-Diaz, M.T.; Martinez, J. L., *J. Phys. Condens. Matter* 1991, 3, 3215.

239. Wochner, P.; Tranquada, J. M.; Buttrey, D. J.; Sachan, V., *Phys. Rev. B* 1998, 57, 1066.

240. Yang, B. X.; Mitsuda, S.; Shirane, G.; Yamaguchi, Y.; Yamauichi, H.; Syono, Y., *J. Phys. Soc. Jpn.* 1987, 56, 2283.

241. Jorgensen, J. D.; Schuttler, H.-B.; Hinks, D. G.; Capone II, D.W.; Zhang, K.; Brodsky, M., *Phys. Rev. Lett.* 1987, 58, 1024.

242. Gnezdilov, V. P.; Yu. Pashkevich, G.; Tranquada, J. M.; Lemmens, P.; Guntherodt, G.; Yeremenko1, A. V.; Barilo, S. N.; Shiryaev, S. V.; Kurnevich, L. A.; Gehrin7, P. M., *Phys. Rev. B* 2004, 69, 174508.

243. Hayward, M. A.; Green, M. A.; Rosseinsky, M. J.; Sloan, J., *J. Am. Chem. Soc.* 1999, 121, 8843.

244. Tassel, C.; Seinberg, L.; Hayashi, N.; Ganesanpotti, S.; Ajiro, Y.; Kobayashi, Y.; Kageyama, H., *Inorg. Chem.* 2013, 52, 6096.

245. Tsujimoto, Y.; Tassel, C.; Hayashi, N.; Watanabe, T.; Kageyama, H.; Yoshimura, K.; Takano, M.; Ceretti, M.; Ritter, C.; Paulus, W., *Nature* 2007, 450, 1062.

246. Romero, F. D.; Leach, A.; Moller, J. S.; Foronada, F.; Blundell, S. J.; Hayward, M. A., *Angew. Chem. Int. Ed.* 2014, 53, 7556.

247. Bang, J.; Matsuishi, S.; Maki, S.; Yamaura, J.-I.; Hiarishi, M.; Takeshita, S.; Yamauichi, I.; Kojima, K.; Hosono, H., *Phys. Rev. B* 2015, 92, 064414.

3

Ultraviolet and Deep-Ultraviolet Nonlinear Optical Materials

P. Shiv Halasyamani

Department of Chemistry, University of Houston, 112 Fleming Building,
Houston, TX 77204-5003, USA

Introduction and Background

Inorganic materials capable of generating coherent radiation through second-harmonic generation (SHG) processes have been used for years to produce radiation at wavelengths where laser sources are not conveniently available.[1] Frequency doubling, or SHG, is a nonlinear optical (NLO) process whereby a specific wavelength of light is converted to half its original, $\lambda \to 1/2\lambda$, or with respect to frequency, $\omega \to 2\omega$.[2] The generation of coherent radiation from the ultraviolet (UV) to the infrared (IR) is accomplished through SHG. Materials capable of generating coherent radiation in the UV, < 300 nm, and deep UV, < 200 nm, are of academic and technological interest. Materials capable of generating these wavelengths are used in attosecond pulse generation, laser systems, photolithography, semiconductor manufacturing, and advanced instrument development.[3–6] In the deep UV, coherent light is also needed for laser-based ultrahigh resolution photoemission spectrometry and photoelectron emission microscopy. Coherent radiation in the deep UV is possible with excimer lasers, e.g. ArF excimer at 193 nm and F_2 excimer at 157 nm. However, attributable to their ease in handling, narrow bandwidth, energy density, peak power density, and tunability, solid-state lasers are often preferred at these wavelengths.

The focus of this chapter is on the design requirements, birefringence and phase-matching, characterization, and specific inorganic materials with NLO applications in the UV and deep UV. Throughout this chapter, the fundamental laser wavelength will be 1064 nm (Nd:YAG). Thus, UV and deep-UV radiation, 266 nm and 177.3 nm, respectively, may be generated through cascaded frequency generation. That is fourth harmonic generation (FOHG) — 1064 nm/4 = 266 nm, 1064 nm <u>SHG</u> 532 nm <u>SHG</u> 266 nm, and sixth harmonic generation (6thHG), 1064 nm <u>SFG</u> 355 nm <u>SHG</u> 177.3 nm where SFG is sum frequency generation using 1064 nm and 532 nm radiation.

Current State of the Field — 266 nm and 177.3 nm Radiation Generation

For the generation of 266 nm radiation, four materials have been reported: $YAl_3(BO_3)_4$ (YAB),[7] $Li_2B_4O_7$ (LB4),[8,9] $CsLiB_6O_{10}$ (CLBO),[10,11] and β-BaB_2O_4 (β-BBO).[12] YAB is reported to have a large SHG coefficient, $d_{11} = 1.7$ pm/V;[13] however, its low transmittance in the 200–320 nm range prohibits its practical application.[14] With LB4, high-quality crystals have been grown[9]; however, the SHG efficiency is very small, i.e. $d_{31} = 0.15$ pm/V.[15] Thus, CLBO and β-BBO remain as viable materials for the generation of 266 nm radiation in both academia and industry. High-quality crystals of both are commercially available through a variety of sources. However, both materials have their drawbacks. $CsLiB_6O_{10}$ is hygroscopic, whereas β-BaB_2O_4 has a large birefringence resulting in walk-off issues that lead to a reduction in the SHG conversion efficiency. With 177.3 nm generation, the situation is even more limiting. Only two materials, $KBe_2BO_3F_2$ (KBBF) and $RbBe_2BO_3F_2$ (RBBF), have been reported to generate coherent radiation at 177.3 nm.[16,17] There are, however, two major issues with both materials. First, toxic BeO must be used in the synthesis. Second, and more importantly, a layered crystal structure is exhibited by both materials with weak A^+–F^- interactions. Attributable to the layering and weak interactions, single crystals no larger than 4 mm along the optical axis have been grown.[16,18]

In order to place the aforementioned materials in context, the spectrum from approximately 225 nm to 125 nm is shown in Figure 1. Included in this figure are those mentioned previously, as well as a few others of note, as well as recently discovered materials that may have UV NLO applications. All of the materials shown in Figure 1 will be discussed later in this chapter. Also included in the figure are the energies for the

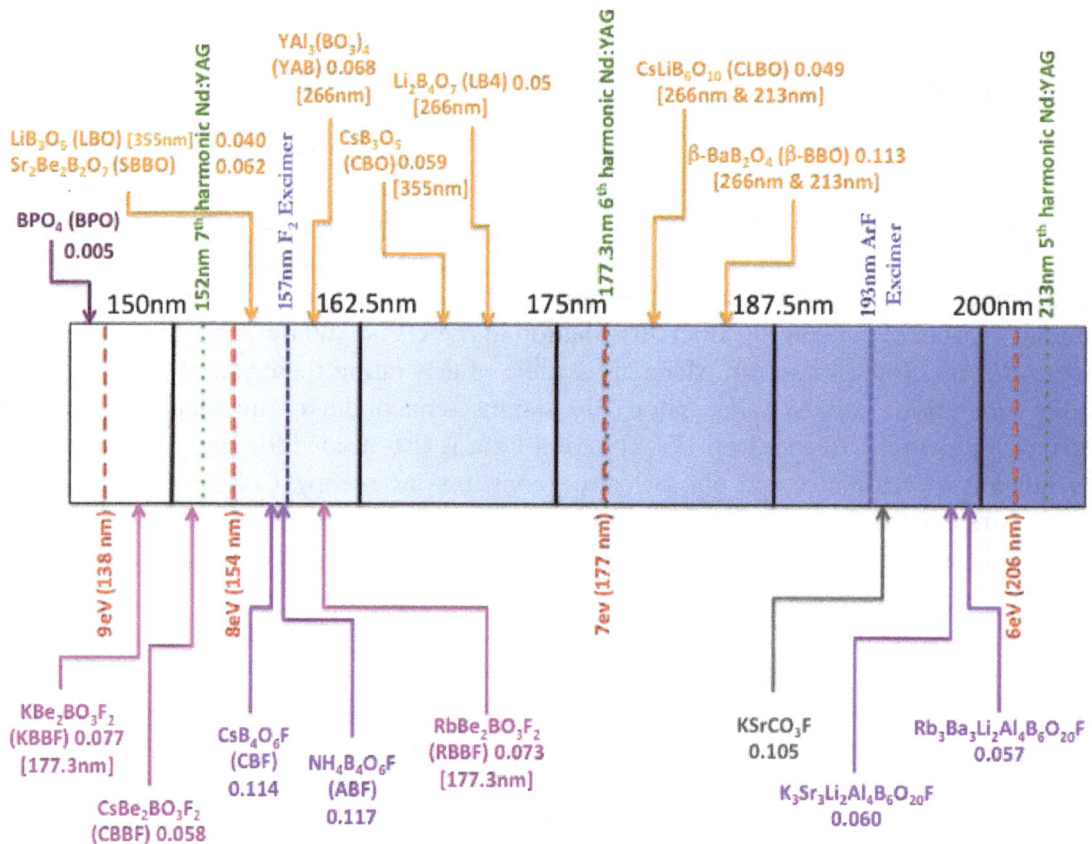

Figure 1. Deep-UV spectrum from approximately 225 nm to 125 nm showing the absorption edges of some UV and deep-UV NLO materials, along with their birefringence, reported lasing wavelengths (in square brackets), as well as excimer wavelengths (F2 and ArF), fifth, sixth, and seventh harmonics of Nd:YAG (1064 nm), and the 6, 7, 8, and 9 eV energies. Adapted from Ref. 143.

fifth, sixth, and seventh harmonic generation from 1064 nm radiation (dotted green lines), as well as the ArF and F_2 excimer wavelengths (dashed blue lines). The wavelengths for 6, 7, 8, and 9 eV (dashed red lines) are also shown. The arrows denote the absorption edge for each material, whereas the numbers in square brackets are the wavelengths where coherent radiation has been reported from a fundamental 1064 nm source.

Design Requirements

There are six fundamental requirements to design a new UV or deep-UV NLO material.

1. Crystallographic Non-centrosymmetry (NCS)

FOHG and 6^{th}HG are only possible in materials that lack a center of symmetry, or an inversion center. In other words, the material must crystallize in an acentric crystal class. A Venn diagram depicting the 21 NCS crystal classes and their relationships is shown in Figure 2. The NCS crystal classes that have the correct symmetry for SHG and higher-order harmonic generation are shown in the bottom oval. All of the NCS crystal classes exhibit the correct symmetry for SHG except for the cubic crystal class 432. As shown previously,[19] in crystal class 432, all of the SHG coefficients, d_{ij}'s are zero. It should be noted that although the cubic crystal classes -43 m (T_d) and 23 (T) exhibit the correct symmetry for SHG, phase-matching is not possible in materials that are found in these crystal classes. We will be discussing phase-matching in much greater detail later in this chapter.

2. Wide Transparency Range

A wide transparency range is required for FOHG and 6^{th}HG, as 266 nm and 177.3 nm radiation is generated. Given the required absorption edge, i.e. for FOHG $\lambda \leq 250$ nm ($E_g > 5.8$ eV) and for 6^{th}HG $\lambda \leq 170$ nm ($E_g > 7.3$ eV), the material in question must be colorless. Thus, d–d or f–f transitions are not possible. As such, most transition metals and lanthanide cations cannot be incorporated. Also, many main group

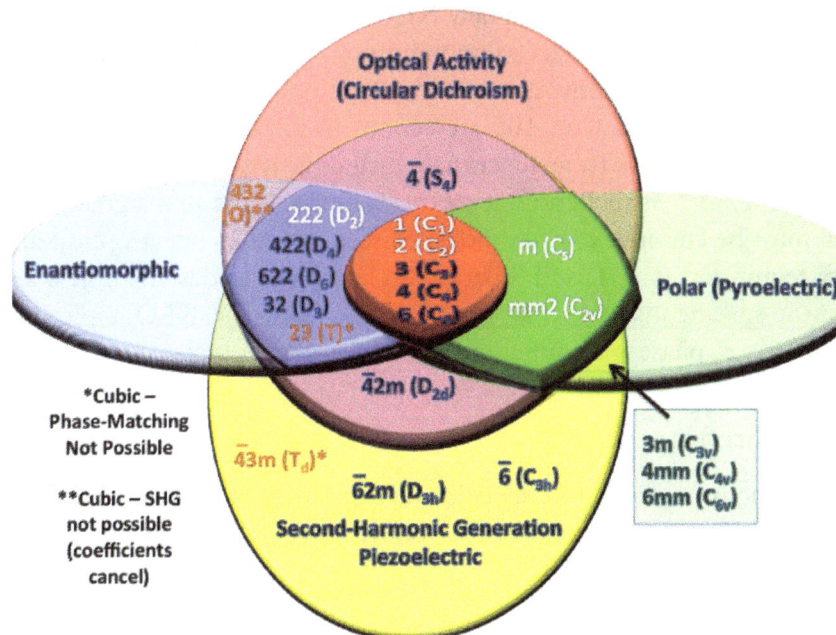

Figure 2.   Venn diagram depicting the relationships between the non-centrosymmetric crystal classes. Note that although SHG is symmetry allowed in crystal classes 23(T) and -43m(T_d), phase-matching is not as both crystal classes are cubic. Uniaxial crystal classes are shown in white, biaxial in blue, and cubic in orange. Adapted from Ref. 144.

cations have absorption edges above 250 nm. Even with these limitations, there are a host of cations and anion groups that may be used. These include alkali and alkaline-earth cations, borates, borate fluorides, borate phosphates, and carbonate fluorides.

3. Large SHG Coefficients — d_{ij}

Directly related to the magnitude of the SHG coefficients, d_{ij}, is the SHG conversion efficiency. The larger the d_{ij} value(s), the larger the SHG conversion efficiency. In the UV and deep-UV regimes, a $d_{ij} > 0.39$ pm/V is required. Notably, 0.39 pm/V is the d_{36} value for KH_2PO_4 (KDP),[20] that is the standard reference used in these measurements. Individual d_{ij} coefficients may be measured using the Maker fringe technique on parallel-plate single crystals that have been cut, indexed, and polished.[21,22] A detailed description of the technique will be discussed later in this chapter.

4. Moderate Birefringence

One of the most important requirements for a technologically viable UV or deep-UV NLO material is a moderate birefringence, i.e. $\Delta n = 0.070$–0.100. A moderate birefringence is required for index phase-matching — an optical direction in the material where the refractive indices are equal, i.e. where $n(\omega) = n(2\omega)$.[23] In other words, where the fundamental and second-harmonic waves have the same propagation speed. With this condition, all materials crystallizing in the cubic NCS classes, 23 (T) and –43 m (T_d), are excluded from index phase-matching applications attributable to their isotropic nature, i.e. $\Delta n = 0$. A moderate birefringence and index phase-matching will be discussed in greater detail later in the chapter.

5. Chemical Stability and Large Laser Damage Threshold (LDT)

Perhaps the most obvious requirement is chemical stability for the NLO material in question. To have any NLO applications, the material must be air- and moisture-stable and not require any toxic reagents for its synthesis. With respect to the laser damage threshold, a single crystal of the material should be able to withstand 5 GW/cm^2 for a single nanosecond pulse at 1064 nm.

6. Growth of Large High-Quality Single Crystals

The large high-quality single crystal growth of a new NLO material is likely the greatest obstacle toward any technological applications. With respect to size, a minimum of 5 mm in all three dimensions is necessary, and for high quality, a rocking curve measurement around a Bragg reflection should generate a peak with a full-width at half-maximum (fwhm) of less than 100 arcseconds (0.0278°), and ideally less than 50 arcseconds (0.0139°). There are a variety of ways to grow crystals including top-seeded solution growth, Czochralski, Bridgman, and floating zone.[24–30] As such, no one method is suitable for all materials. In addition to growing the crystal, the crystal must be cut, indexed, and polished to measure its linear-optical and NLO properties.

In order to satisfy requirements 3, 4, and 5, large single crystals as described in 6 are required.

Before describing the specific materials that have UV and deep-UV NLO applications, a discussion on refractive index, birefringence, phase-matching, and the Maker fringe method is necessary.

Refractive Index and Birefringence

The difference in the refractive indices at a specific wavelength is defined as the birefringence, and only occurs in anisotropic crystal systems, i.e. non-cubic crystal classes. The non-cubic crystal classes can be divided into two groups: uniaxial and biaxial.

Uniaxial	Biaxial
Hexagonal	Orthorhombic
Tetragonal	Monoclinic
Trigonal	Triclinic

In uniaxial systems, a beam of light is split into ordinary (*o*) and extraordinary (*e*) beams, i.e. n_o and n_e. The value of the birefringence is the difference between n_o and n_e, $\Delta n = |n_o - n_e|$. When $n_e > n_o$ ($n_e < n_o$), the material is termed a positive (negative) uniaxial crystal.[31,32] With biaxial crystal systems, the situation is somewhat more complex. There are three refractive indices, n_x, n_y, and n_z, along the *x*-, *y*-, and *z*-axis, respectively. By convention, $n_x < n_y < n_z$. The value of the birefringence is taken to be the difference between the maximum values of n_z and n_x. If $n_z - n_y > n_y - n_x$ ($n_z - n_y < n_y - n_x$), the material is termed positive (negative) biaxial optical crystal.

Refractive Index Measurements

In order to determine the birefringence, the refractive index at a variety of wavelengths must be measured. One of the most accurate methods for measuring the refractive index on a solid is the minimum deviation technique[33] on a large single crystal. With this method, a crystal is grown and cut as a wedge or prism, polished, and indexed. Monochromatic sources from the UV to the IR are used to measure the refractive indices at several wavelengths.[34–37] An example was recently reported for $Ba_3(ZnB_5O_{10})PO_4$ (BZBP).[38] The polished and indexed wedge shaped crystal for BZBP is shown in Figure 3(a). As seen, a relatively large crystal, several milli meters in all three dimensions, is needed for the measurement. In addition, the vertex angle is inversely related to the birefringence, i.e. the smaller the birefringence, the larger the vertex angle needs to be to obtain the refractive index values at a range of wavelengths. The refractive index data, from

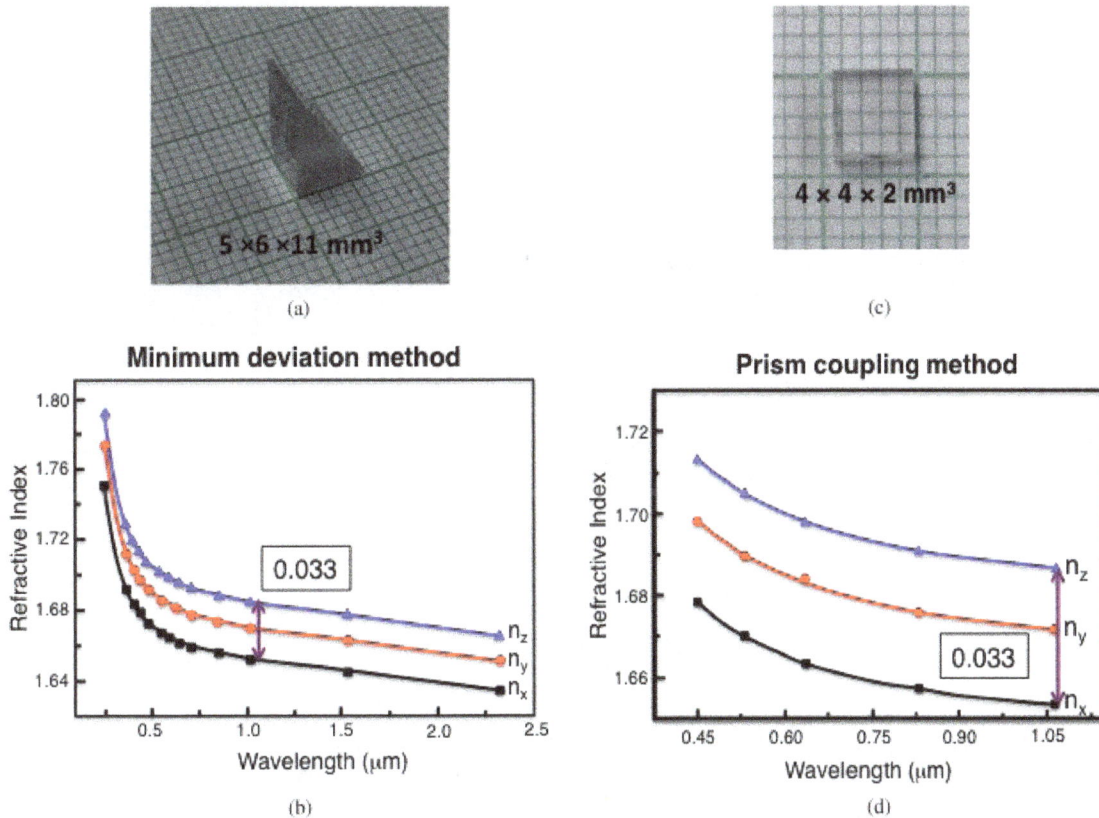

Figure 3. (a) Single crystal of $Ba_3(ZnB_5O_{10})PO_4$ (BZBP) cut into a wedge, polished, and indexed for refractive index measurements by the (b) minimum deviation method. (c) Single crystal of BZBP cut, polished, and indexed for refractive index measurements by the (d) prism coupling method. In both (b) and (d), the birefringence at 1064 nm is determined to be 0.033. The curves are fit to the refractive index data from the Sellmeier equations.

the UV to the IR, for BZBP using the minimum deviation technique are shown in Figure 3(b). As BZBP is biaxial, three refractive indices, n_x, n_y, and n_z are measured. Also since $n_z-n_y < n_y-n_x$, BZBP is a negative biaxial crystal. It is also possible to determine the refractive indices using smaller crystals. A prism coupler sold by Metricon Co. (http://www.metricon.com; no endorsement is implied) enables one to determine the refractive indices at up to five wavelengths using a smaller crystal, 2–4 mm (see Figure 3(c). With this technique, the crystal must also be cut, polished, and indexed. Using these crystals and the prism coupling method, the refractive index data for BZBP was also acquired (see Figure 3(d)). With both measurements, the refractive index data may be fit as a function of wavelength to the Sellmeier equation[39]:

$$n_i^2 = A + \frac{B}{\lambda^2 - C} - D\lambda^2$$

where λ is the wavelength in μm, and A, B, C, and D are the Sellmeier parameters. As there are four Sellmeier parameters to be determined, at least a minimum of five wavelengths are required. Thus, the minimum deviation technique is more accurate compared with the prism coupling method. The solid lines in Figures 3(b) and 3(d) are the fits to the refractive index data for BZBP using the Sellmeier equation. Once the Sellmeier equations are known, it is possible to calculate the refractive index at any wavelength, and subsequently allows one to determine the phase-matching wavelength range.

Phase-Matching in Uniaxial and Biaxial Systems

Phase-matching, or the material's ability to be phase-matched, is of crucial importance for NLO applications. We will exclusively be discussing angle phase-matching, which is sometimes called index matching, angle matching, or birefringence matching, but hereafter simply phase-matching. Our discussion on phase-matching is distinct from quasi-phase-matching[40–42] and non-critical phase-matching methods.[43,44] Also in our discussion, we are going to assume the fundamental wavelength, with angular frequency ω, is 1064 nm; therefore, the SHG wavelength, with angular frequency 2ω, is 532 nm. In the SHG process, there are fundamental and second-harmonic wave vectors, $k(\omega)$ and $k(2\omega)$, respectively, propagating through the crystal. The wave vectors, $k(\omega)$ and $k(2\omega)$, may be defined as follows[45]:

$$k(\omega) = \omega \times n(\omega)/c \qquad k(2\omega) = 2\omega \times n(2\omega)/c$$

with $n(\omega)$ and $n(2\omega)$ defined as the refractive index at 1064 nm, and 532 nm, respectively, and c is the speed of light. In general $n(\omega) \neq n(2\omega)$, and there is phase-mismatch, Δk, between the two waves, i.e. $\Delta k = k(2\omega) - 2k(\omega) = (4\pi/\lambda)[n(2\omega) - n(\omega)]$.

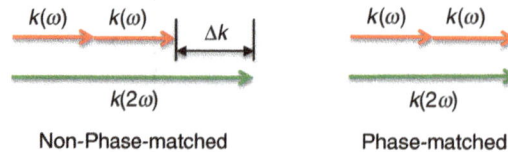

Non-Phase-matched   Phase-matched

The most efficient SHG will occur with $\Delta k = 0$. This is known as the phase-matching condition.[46] In real materials, k is a vector quantity, and the direction of propagation in the material where $n(\omega) = n(2\omega)$, i.e. the refractive index of the fundamental equals the refractive index of the second harmonic, should be considered. Phase-matching occurs in both uniaxial and biaxial systems, and two types may occur — Type I and Type II. Both will be discussed. A convenient manner in which to understand phase-matching is to begin by representing the 3D index ellipsoid as a 2D projection of the refractive index.

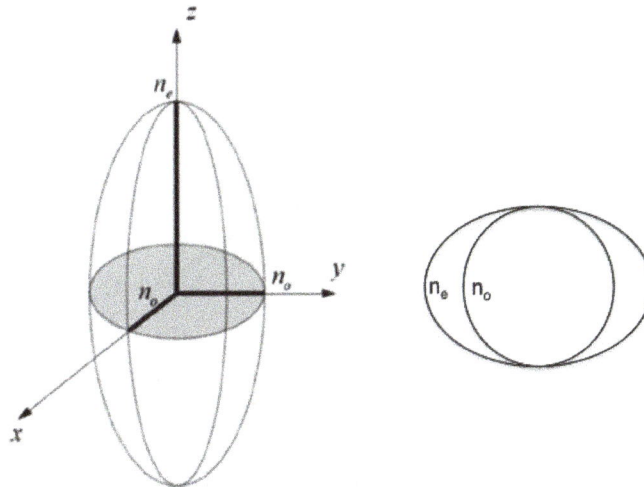

Uniaxial

Type I Phase-Matching — Positive and Negative Uniaxial: The projections of the refractive indices in the *xz*-plane at ω and 2ω, 1064 nm and 532 nm, respectively, are shown in Figures 4(a) and 4(b). The representation is for a positive uniaxial crystal, $n_e > n_o$. A tetragonal crystal system is assumed, and thus the optic axis is parallel to the C_4 axis. When Figures 4(a) and 4(b) are 'combined', the phase-matching angle can be determined (Figure 4(c)). Recall, that for phase-matching $n(\omega) = n(2\omega)$. As seen in Figure 4(c), this occurs when $n_e(\omega)$ (dashed red ellipse) intersects with $n_o(2\omega)$ (solid green circle). This is shown by a green dot. Another way to describe $n(\omega) = n(2\omega)$ is by $e(\omega) + e(\omega) \rightarrow o(2\omega)$ (eeo), i.e. two extraordinary waves at the fundamental wavelength combine to produce an ordinary wave at the second harmonic, i.e. Type I phase-matching for positive uniaxial system is often described as eeo. The Type I phase-matching angle, $\theta_m(I)$, is taken from the optic axis to the green arrow. The phase-matching 'point' in 3D forms symmetrical cones (Figure 5). The phase-matching angle, $\theta_m(I)$, can be calculated as follows[47]:

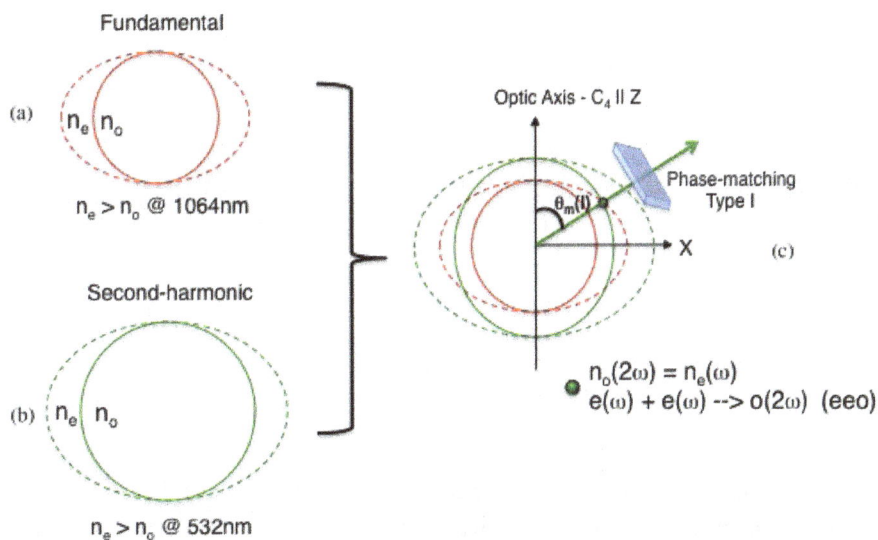

Figure 4. Refractive index projections at 1064 nm (red) and 532 nm (green) are shown for a positive uniaxial crystal. These curves are combined and a Type I phase-matching angle, $\theta_m(I)$, is determined. Type I phase-matching occurs when $e(e) + e(\omega) \rightarrow o(2\omega)$ (eeo). Adapted from Ref. 23.

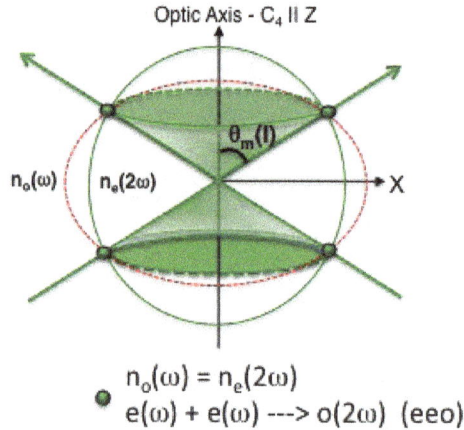

Figure 5.   Symmetric cones, shown in green, for Type I phase-matching for positive uniaxial crystals. Adapted from Ref. 23.

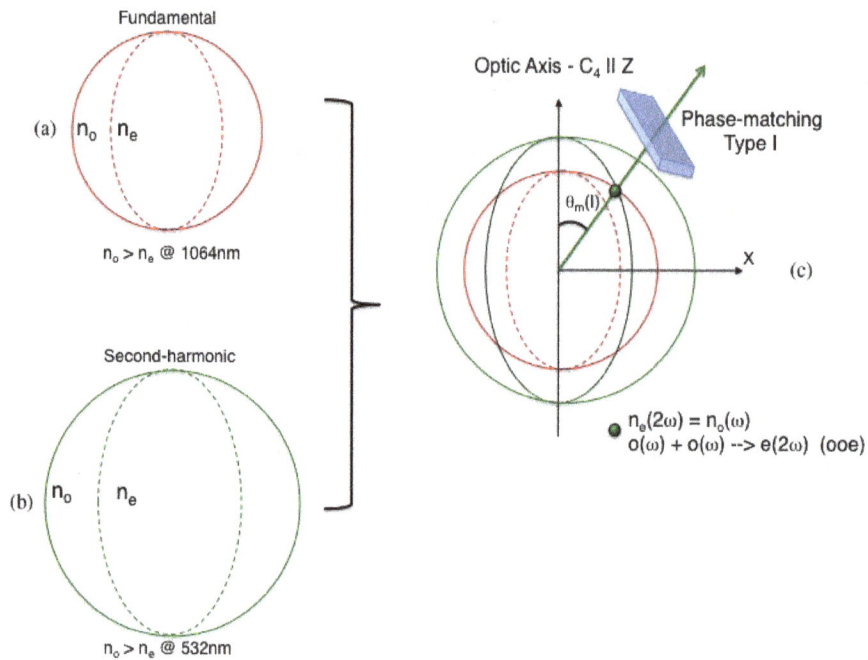

Figure 6.   Refractive index projections at 1064 nm (red) and 532 nm (green) are shown for a negative uniaxial crystal. These curves are combined and a Type I phase-matching angle, $\theta_m(I)$, is determined. Type I phase-matching occurs when $o(\omega) + o(\omega) \rightarrow e(2\omega)$ (*ooe*). Adapted from Ref. 23.

$$\sin^2(\theta_m) = \left(\frac{n_e(\omega)}{n_0(2\omega)} \right)^2 \frac{n_0^2(\omega) - n_0^2(2\omega)}{n_0^2(\omega) - n_0^2(\omega)}$$

where refractive indices $n_e(\omega)$, $n_o(\omega)$, and $n_o(2\omega)$ are determined experimentally as described previously.

For a tetragonal negative uniaxial crystal, the situation is similar (see Figure 6). As before, the solid and dashed red and green curves represent the ordinary and extraordinary waves at 1064 and 532 nm, respectively. Recall that for a negative uniaxial crystal, $n_o > n_e$. With this situation, $n(\omega) = n(2\omega)$ when $n_o(\omega) = n_e(2\omega)$, i.e. the solid red circle intersects with the green dashed ellipse, or $o(\omega) + o(\omega) \rightarrow e(2\omega)$ (*ooe*) — the green dot in Figure 6. That is, two ordinary waves at the fundamental wavelength combine to produce an

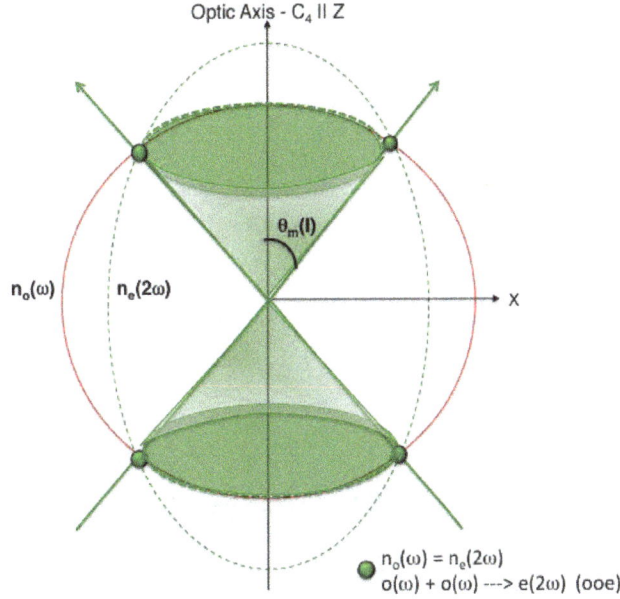

Figure 7. Symmetric cones, shown in green, for Type I phase-matching for negative uniaxial crystals. Adapted from Ref. 23.

extraordinary wave at the second harmonic. Analogous to the positive uniaxial situation, symmetrical phase-matching cones are also observed (see Figure 7). The phase-matching angle, $\theta_m(I)$, for a negative uniaxial crystal can be calculated as follows[47]:

$$\sin^2(\theta_m) = \left(\frac{n_e(2\omega)}{n_0(\omega)}\right)^2 \frac{n_0^2(2\omega) - n_0^2(\omega)}{n_0^2(2\omega) - n_0^2(2\omega)}$$

where the refractive indices $n_e(2\omega)$, $n_0(\omega)$, and $n_0(2\omega)$ are determined experimentally.

Type II Phase-Matching: In some NLO materials, a second symmetrical phase-matching cone is observed. With Type II phase-matching, phase-matching occurs when $[n_e(\omega) + n_0(\omega)]/2 = n_e(2\omega)$ or $[n_e(\omega) + n_0(\omega)]/2 = n_0(2\omega)$. The shape of the $[n_e(\omega) + n_0(\omega)]/2$ curve for both the positive and negative uniaxial situations is an ellipse.

Positive Uniaxial: For positive uniaxial crystals, Type II phase-matching occurs when $[n_e(\omega) + n_0(\omega)]/2 = n_0(2\omega)$ (see Figure 8). A tetragonal system is assumed, and as before, the solid and dashed red and green curves represent the ordinary and extraordinary waves, respectively, at 1064 nm (red) and 532 nm (green). With Type II phase-matching, an extraordinary and ordinary wave from the fundamental wavelength combine to produce an ordinary wave at the second-harmonic, i.e. $e(\omega) + o(\omega) \rightarrow o(2\omega)$ (eoo). The calculation of the Type II phase-matching angle is more complicated as $[n_e(\omega) + n_0(\omega)]/2 = n_0(2\omega)$. In this situation,[47]

$$\frac{1}{2}\left(\frac{n_0(\omega)n_e(\omega)}{\sqrt{n_0^2(\omega)\sin^2\theta_m + n_e^2(\omega)\cos^2\theta_m}} + n_0(\omega)\right) = n_0(2\omega)$$

assuming $n_0(\omega)$, $n_e(\omega)$, and $n_0(2\omega)$ are determined experimentally, one can numerically solve this nonlinear equation for θ_m.

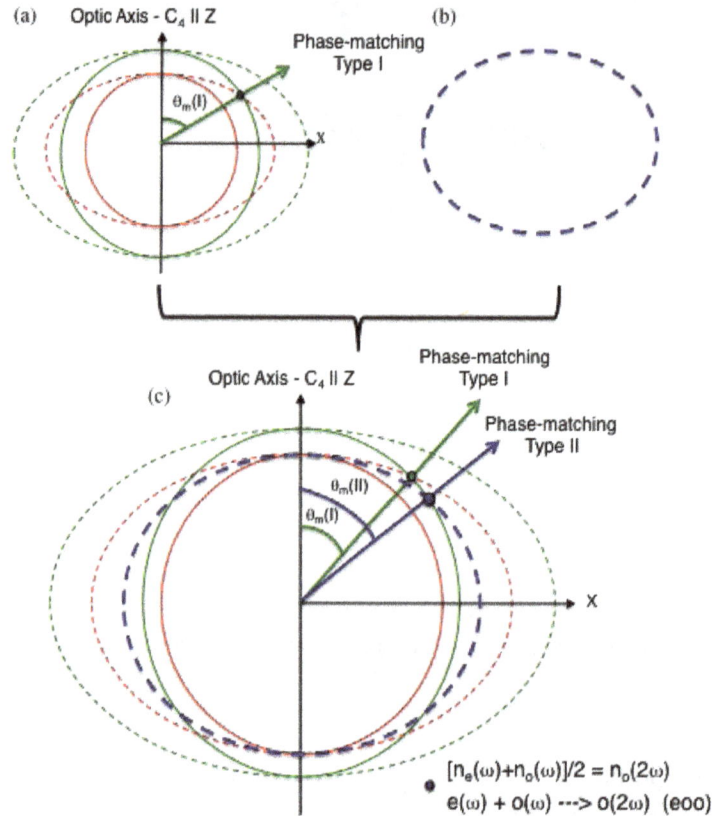

Figure 8. (a) Type I phase-matching for a positive uniaxial crystal and (b) an ellipse representing $[n_e(\omega) + n_o(\omega)]/2$. These curves are combined in (c), and the Type II phase-matching angle, $\theta_m(II)$, is determined (blue curve and arrow). The Type I phase-matching angle, $\theta_m(I)$, is also shown (green curve and line). Type II phase-matching occurs when $e(\omega) + o(\omega) \to o(2\omega)$ (*eoo*). Adapted from Ref. 23.

Negative Uniaxial: For negative uniaxial crystals, Type II phase-matching occurs when $[n_e(\omega) + n_o(\omega)]/2 = n_e(2\omega)$ (see Figure 9). That is, an extraordinary and ordinary wave from the fundamental wavelength combine to produce an extraordinary wave at the second harmonic, i.e. $e(\omega) + o(\omega) \to e(2\omega)$ (*eoe*). Similar to positive uniaxial situation, calculating theta for Type II phase-matching in a negative uniaxial crystal is possible. One can determine $\theta_m(II)$ by solving the following equation assuming that $n_o(\omega)$, $n_e(\omega)$, $n_o(2\omega)$, and $n_e(2\omega)$ have been determined experimentally.

$$\frac{1}{2}\left(\frac{n_0(\omega)n_e(\omega)}{\sqrt{n_0^2(\omega)\sin^2\theta_m + n_e^2(\omega)\cos^2\theta_m}} + n_0(\omega) \right) = \frac{n_0(2\omega)n_e(2\omega)}{\sqrt{n_0^2(2\omega)\sin^2\theta_m + n_e^2(2\omega)\cos^2\theta_m}}$$

As with Type I phase-matching, for Type II phase-matching symmetrical cones are observed. The Type II phase-matching cones (blue) along with the Type I phase-matching cones (green) are shown in Figure 10.

A summary of the phase-matching conditions and angle calculations for uniaxial systems is given in Table 1.

Biaxial

In biaxial systems, the situation is more cumbersome compared with uniaxial materials owing to the three independent refractive indices from which phase-matching can be achieved. The phase-matching condition for biaxial crystals is, technically, similar to uniaxial crystals, i.e. the intersection of the refractive index

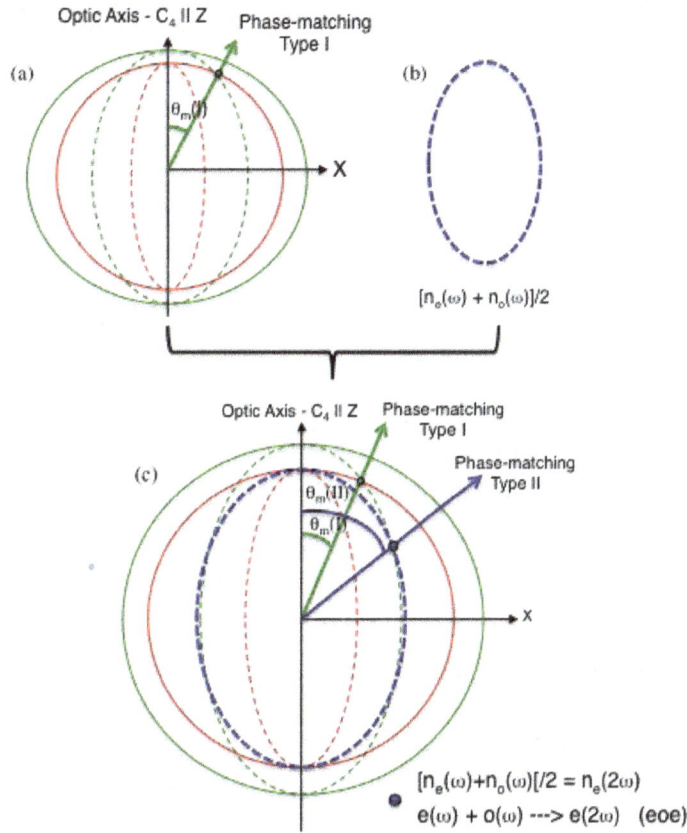

Figure 9. (a) Type I phase-matching for a negative uniaxial crystal and (b) an ellipse representing $[n_e(\omega) + n_o(\omega)]/2$. These curves are combined in (c), and the Type II phase-matching angle, $\theta_m(II)$, is determined (blue curve and arrow). The Type I phase-matching angle, $\theta_m(I)$, is also shown (green curve and line). Type II phase-matching occurs when $e(\omega) + o(\omega) \to e(2\omega)$ (eoe). Adapted from Ref. 23.

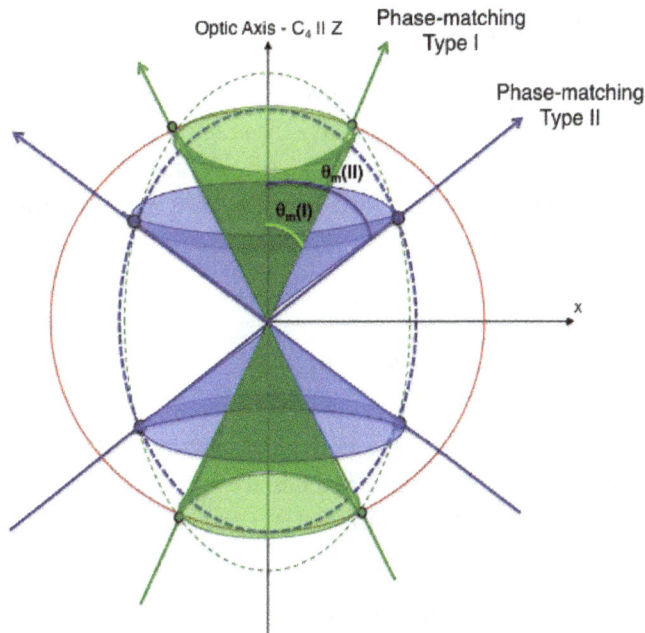

Figure 10. Type I (green) and Type II (green) negative uniaxial phase-matching cones are shown. The phase-matching angles $\theta_m(I)$ and $\theta_m(II)$ are also given. Adapted from Ref. 23.

Table 1. Phase-matching conditions and angle calculations for uniaxial systems.

	Uniaxial	
	Phase-matching condition	Phase-matching angle (θm)
Type I — Positive	$n_e(\omega) = n_o(2\omega)$ $e(\omega) + e(\omega) \rightarrow o(2\omega)$ *eeo*	$\sin^2(\theta_m) = \left(\dfrac{n_e(\omega)}{n_o(2\omega)}\right)^2 \dfrac{n_o^2(\omega) - n_o^2(2\omega)}{n_e^2(\omega) - n_o^2(\omega)}$
Type I — Negative	$n_o(\omega) = n_e(2\omega)$ $o(\omega) + o(\omega) \rightarrow e(2\omega)$ *ooe*	$\sin^2(\theta_m) = \left(\dfrac{n_e(2\omega)}{n_o(\omega)}\right)^2 \dfrac{n_o^2(2\omega) - n_o^2(\omega)}{n_e^2(2\omega) - n_o^2(2\omega)}$
Type II — Positive	$[n_e(\omega) + n_o(\omega)]/2 = n_o(2\omega)$ $e(\omega) + o(\omega) \rightarrow o(2\omega)$ *eoo*	$\dfrac{1}{2}\left(\dfrac{n_o(\omega)n_e(\omega)}{\sqrt{n_o^2(\omega)\sin^2\theta_m + n_e^2(\omega)\cos^2\theta_m}} + n_o(\omega)\right) = n_o(2\omega)$
Type II — Negative	$[n_e(\omega) + n_o(\omega)]/2 = n_e(2\omega)$ $e(\omega) + o(\omega) \rightarrow e(2\omega)$ *eoe*	$\dfrac{1}{2}\left(\dfrac{n_o(\omega)n_e(\omega)}{\sqrt{n_o^2(\omega)\sin^2\theta_m + n_e^2(\omega)\cos^2\theta_m}} + n_o(\omega)\right) = \dfrac{n_o(2\omega)n_e(2\omega)}{\sqrt{n_o^2(2\omega)\sin^2\theta_m + n_e^2(2\omega)\cos^2\theta_m}}$

surfaces at the fundamental and second-harmonic frequencies, $n(\omega) = n(2\omega)$. Although three refractive indices exist in a biaxial crystal, we will focus on the smallest and largest. The beam with the smaller refractive index is termed fast (*f*), whereas the beam with the larger refractive index is termed slow (*s*). As the velocity of light equals the speed of light divided by the refractive index, i.e. $v = c/n$, the terms 'slow' and 'fast' should be evident. The slow and fast beams from the fundamental light combine to generate the second-harmonic radiation (see Figure 11). Figures 11(a) and 11(b) show the refractive index curves at 1064 nm and 532 nm, respectively. When Figures 11(a) and 11(b) are combined in Figure 11(c), phase-matching is observed when $n(\omega) = n(2\omega)$. Type I phase-matching occurs when two fundamental slow beams, $n_s(\omega)$, combine to generate a second-harmonic beam, $n_f(2\omega)$, i.e. where the solid red, $n(\omega)$, and solid green, $n(2\omega)$, curves cross. Type II phase-matching occurs when a fundamental slow beam, $n_s(\omega)$, and fundamental fast beam, $n_f(\omega)$, combine to generate a second-harmonic beam, $n_f(2\omega)$, i.e. where the blue dashed curve $(n_s(\omega) + n_f(\omega))/2$ crosses the solid green curve, $n(2\omega)$.

The phase-matching conditions for biaxial systems have been described extensively by Hobden.[47] He determined 13 phase-matching conditions for biaxial crystals. The list of these conditions and phase-matching loci are beyond the scope of this chapter. Calculating the phase-matching angles, θ_m and ϕ_m, in biaxial systems are more complicated compared with uniaxial systems. One must solve the following equation for θ and ϕ:

$$\frac{k_x^2}{n^{-2} - n_x^{-2}} + \frac{k_y^2}{n^{-2} - n_y^{-2}} + \frac{k_z^2}{n^{-2} - n_z^{-2}} = 0$$

where $k_x = \sin\theta\cos\phi$, $k_y = \sin\theta\sin\phi$, and $k_z = \cos\theta$. θ is the angle from the z-axis, and ϕ is the angle from the x-axis in the xy-plane. The three principle refractive indices are n_x, n_y, and n_z. Explicit solutions for the above equation for Type I and Type II phase-matching in biaxial systems is given by Yao *et al.*[48]

Phase-Matching Angle and Wavelength Range

In addition to the phase-matching angles, θ_m (uniaxial), and θ_m and ϕ_m (biaxial), there are phase-matching wavelength ranges that arise from the dispersion of the refractive indices with wavelength.

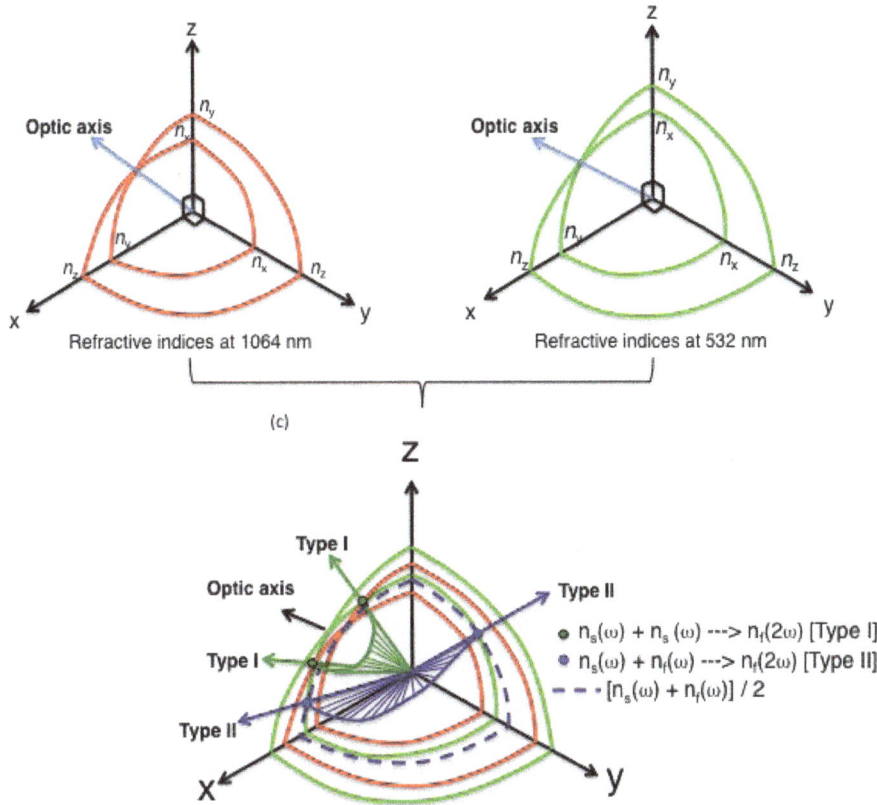

Figure 11.   Projections of the refractive indices at 1064 nm (red) and 532 nm (green) are shown. These curves are combined and the Type I (green dots) and Type II (blue dots) phase-matching conditions are revealed. Adapted from Ref. 23.

These are the wavelengths where phase-matching is possible, and these are distinct for Type I and Type II phase-matching. Specific uniaxial and biaxial materials will be used to describe the phase-matching wavelengths.

As a uniaxial example, we will use $K_3B_6O_{10}Cl$ (KBOC), a positive uniaxial NLO material.[49,50] KBOC was reported in 2010 as an NLO material with a deep-UV absorption edge of 180 nm. A large crystal of KBOC was cut into a wedge, and the refractive indices were measured by the minimum deviation method at several wavelengths between 200 nm and 2000 nm. The birefringence is 0.0464 at 1064 nm (see Figure 12). The fits to the Sellmeier equations are shown in Figure 12 — solid blue and black lines. Using the Type I and Type II positive uniaxial equations given in Table 1, we can calculate θ_m as a function of wavelength, λ. This data is shown in Figure 13. A great deal of information is provided in Figure 13. First, we can determine the Type I and Type II phase-matching wavelength ranges for fundamental radiation, by extrapolating the blue (Type I) and red (Type II) curves to the y-axis. KBOC is Type I phase-matchable from 544 nm to 3772 nm, and Type II phase-matchable from 748 nm to 2848 nm. Also, from the data shown in Figure 13, we can determine the exact angle to cut the crystal for phase-matching at a specific wavelength. The black squares in Figure 13 represent 1064 nm (the black arrow indicates 1064 nm on the left y-axis). Reading off the x-axis, we can see that for Type I phase-matching the propagation direction is 33.24° with respect to the optic axis (z-axis), whereas for Type II phase-matching the propagation direction is 51° with respect to the optic axis. Both are perpendicular to the cut crystal wafer.

For a biaxial NLO material, we will use $Ba_3(ZnB_5O_{10})PO_4$ (BZBP) as an example.[38,51] The refractive index data and fits to the Sellmeier equations for BZBP have been shown in Figure 3(b). With a biaxial material, there are two phase-matching angles, θ_m and ϕ_m. θ_m represents the phase-matching angle with

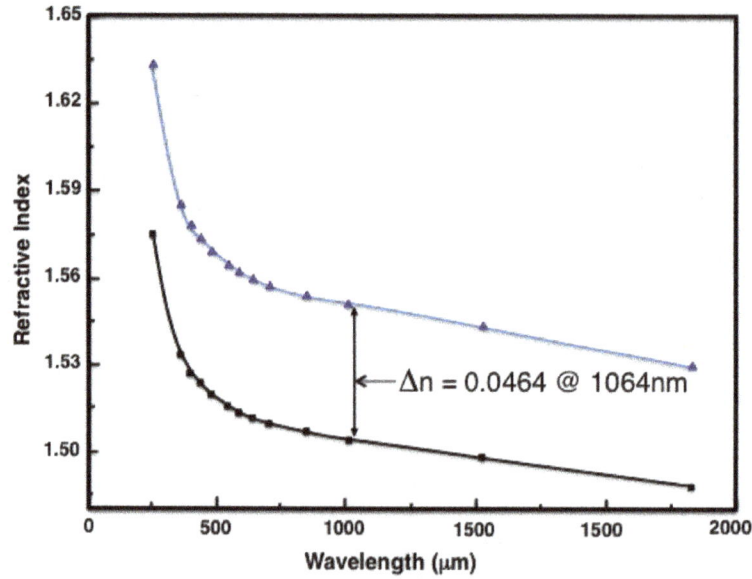

Figure 12. Refractive index data using the minimum deviation method for $K_3B_6O_{10}Cl$ (KBOC) are shown. A birefringence of 0.0464 at 1064 nm is determined, with the curves being fits from the Sellmeier equations. Adapted from Ref. 23.

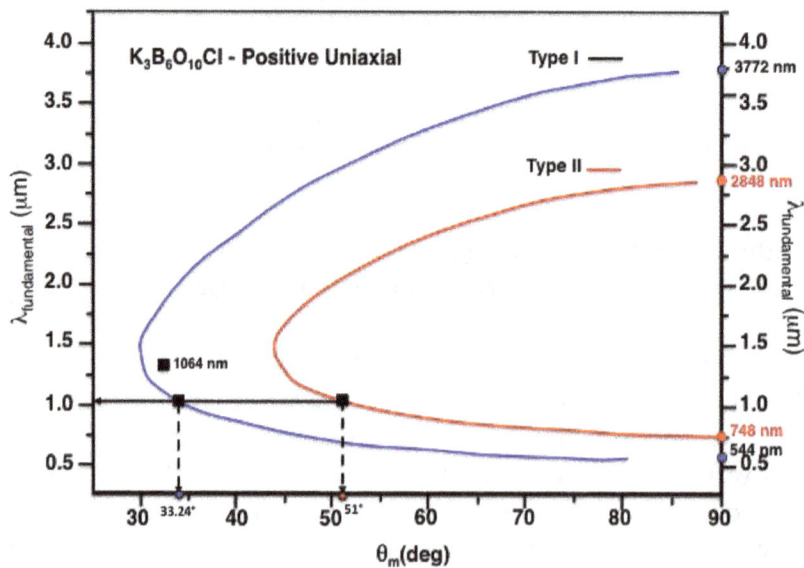

Figure 13. The Type I (blue curve) and Type II (red curve) phase-matching angles and wavelength ranges are shown. At 1064 nm, a Type I phase-matching angle of 33.24° is determined, whereas a Type II phase-matching angle of 51.0° is found. Both are indicated by the black squares on the curves and black arrows pointing to the x-axis. Adapted from Ref. 23.

respect to the z-axis, whereas ϕ_m represents the phase-matching angle in the xy-plane with respect to the x-axis. The equations necessary to solve for θ_m and ϕ_m with respect to the fundamental wavelength have been given earlier. For BZBP, the phase-matching angles (θ_m, ϕ_m) with respect to the fundamental wavelength are shown in Figure 14. It is possible to determine the Type I and Type II phase-matching wavelength ranges by extrapolating the maxima and minima of the blue (Type I) and red (Type II) curves to the y-axis — blue (Type I) and red (Type II) arrows. For BZBP, the Type I (Type II) phase-matching wavelength ranges are

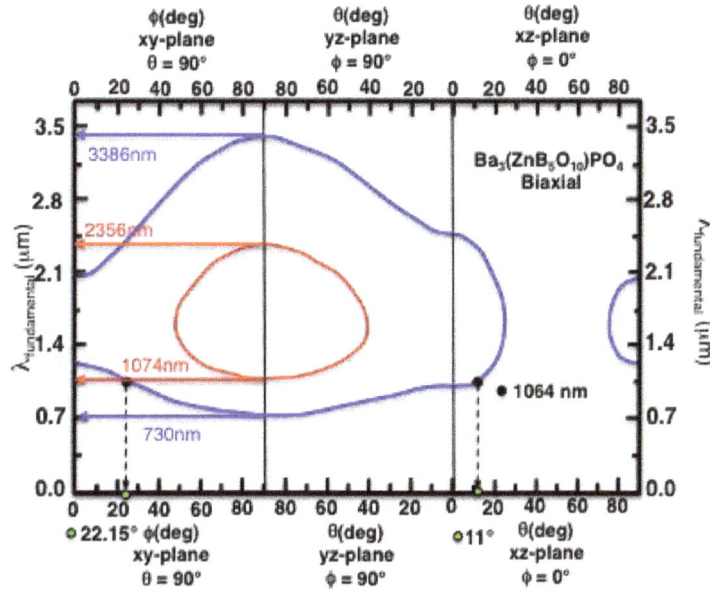

Figure 14. Phase-matching angles in the principle planes, q and f, with respect to the fundamental wavelength for $Ba_3(ZnB_5O_{10})$ PO_4 (BZBP). The Type I (Type II) phase-matching curves are shown in blue (red). The Type I (Type II) phase-matching wavelength range is 730–3386 nm (1074–2356 nm). At 1064 nm (black dots), phase-matching angles of 22.15° and 11° are determined in the *xy*- and *xz*-plane, respectively. Adapted from Ref. 23.

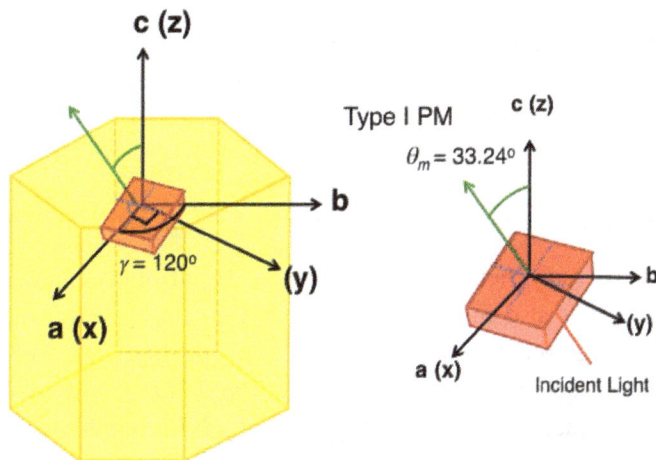

Figure 15. An ideal (yellow) $K_3B_6O_{10}Cl$ crystal is shown on the right, where a Type I phase-matching angle of 33.24° has been determined. When the crystal (red) is cut at this phase-matching angle and placed perpendicular to the propagation direction of the incident beam, phase-matching is achieved. Adapted from Ref. 23.

730–3386 nm (1074–2356 nm). From Figure 14, we can also determine the phase-matching angles at 1064 nm. In Figure 14, 1064 nm are shown by black circles. Drawing a vertical line down to the *x*-axis shows that for $\theta = 90°$, $\phi_m = 22.15°$, whereas for $\phi = 0°$, $\theta_m = 11.00°$. For both KBOC and BZBP, ideal crystal examples should prove illustrative.

An ideal crystal of KBOC is shown in Figure 15 (yellow crystal). As stated earlier, a Type I phase-matching angle of 33.24° has been calculated (Figure 13) and is the angle from the optic axis (*z*-axis). A crystal wafer is cut at this phase-matching angle (red crystal), and phase-matching is achieved when the crystal is placed

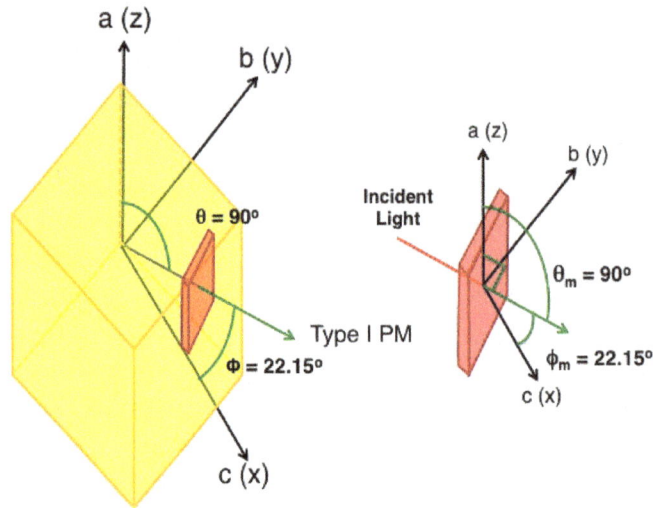

Figure 16. An ideal (yellow) $Ba_3(ZnB_5O_{10})PO_4$ crystal is shown left, where Type I phase-matching angles of $\theta_m = 90°$ and $\phi_m = 22.15°$ have been determined. When the crystal (red) is cut at these phase-matching angles and placed perpendicular to the propagation direction of the incident beam, phase-matching is achieved. Adapted from Ref. 23.

perpendicular to the propagation direction of the incident beam. With Type II phase-matching, the crystal would be cut at 51° with respect to the optic axis. For BZBP, an idea crystal is shown in Figure 16 (yellow crystal). As stated earlier, Type I phase-matching is achieved with $\phi = 90°$ and $\theta = 22.15°$. Recall that θ is the angle from the optic axis (z-axis), whereas ϕ is the angle in the xy-plane. A crystal wafer (red crystal) is cut at these phase-matching angles, and phase-matching is achieved when the crystal is placed perpendicular to the propagation direction of the incident beam.

Maker Fringe Measurements

In addition to refractive index measurements, determining the birefringence and subsequent phase-matching angles and wavelength ranges, it is also important to determine the individual SHG coefficients, i.e. $d_{ijk's}$. This is done through Maker fringe measurements[21,22] on cut, polished, and indexed single crystals that are on the order of $4 \times 4 \times 2$ mm. A schematic of the Maker fringe system is shown in Figure 17, along with a picture of a crystal used in the measurement. Data from the d_{36} NLO coefficient of KDP (KH_2PO_4) is used as a reference.

According to Maker fringe theory, the generated second-harmonic power, $P_{2\omega}$, may be expressed as

$$P_{2\omega}(\theta) = \frac{512\pi^2}{cw^2(n_\omega^2 - n_{2\omega}^2)^2} d^2 P_\omega^2 t_\omega^4 T_{2\omega} R(\theta)\beta(\theta)\sin^2\psi$$

where

$$\psi = \frac{2\pi L}{\lambda}(n_\omega \cos\theta_\omega - n_{2\omega} \cos\theta_{2\omega})$$

In these equations, c is the speed of light in air, w is the radius of light beam, n_ω and $n_{2\omega}$ are the refractive indices at the fundamental and harmonic wavelengths, respectively, d is the NLO coefficient, P_ω is the

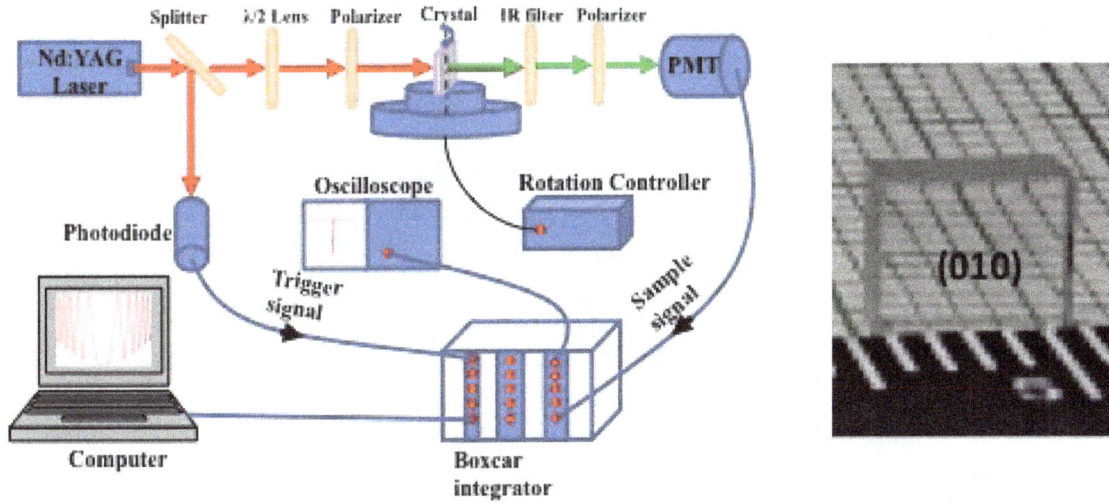

Figure 17. Schematic of the Maker fringe measurement setup with a cut, polished, and indexed crystal needed for the measurement.

fundamental beam power, t_ω and $T_{2\omega}$ are the transmission coefficients of the fundamental wave and the harmonic wave, respectively; $R(\theta)$ is the incident multiple-reflection correction, $p(\theta)$ is the projection factor, $\beta(\theta)$ is the beam size correction, L is the sample thickness, and λ is the wavelength of the fundamental beam in air, respectively.

If

$$f(\mathrm{n},\theta) = \frac{1}{(\mathrm{n}_\omega^2 - \mathrm{n}_{2\omega}^2)} t_\omega^4 T_{2\omega} R(\theta) p^2(\theta) \beta(\theta)$$

then $P_{2\omega}$ can be simplified to

$$P_{2\omega}(\theta) = \frac{512\pi^2 d^2 p_\omega^2}{cw^2} f(\mathrm{n},\theta) \sin^2 \psi$$

The $f(n,\theta)$ describes the envelope of the Maker fringe, whereas $\sin^2\psi$ determines the minima oscillating position. By fitting the calculated and measured Maker fringes, a constant, $C = 512\pi^2 d^2 p_\omega^2/(cw^2)$, can be obtained. The magnitude of the second-order NLO coefficient of a crystal can be determined relative to d_{36} of a KDP crystal, and the coefficient equation can be expressed as follows:

$$d_{\mathrm{sample}} = \sqrt{\frac{C_{\mathrm{sample}}}{C_{\mathrm{KDP}}}} \times d_{36}(\mathrm{KDP})$$

The Maker fringes for the NLO material $Ba_4B_{11}O_{20}F$ (BBOF) and d_{36} KDP are shown in Figure 18,[52] along with pictures of the crystals used in the measurements.

NLO Materials

As stated earlier, attributable to the required wide transparency range for UV and deep-UV NLO applications, the material in question must be colorless. Thus, *d–d* or *f–f* transitions are not possible. Thus, most

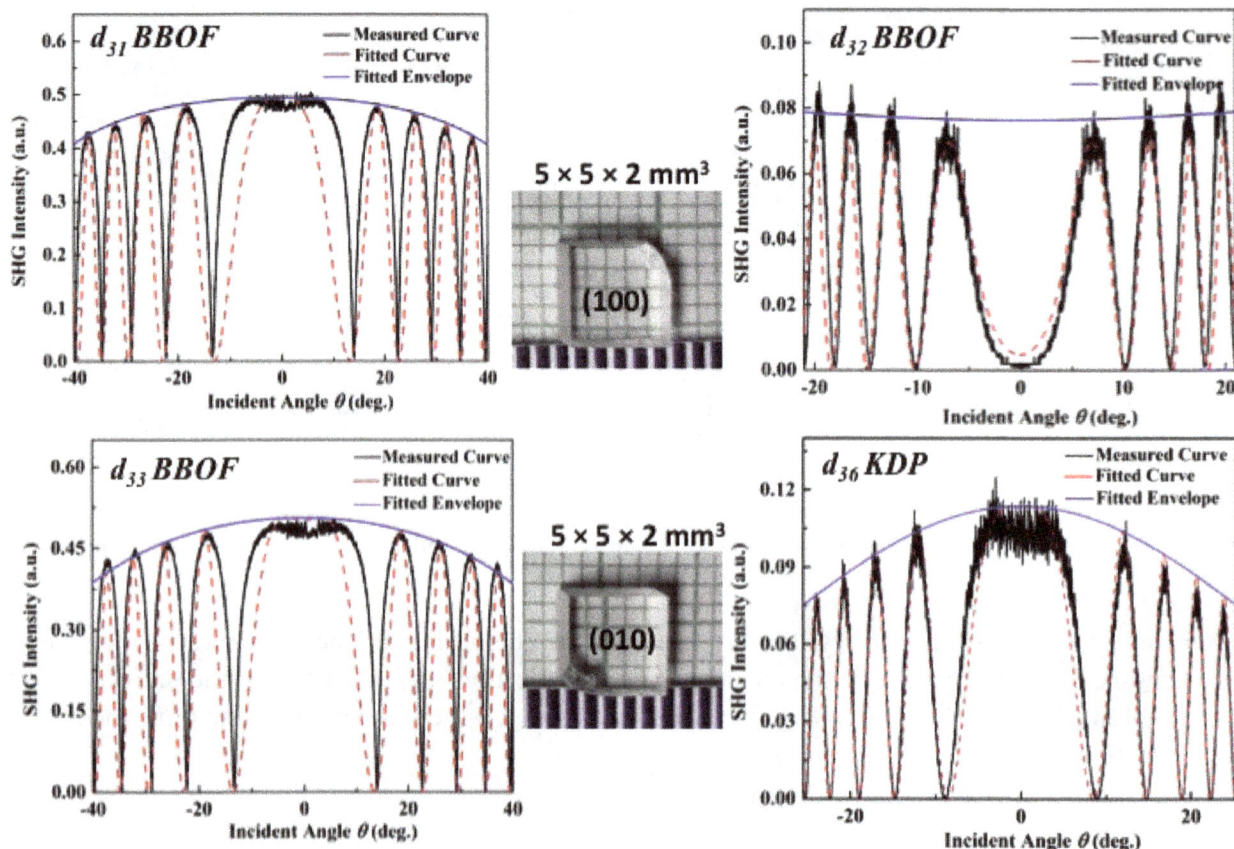

Figure 18. Experimental (black solid lines) and theoretical (red dashed lines) Maker fringes for d_{31}, d_{32}, and d_{33} Ba$_4$B$_{11}$O$_{20}$F (BBOF) and d_{36} KDP. The BBOF crystals used for the Maker Fringe experiments are also shown. Adapted from Ref. 53.

transition metals and lanthanide cations cannot be incorporated. In addition, many main group cations have absorption edges above 250 nm. Even with these limitations, there are a host of cations and anions than may be incorporated in a UV or deep-UV NLO material.

Borates

Dominating the deep-UV NLO group of materials, and NLO materials as a whole are borates.[53,54] An issue with using borates in UV and deep-UV NLO applications is the dangling bond or terminal oxygen problem. Most of the borate materials described herein are composed of $(BO_3)^{3-}$, $(B_3O_6)^{3-}$, or $(B_3O_7)^{5-}$ groups. Chen *et al*. calculated the orbital energies of each group (see Figure 19).[55–57] If the dangling bond or terminal oxygen can be removed, i.e. all of the oxygen atoms bond to another atom in addition to boron, the energy gap is increased — blueshifted.

BPO$_4$ (BPO): BPO was reported in 1934[58] with the binary phase diagram, BPO$_4$–Li$_4$P$_2$O$_7$, reported in 1961.[59] The material exhibits a 3D structure of corner linked BO$_4$ and PO$_4$ tetrahedra.[60] BPO exhibits the shortest absorption edge of all the known NCS borates — 134 nm (9.25 eV).[61] Large crystals have been grown by top-seeded solution growth methods using Li$_2$O–Li$_4$P$_2$O$_7$ ($15 \times 10 \times 12$ mm^3) (see ref. 62) and Li$_2$O–MoO$_3$ fluxes ($31 \times 18 \times 16$ mm^3).[63] In addition, Maker fringe measurements revealed a $d_{36} = 0.76$ pm/V.[61] Unfortunately, the birefringence of BPO is extremely small, 0.0052 @ 1014 nm and 0.0056 @ 589 nm.[61] As such, BPO cannot be used in NLO devices.

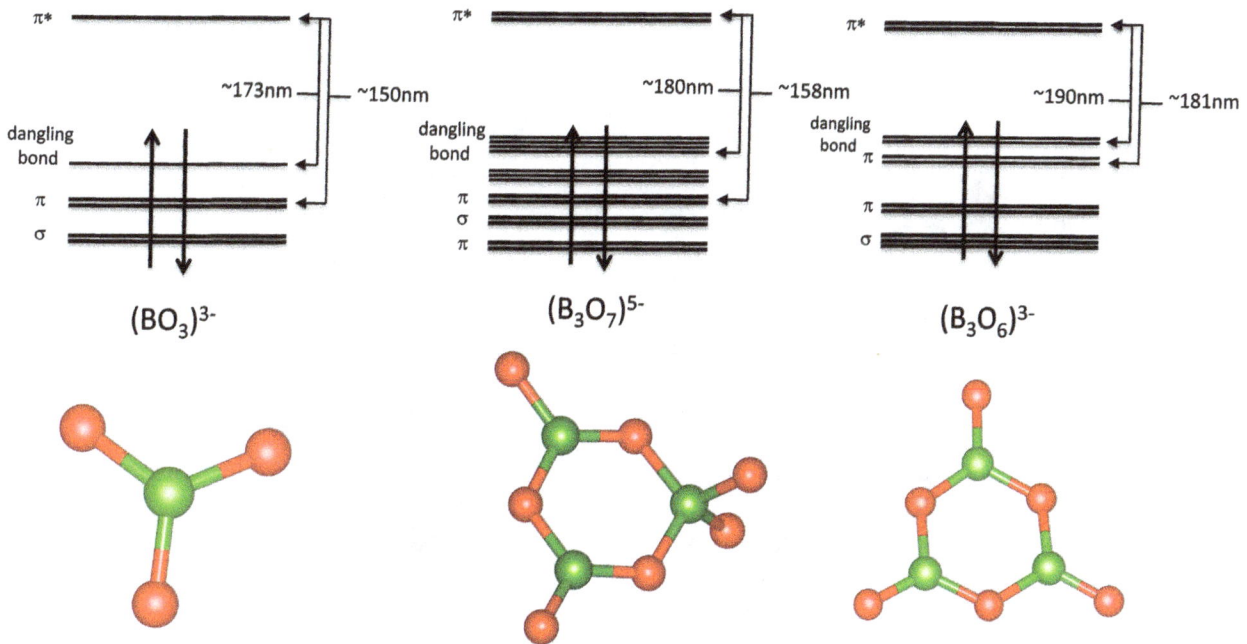

Figure 19. Energy diagrams for $(BO_3)^{3-}$, $(B_3O_7)^{5-}$, and $(B_3O_6)^{3-}$ groups. The removal of the dangling bond that results in full connectivity of the borate group increases the band gap energy.[55–57] Adapted from Ref. 143.

LiB_3O_5 (LBO): LBO was first reported in 1955[64,65] with the phase diagram of the pseudo-binary Li_2O–B_2O_3 system reported in 1958.[66] The crystal structure of LBO was reported in 1978 by Konig *et al.*,[67] and confirmed in 1980 by Ihara *et al.*[68] Variable temperature, 20°C, 227°C, and 377°C, single-crystal data were reported by Shepelev *et al.*,[69] with no observed phase-transition to at least 377°C. The material exhibits helices of fully connected $(B_3O_7)^{5-}$ groups separated by Li^+ cations (see Figure 20). However, it was not until 1989 that LBO was recognized by Chen's group as an NLO material.[35] LBO is reported to have an absorption edge of 155 nm (8 eV), attributable to the π–π^* transition of the localized molecular orbitals of the $(B_3O_7)^{5-}$ group (see Figure 19). This value is in good agreement with the prediction from the orbital energy calculations. With respect to crystal growth, large single crystals have been grown despite the fact that LBO melts incongruently.[70–87] It has been determined that the addition of MoO_3 to the Li_2O-B_2O_3 flux reduces the viscosity and enables the growth of optically perfect LBO crystals.[88,89] Using a Li_2O–B_2O_3–MoO_3 flux, crystals larger than 5 cm^3 have been grown.[90,91] As optically perfect LBO crystals have been grown, researchers have been able to investigate the linear and nonlinear optical properties of the material in great detail.[1] SHG coefficients have been measured and are reported to be $d_{31} = 0.83$ pm/V, $d_{32} = 0.88$ pm/V, and $d_{33} = 0.049$ pm/V.[35] In principle, with an absorption edge of 155 nm (8 eV), LBO should be useful in deep UV, i.e. sixth harmonic generation at 177.3 nm. The issue however is the small birefringence, $\Delta n \sim 0.045$, thought to be attributable to the helical arrangements of the infinite (B_3O_7) chains.[35] Attributable to this small birefringence, SHG below 200 nm is not possible.

$Sr_2Be_2B_2O_7$ (SBBO): SBBO was reported by Chen *et al.* in 1995 with an absorption edge of 155 nm (8 eV).[92,93] The structure is comprised of layers of connected BeO_4 and BO_3 polyhedra that are separated by Sr^{2+} cations (see Figures 21 and 22). The lack of a dangling oxygen bond in the $(BO_3)^{3-}$ network has been suggested for the blueshift in the absorption of the material and is consistent with the proposed orbital energy diagram (see Figure 19).[1] Crystals have been grown by the top-seeded solution method (7 mm × 7 mm × 3 mm),[92] as well as hydrothermally[94] using $SrCO_3$, BeO, and H_3BO_3 as reagents. An SHG coefficient,

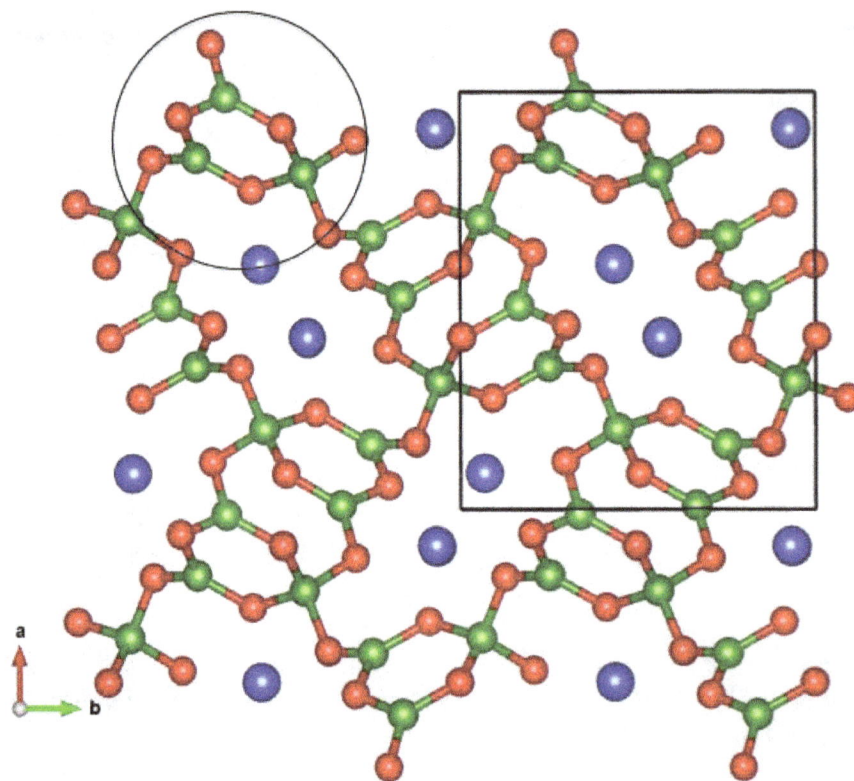

Figure 20.　Ball-and-stick representation of LiB_3O_5 (LBO). The $(B_3O_7)^{5-}$ group is circled.[143]

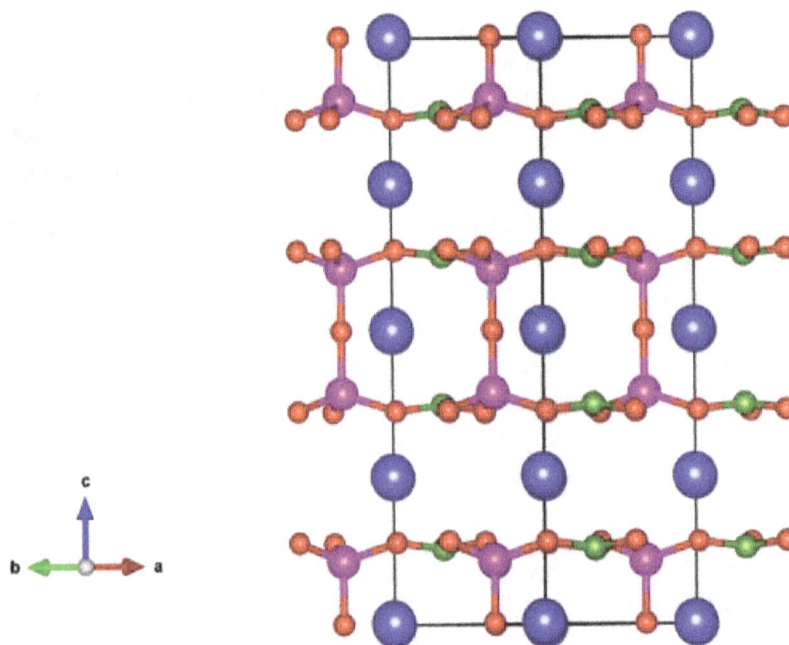

Figure 21.　Ball-and-stick representation of $Sr_2Be_2B_2O_7$ (SBBO). Connected BeO_4 and BO_3 groups comprise the layers that are separated by the Sr^{2+} cations.[143]

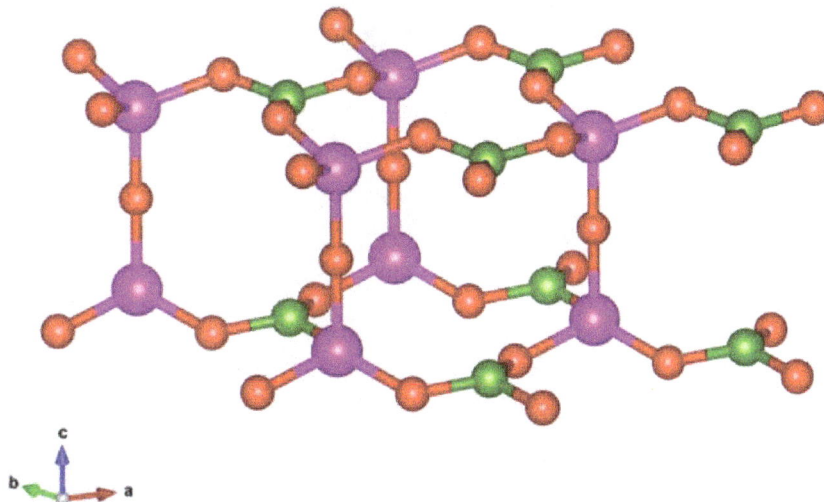

Figure 22. The BeO_4–BO_3 connectivity in $Sr_2Be_2B_2O_7$.[143]

d_{22} = 2.00–2.48 pm/V, has been reported,[92] as well as a birefringence of 0.062 at 589 nm.[93] Thus, SBBO seems to be a good candidate for deep-UV NLO applications as the absorption edge is 155 nm (8.0 eV). To date, however, the largest crystal reported grown is from the original report in 1995, i.e. $7 \times 7 \times 3$ mm^3. In the original report, it is stated that large crystals are 'easy to grow'.[92] Similar size crystals have been grown by Kolis *et al.*[94] A major issue with SBBO is the use of BeO. BeO is highly toxic, and inhalation can cause pneumonia-like symptoms or cancer.

$YAl_3(BO_3)_4$ (YAB): YAB was first reported in 1962 by Ballman[95] and was shown to be NLO active in 1974 by Filimonov.[96] YAB exhibits a 3D crystal structure consisting of AlO_6 octahedra that link corners with BO_3 groups (see Figure 23). Detailed NLO properties were reported by Rytz *et al.* in 2008 and suggest the material may have uses with the generation of 266 nm radiation.[13] In fact, 240 kW peak power at 266 nm has been observed.[59] YAB is reported to have an absorption edge of 160 nm or 170 nm depending on the quality of the crystal.[13,97] One issue with YAB is the non-trivial absorption in the UV caused by Cr^{3+} and Fe^{3+} impurities.[98] Nonetheless, high-quality crystals $30 \times 30 \times 25$ mm^3 have been grown using a TSSG technique.[97] A d_{11} = 1.7 pm/V has been reported with a birefringence of 0.068 at 1064 nm.[13]

CsB_3O_5 (CBO): CBO was discovered in 1958 by Krogh-Moe who initially determined the lattice parameters.[99] The crystal structure was determined 16 years later in 1974, also by Krogh-Moe.[8] The nonlinear optical capabilities of CBO were recognized by Chen *et al.*, who also calculated the SHG coefficients.[35] These were experimentally confirmed in 1993 by Wu *et al.*[36] CBO also contains $(B_3O_7)^{5-}$ groups similar to LBO (see Figure 24) in a 3D network, which is reported to have an absorption edge of 167 nm (7.42 eV), with an experimentally determined d_{14} = 0.75 pm/V. CBO melts congruently, and as such large single crystals have been grown through melt, Czochralski, and Kyropoulos techniques.[100–102] Similar to LBO and as we will show CLBO, the small birefringence of CBO, $\Delta n \sim 0.059$, precludes the material from being used for SHG below 200 nm.

$Li_2B_4O_7$ (LB4): This lithium borate was reported by the Japanese in early 1990s.[15] Unfortunately, initially denoted 'LBO', $Li_2B_4O_7$ (LB4) is often confused with LiB_3O_5, i.e. LBO. LB4 has an absorption edge of 170 nm (7.3 eV), and very large high optical quality single crystals, 80–90 mm in diameter and 100–140 mm in length, have been grown by Czochralski methods.[103] Maker fringe measurements have been performed and a d_{13} = 0.15 pm/V has been measured.[104] Finally, LDT of 40 GW/cm^2 (see ref. 103) has been reported, and a birefringence of 0.055 at 1014 nm has been measured.[104] In spite of this moderate birefringence,

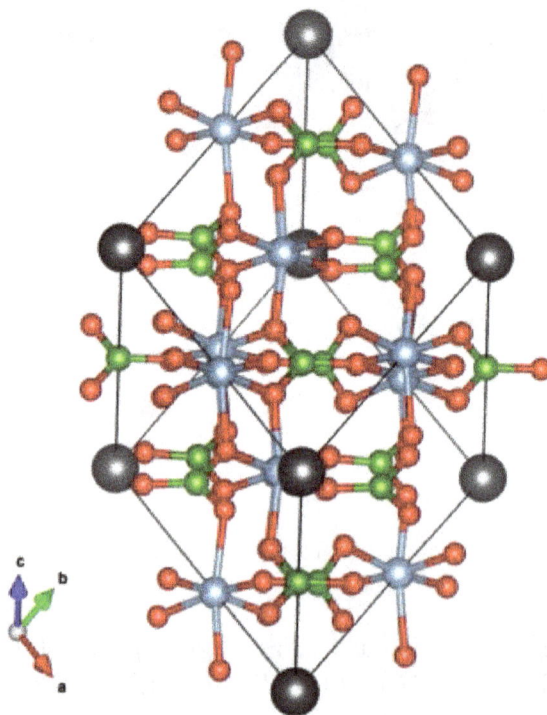

Figure 23.   Ball-and-stick representation of $YAl_3(BO_3)_4$. The 3D network is formed by the connectivity between the AlO_6 octahedra and BO_3 groups. [143]

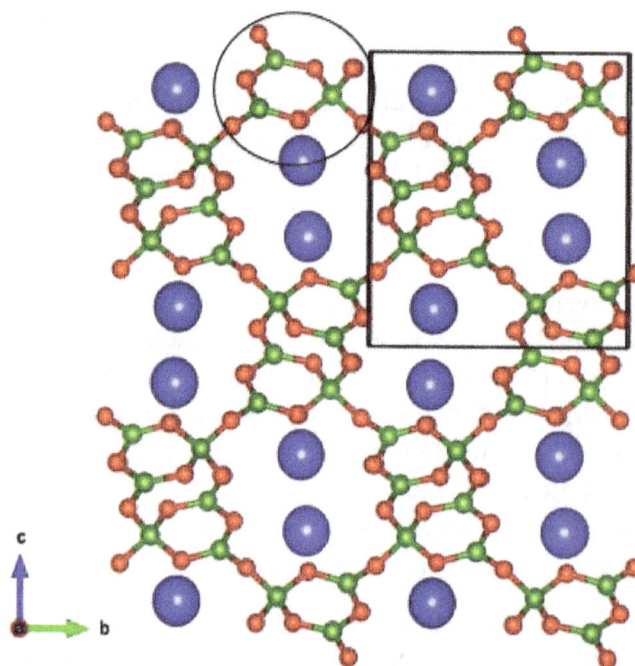

Figure 24.   Ball-and-stick representation of CsB_3O_5 (CBO). Similar to LiB_3O_5 (LBO), a $(B_3O_7)^{5-}$ group (circled) is observed.[143]

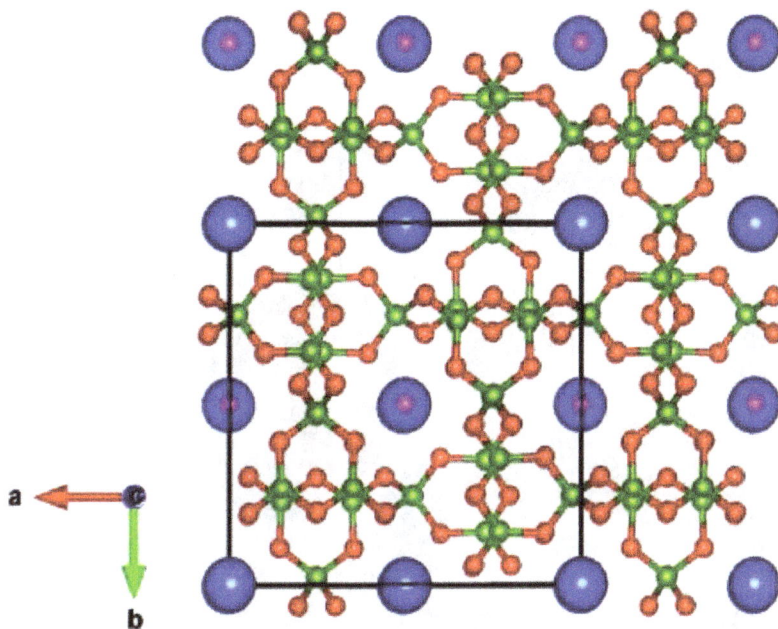

Figure 25. Ball-and-stick representation of $CsLiB_6O_{10}$ (CLBO). In CLBO, the $(B_3O_7)^{5-}$ group is connected in all three dimensions.[143]

sum frequency generation of tunable vacuum-UV femtosecond pulses from 185–170 nm has been demonstrated.[105] It remains to be seen if LB4 can be used in deep-UV technological applications.

$CsLiB_6O_{10}$ (CLBO):[106] CLBO was reported in Japan in 1995 and was immediately shown to be a promising NLO material.[106] CLBO, similar to LBO and CBO, contains a network of 3D connected $(B_3O_7)^{5-}$ groups that have no dangling oxygen atoms (see Figure 25). CLBO has an absorption edge of 180 nm (6.89 eV). Extremely large crystals, $14 \times 11 \times 11$ cm^3, have been grown from a Cs_2CO_3–Li_2CO_3–B_2O_3 melt with a 1:1:5.5 molar ratio over 45 days. PSHG measurements revealed an SHG efficiency of approximately 90 × α-SiO_2. Through single-crystal NLO measurements, $d_{36} = 0.95$ pm/V,[106] 0.74 pm/V,[107] and 0.67 pm/V[108] have been reported. The LDT for CLBO is 29 GW/cm^2 at 1064 nm.[5] Similar to LBO and CBO, the small birefringence of CLBO ($\Delta n \sim 0.049$) does not permit the material to be used to generate coherent radiation below 200 nm.[106]

β-BaB_2O_4 (BBO):[34,109–111] β-BBO is one of the most well-investigated NLO materials and is commercially available. As such, only a brief description of the material will be given here. BBO has a α-high temperature and β-low temperature form, and only the low-temperature form is NCS and SHG active.[111] β-BBO was initially thought to crystallize in space group R3,[84] but this was later corrected to *R3c*.[109,110] The material contains isolated $(B_3O_6)^{3-}$ groups that are separated by Ba^{2+} cations (see Figure 26). The borate was recognized as a potential NLO material by Chen in 1985.[111] The melting point of BBO is 1095 ± 5°C and the α–β phase transition is at 925 ± 5°C.[111] Thus, flux growth techniques are necessary for crystal growth. A number of techniques have been used over the past 30 years including flux methods involving BBO–$NaBaBO_3$,[112] BBO–NaF,[113,114] TSSG,[115,116] Czochralski,[117–120] and immersion seeded methods.[121] High-quality optical crystals of 50 mm in diameter and 30 mm in height have been grown. β-BBO has an absorption edge of 185 nm (6.7 eV), consistent with the full connectivity of the $(B_3O_6)^{3-}$ ring groups. Powder SHG measurements indicate Type I phase-matching behavior with an efficiency of approximately 164 × α-SiO_2. As high-quality crystals have been grown, Maker fringe measurements have been performed.

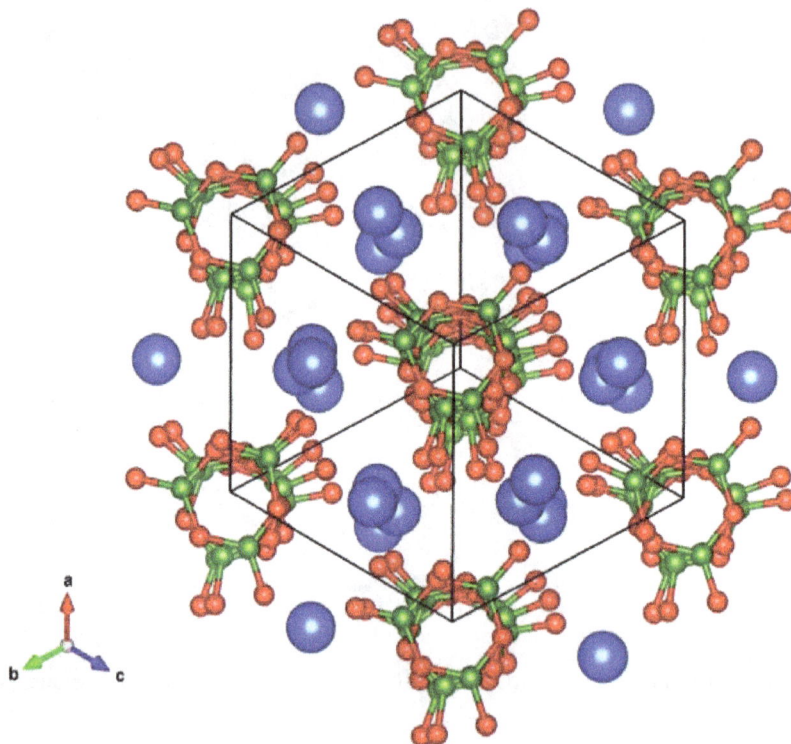

Figure 26.   Ball-and-stick representation of β-BaB$_2$O$_4$ (β-BBO). The (B$_3$O$_6$)$^{3-}$ groups are stacked between the Ba^{2+} cations.[143]

β-BBO has one of the largest reported d_{ij} coefficients, i.e. $d_{22} = 1.60 \pm 0.05$ pm/V (see ref. 121) and $d_{22} = 2.20 \pm 0.05$ pm/V.[122] Depending on the crystal quality, LDTs have been reported ranging from 10 to 50 GW/cm^2.[1] Finally, a birefringence of 0.1127 has been measured at 1064 nm. Attributable to other optical issues, i.e. too large walk-off angle and too small bandwidth, β-BBO is not used for applications below 200 nm. β-BBO is the material of choice for fourth and fifth harmonic generation, i.e. 266 nm and 213 nm, for Nd:YAG laser systems.

Borate fluorides

Another group of materials that have potential applications for deep-UV NLO technologies are metal borate fluorides.[123] With these materials, it is thought that the addition of fluoride to a fully linked borate framework will blueshift the absorption edge.

ABe$_2$BO$_3$F$_2$ (A = K, Rb, or Cs: KBBF, RBBF, and CBBF)[110–117]: These three isostructural alkali beryllium borate fluorides were discovered in the former Soviet Union in the 1970s.[124,125] Initial structural characterization indicated the materials were monoclinic and crystallized in space group *C2*. Chen *et al.* determined this to be incorrect and showed that all three materials are trigonal — space group *R32*.[126] The materials exhibit some of the deepest absorption edges, i.e. 147–160 nm (8.2–7.8 eV). Crystals of these materials have been notoriously difficult to grow, not only attributable to the toxicity of BeO, but also the tendency for the crystals to layer. The material consists of $^2_\infty$[(BeO$_3$F)(BO$_3$)] layers that are separated by K$^+$ cations (see Figures 27 and 28). The fluoride anions on the Be^{2+} cations alternate directions. Crystals have been grown by self-flux methods, i.e. ABBF + AF + B$_2$O$_3$,[1] as well as seeded hydrothermal methods using AF as a mineralizer.[127,128] There is however some debate on which technique produces the best crystals for optical applications.[16] At present, however, crystals with high optical quality have not been grown larger than 4 mm along the *c*-axis direction.

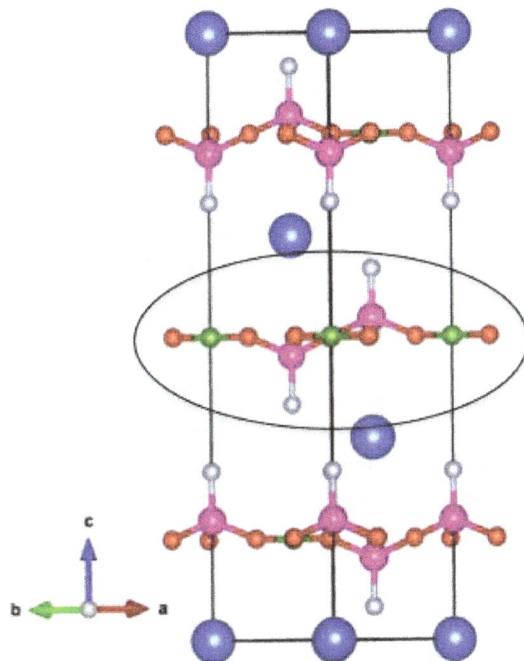

Figure 27. Ball-and-stick representation of (a) $KBe_2BO_3F_2$ (KBBF). The (b) $^2_\infty[(BeO_3F)(BO_3)]$ layer is circled.[143]

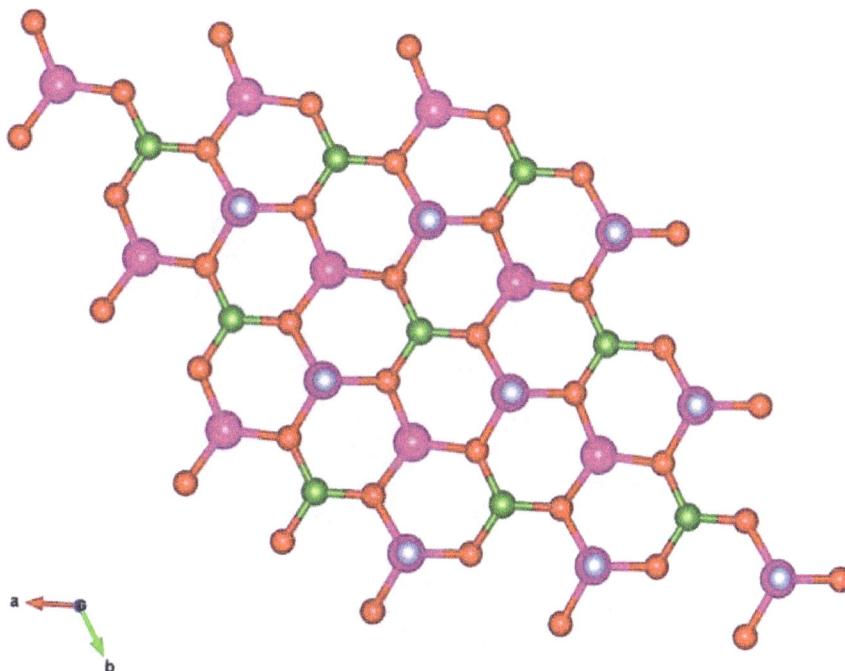

Figure 28. Ball-and-stick representation of the $^2_\infty[(BeO_3F)(BO_3)]$ layer in $KBe_2BO_3F_2$.[143]

Powder SHG measurements revealed Type I phase-matching behavior and efficiencies of approximately 50 × α-SiO_2.[127] SHG coefficients have been determined for all three materials and are $d_{11} = 0.47 \pm 0.01$ pm/V (KBBF), 0.45 ± 0.01 pm/V (RBBF), and 0.5 pm/V (CBBF). The LDT for KBBF has been reported as high as 900 GW/cm^2. The LDTs are not known for RBBF or CBBF. Birefringence values are 0.08 (KBBF — 1064 nm), 0.073 (RBBF — 694.3 nm), and 0.058 (CBBF — 1014 nm).[1]

Carbonate fluorides

Recently a number of NLO active mixed-metal carbonate fluorides have been discovered with absorption edges below 200 nm. Mixed-metal carbonate fluorides, $AMCO_3F$ (A = alkali metal and M = alkaline-earth cation), exhibit (i) fully bonded $(CO_3)^{2-}$ groups, that is there is (are) no terminal oxygen atom(s). As stated earlier, in borates it has been demonstrated that dangling or terminal oxygen atom(s) redshift the absorption edge. And (ii) a blueshift of the absorption edge is attributable to the inclusion of fluoride. These materials are synthetically accessible through both hydrothermal and ceramic methods.

$KCaCO_3F$, $KSrCO_3F$, $RbCaCO_3F$, $RbSrCO_3F$, $CsCaCO_3F$[129]: The initial report on carbonate fluorides as possible UV and/or deep-UV NLO materials was from Zou *et al.* in 2011.[129] $ASrCO_3F$ (A = K, Rb) and $KCaCO_3F$ are isostructural, as are $ACaCO_3F$ (A = Rb, Cs), and consist of $CaCO_3F_2$ or $SrCO_3F_2$ polyhedra separated by the alkali cation (see Figures 29 and 30). All five materials are reported to exhibit an absorption edge below 200 nm (6.2 eV), but no exact value was given. PSHG measurements indicate Type I phase-matching behavior with efficiencies ranging from $45 \times \alpha\text{-}SiO_2$ ($RbCaCO_3F$ and $CsCaCO_3F$) to $130 \times \alpha\text{-}SiO_2$ ($KSrCO_3F$ and $RbSrCO_3F$) and $145 \times \alpha\text{-}SiO_2$ ($KCaCO_3F$).

$RbMgCO_3F$[130]: This material was reported in 2015 by Tran *et al.*[130] The material exhibits a 3D structure consisting of corner-shared $Mg(CO_3)_3F_2$ polyhedra. The Mg^{2+} cations are connected to carbonate groups in the *ab*-plane and a bridging fluoride along the *c*-axis direction (see Figure 31). An absorption edge below 190 nm (6.53 eV) has been reported. PSHG measurements with 1064 nm and 532 nm radiation revealed SHG efficiencies of approximately $160 \times \alpha\text{-}SiO_2$ and $0.6 \times \beta\text{-}BBO$, respectively. Type I phase-matching behavior was also observed. Large single crystals have yet to be grown, and as such SHG coefficients, LDT, and birefringence data have not been reported.

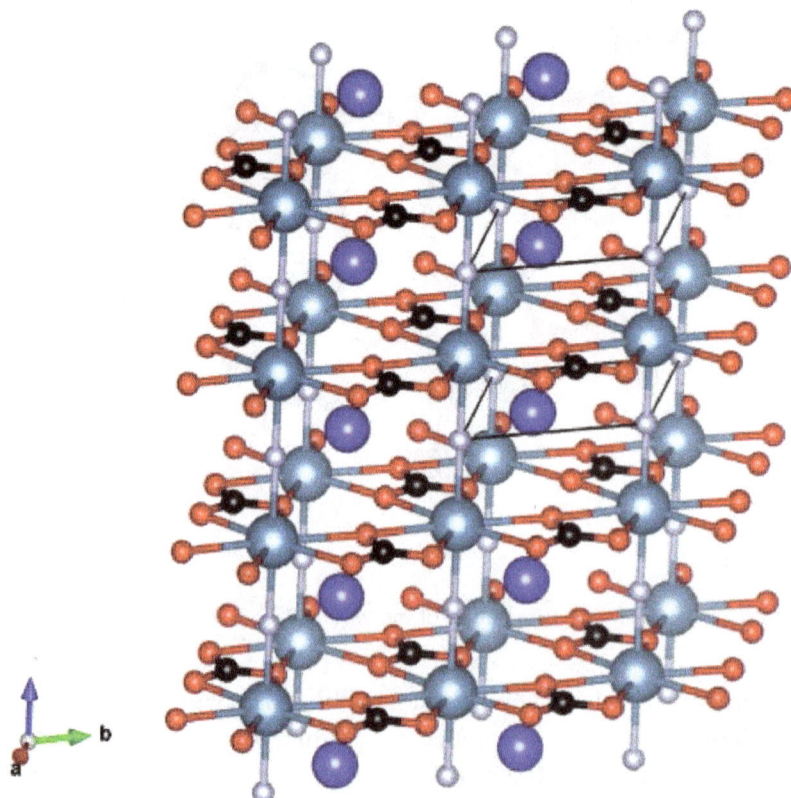

Figure 29. Ball-and-stick representation of $KCaCO_3F$ showing the $CaCO_3F_2$ groups that are separated by the K^+ cations.[143]

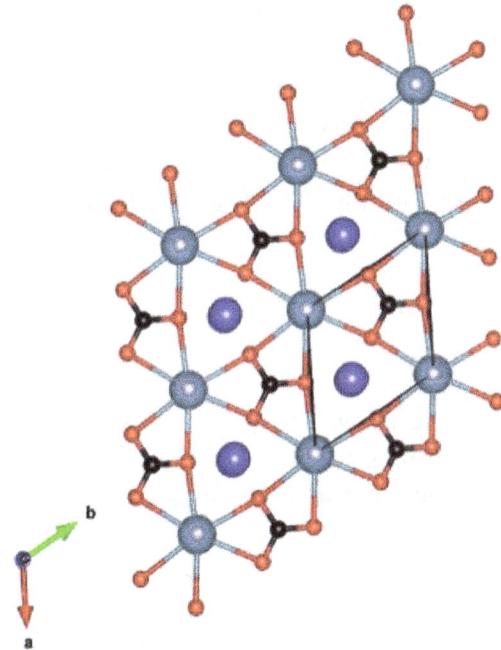

Figure 30. Ball-and-stick representation of one layer of $KCaCO_3F$.[143]

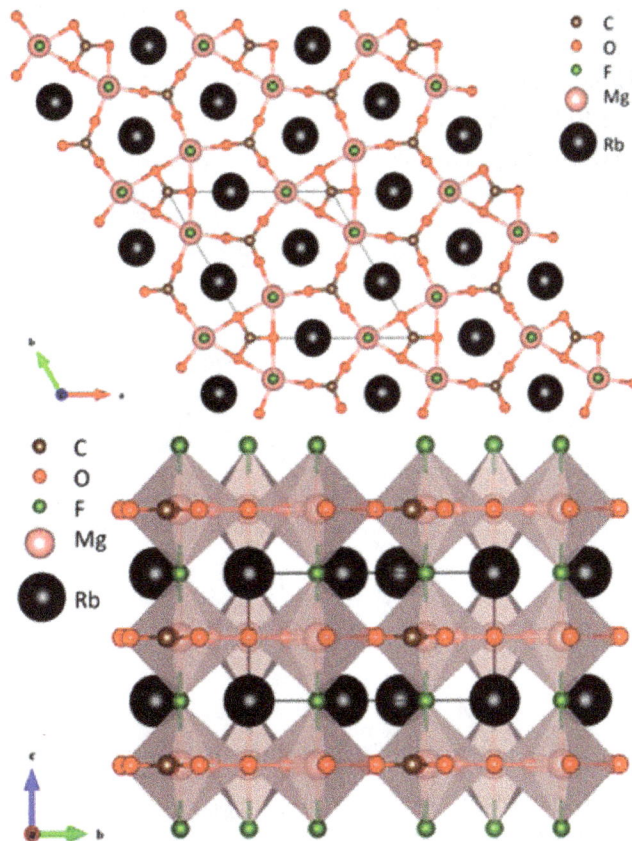

Figure 31. Ball-and-stick and polyhedral representation of $RbMgCO_3F$. In the top figure, the Rb cations reside in the pentagonal 'tunnels' created by the $MgCO_3F$ polyhedra, whereas in the bottom figure, F atoms connect the $MgCO_3F$ polyhedra along the c-axis direction.[130]

KSrCO₃F (crystal growth)[131]: Very recently, Zhang and Halasyamani, reported on the successful crystal growth of $KSrCO_3F$.[131] Using a top-seeded solution growth method, a $KF–Li_2CO_3–SrCO_3$ flux with a molar ratio of 7:10:3 was used to grow relatively large crystals. The crystal growth process is somewhat slow, with a $5 \times 3 \times 2$ mm³ crystal grown over 35 days. These crystals are large enough for refractive index and Maker fringe measurements. A birefringence of 0.1049 @ 1064 nm was determined, and a $d_{22} = 0.50$ pm/V was measured. In addition, the SHG phase-matching wavelength limit was calculated to be 200 nm. All of this suggests that $KSrCO_3F$ is a viable NLO material for FOHG generation — 266 nm.

Structure–Property Relationships

Recently, Tran *et al.* examined all of the known NLO active mixed-metal carbonate fluoride materials in order to better understand the relationships between their NCS structures and SHG properties.[132] They plotted the SHG efficiency vs the specific acentric mode displacement (SAMD) parameter vs cation variance in a ternary diagram (see Figure 32). The SAMD enables one to quantify, to an extent, the acentricity in any NCS structure by comparing it to a hypothetical undistorted centrosymmetric structure.[133] The magnitude of the atomic displacements required to relate the two structures is normalized to the unit cell volume that results in a quantification of acentricity. As seen in Figure 32, the known $AMCO_3F$ materials can be divided into four groups. The group with the largest SHG efficiency contains $RbPbCO_3F$ and $CsPbCO_3F$. This is perhaps not too surprising, as the addition of Pb^{2+} with its stereoactive lone pair increases the SHG efficiency of the carbonate fluoride. Including Pb^{2+}, however, redshifts the absorption edge to approximately 300 nm, and as such these materials are not useful in the UV or deep UV. In examining groups 2, 3, and 4, we note that the SHG efficiency decreases, on average, with large A and smaller M cations, or conversely, a greater SHG efficiency is observed for smaller A and larger M cations. It also seems that a smaller SAMD and moderate variance result in a greater SHG efficiency. It should be noted, however, that there are only 14 reported NLO active compounds in the $AMCO_3F$ family; thus, the trends described herein should be taken as provisional. Nonetheless, it is clear that there is a relationship between the SHG efficiency, SAMD, and cation variance.

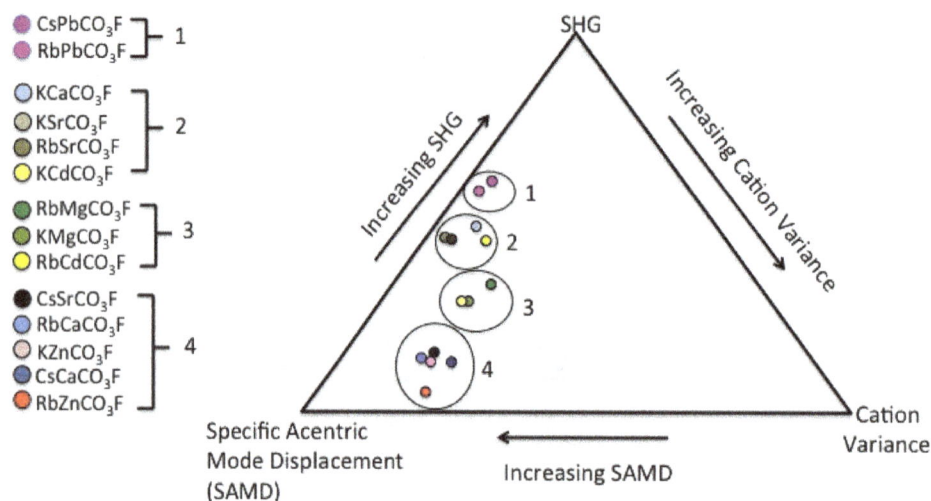

Figure 32.   Relationship between the SHG efficiency, Specific Acentric Mode Displacement (SAMD), and Cation Variance in mixed-metal SHG active carbonate fluorides — $AMCO_3F$ (A = alkali metal, M = alkaline-earth cation, Zn, Cd, or Pb).[132]

Table 2. Known and predicted fluorooxoborates with possible-UV and/or deep-UV NLO applications.

	$Li_2B_3O_4F_3$	$Li_2B_6O_9F_2$	KB_4O_6F–I[a]	KB_4O_6F–II[a]	CsB_4O_6F	$NH_4B_4O_6F$	$BaB_4O_6F_2$[a]
Abs. edge	<200 nm[b]	<200 nm	161 nm[b]	161 nm[b]	155 nm	158 nm	158 nm[b]
PSHG	Unknown	0.9 × KDP	1.0 × KDP[b]	1.0 × KDP[b]	1.9 × KDP	1.2 × KDP	1.0 × KDP[b]
Birefring.	0.05[b]	0.07[b]	0.087[b]	0.098[b]	0.113	0.117	0.065[b]
SHG limit	Unknown	192 nm[b]	161 nm[b]	161 nm[b]	171 nm	158 nm	180 nm[b]

Notes: [a]Predicted Compound
[b]Calculated

Fluorooxoborates

Also recently discovered are a series of fluorooxoborates with potential UV and deep-UV NLO applications. The known and predicted fluorooxoborates are given in Table 2, along with their absorption edge, powder SHG efficiency (PSHG) vs KDP, birefingence, and SHG limit.

Two Li compounds, $Li_2B_3O_4F_3$ and $Li_2B_6O_9F_2$,[134,135] seem to exhibit suitable birefringences for UV NLO applications.[136] However, neither has been grown as large enough crystals for refractive index measurements. Three fluorooxoborates have been predicted to have UV and/or deep-UV NLO applications — KB_4O_6F–I, KB_4O_6F–II, and $BaB_4O_6F_2$.[137] It should be noted that none of these three materials has been synthesized. As seen in Table 2, these three materials have calculated absorbance edges between 158–161 nm, with suitable calculated birefringences, and suggested SHG limits below 200 nm. In fact KB_4O_6F–I and KB_4O_6F–II have calculated SHG limits below 177.3 nm — 6[th]HG.[137] Very recently, two fluorooxoborates have been synthesized that seem to have possible deep-UV NLO applications. These materials are CsB_4O_6F (see ref. 138) and $NH_4B_4O_6F$.[139] Both compounds have very short absorption edges, i.e. 155 nm — Cs and 158 nm — NH_4. In addition, crystals have been grown large enough to determine their birefringence, ~0.11 for both, and SHG limits, 171 nm — Cs and 158 nm — NH_4. Crystals have not been grown large enough to determine their NLO coefficients, i.e. the d_{ij}'s.

$X_3Y_3Li_2Al_4B_6O_{20}F$ (X = alkali cation, Y = alkaline-earth cation)

Another family of materials that has been recently discovered with possible UV NLO applications is the seemingly complex $X_3Y_3Li_2Al_4B_6O_{20}F$ stoichiometry (X = alkali cation, Y = alkaline-earth cation). A schematic of how $X_3Y_3Li_2Al_4B_6O_{20}F$ is determined from $Sr_2Be_2B_2O_7$ is given in Figure 33. Briefly, the $Sr_2Be_2B_2O_7$ stoichiometry is tripled, resulting in '$Sr_6Be_6B_6O_{21}$' (not a compound). 'Sr6' is replaced with 'X_3Y_3', 'Be6' with 'Li_2A_{l4}', and 'O_{21}' with '$O_{20}F$', resulting in $X_3Y_3Li_2Al_4B_6O_{20}F$. One important consequence of the substitutions is the removal of Be from the final compound. The second is that the layering issue observed in $KBe_2BO_3F_2$ is addressed. The structural evolution from $KBe_2BO_3F_2$ is shown in Figure 34. As seen in Figure 34, the layering habit observed in $KBe_2BO_3F_2$ is improved with $Sr_2Be_2B_2O_7$ and $NaCaBe_2(BO_3)_2F$, but Be^{2+} is still present and the materials are still layered. Stronger layer stability is observed in $K_3Sr_3Li_2Al_4B_6O_{20}F$ (see ref. 140) and large single crystals have been grown, but layering still remains. Through the judicious selection of cations, $Rb_3Ba_3Li_2Al_4B_6O_{20}F$ (see ref. 141) was synthesized which exhibits a 3D crystal structure.

With this $X_3Y_3Li_2Al_4B_6O_{20}F$ stoichiometry, three materials have been reported: $K_3Sr_3Li_2Al_4B_6O_{20}F$, $K_3Ba_3Li_2Al_4B_6O_{20}F$, (see ref. 142) and $Rb_3Ba_3Li_2Al_4B_6O_{20}F$ (see Table 3).

As seen in Table 3, large enough crystals have only been grown for $K_3Sr_3Li_2Al_4B_6O_{20}F$ and $Rb_3Ba_3Li_2Al_4B_6O_{20}F$ to determine the birefringence and SHG limit. With both $K_3Sr_3Li_2Al_4B_6O_{20}F$ and

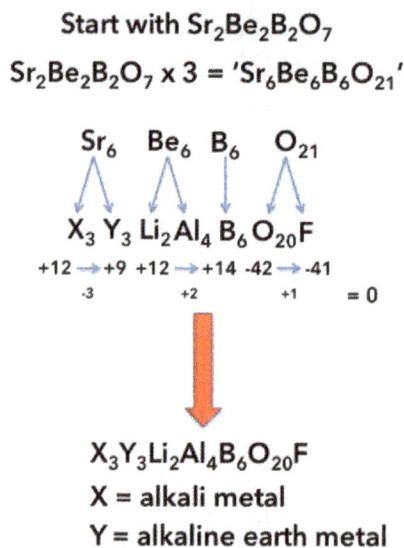

Start with $Sr_2Be_2B_2O_7$

$Sr_2Be_2B_2O_7 \times 3 = \text{'}Sr_6Be_6B_6O_{21}\text{'}$

$Sr_6 \quad Be_6 \quad B_6 \quad O_{21}$

$X_3 \; Y_3 \; Li_2Al_4 \; B_6 \, O_{20}F$

$+12 \rightarrow +9 \quad +12 \rightarrow +14 \quad -42 \rightarrow -41$

$\quad\quad -3 \quad\quad\quad +2 \quad\quad\quad +1 \quad = 0$

$X_3Y_3Li_2Al_4B_6O_{20}F$

X = alkali metal

Y = alkaline earth metal

Figure 33.   Schematic showing how $Sr_2Be_2B_2O_7$ evolves into $X_3Y_3Li_2Al_4B_6O_{20}F$.

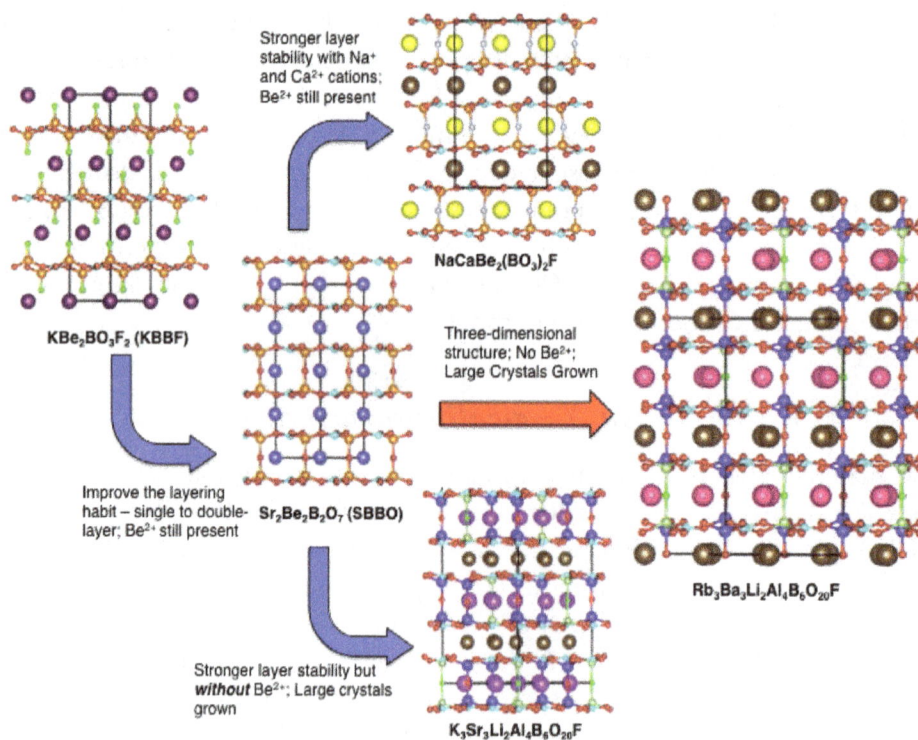

Figure 34.   Structural evolution from $KBe_2BO_3F_2$ to $Rb_3Ba_3Li_2Al_4B_6O_{20}F$. The layering habit of $KBe_2BO_3F_2$ is improved in $Sr_2Be_2B_2O_7$ and $NaCaBe_2(BO_3)_2F$, but Be^{2+} is still present and both lack 3D connectivity. Stronger layer stability is observed in $K_3Sr_3Li_2Al_4B_6O_{20}F$ and large crystals have been grown,[140] but the layered crystal structure remains. By carefully selecting cations, the successful synthesis of $Rb_3Ba_3Li_2Al_4B_6O_{20}F$ was achieved — a material with a 3D structure.[141]

Table 3. Absorption edge, powder SHG intensity, birefringence @ 1064 nm, and SHG limit for $K_3Sr_3Li_2Al_4B_6O_{20}F$, $K_3Ba_3Li_2Al_4B_6O_{20}F$, and $Rb_3Ba_3Li_2Al_4B_6O_{20}F$.

	$K_3Sr_3Li_2Al_4B_6O_{20}F$	$K_3Ba_3Li_2Al_4B_6O_{20}F$	$Rb_3Ba_3Li_2Al_4B_6O_{20}F$
Abs. edge	190 nm	190 nm	198 nm
PSHG	$1.7 \times KDP$	$1.5 \times KDP$	$1.5 \times KDP$
Birefring.	0.060	Unknown	0.057
SHG limit	224 nm	Unknown	243 nm

$Rb_3Ba_3Li_2Al_4B_6O_{20}F$, the SHG limit is below 266 nm, indicating both materials have FOHG (266 nm) NLO applications.

Conclusions and Outlook

The discovery of new UV and deep-UV NLO materials remains an ongoing challenge for chemists, material scientists, and physicists. As the design requirements for these materials are rather strict, viable NLO materials are limited. In addition to the synthesis, single-crystal growth, i.e. crystals larger than 5 mm, is required in order to determine any technological application. This adds another layer of difficulty to an already complex scientific challenge. Despite these restrictions, new materials are being discovered, along with their crystal growth, that have potential UV or deep-UV NLO applications. These include $KSrCO_3F$, CsB_4O_6F, $NH_4B_4O_6F$, $K_3Sr_3Li_2Al_4B_6O_{20}F$, and $Rb_3Ba_3Li_2Al_4B_6O_{20}F$. It remains to be seen if any of these materials can replace $CsLiB_6O_{10}$ (CLBO) or β-BaB_2O_4 (β-BBO) for 266 nm generation, or $KBe_2BO_3F_2$ (KBBF) for 177.3 nm generation.

References

1. Chen, C.; Sasaki, T.; Li, R. K.; Wu, Y.; Lin, Z.; Mori, Y.; Hu, Z.; Wang, J.; Uda, S.; Yoshimura, M.; Kaneda, Y., *Nonlinear optical Borate Crystal: Principals and Applications.* Wiley-VCH: Weinheim, Germany, 2012.
2. Franken, P. A.; Hill, A. E.; Peters, C. W.; Weinreich, G., Generation of Optical Harmonics. *Phys. Rev. Lett.* 1961, 7, 118–120.
3. Savage, N., Ultraviolet lasers. *Nat. Photo.* 2007, 1, 83–84.
4. Cyranoski, D., China's crystal cache. *Nature (London)* 2009, 457, 953.
5. Yao, W.; He, R.; Wang, X.; Lin, Z.; Chen, C., Analysis of Deep-UV Nonlinear Optical Borates: Approaching the End. *Adv. Opt. Mater.* 2014, 2, 411–417.
6. Tressaud, A.; Poeppelmeier, K. R., *Photonic and Electronic Properties of Floride Materials.* Elsevier: Boston, 2016.
7. Liu, Q.; Yan, X.; Gong, M.; Liu, H.; Zhang, G.; Ye, N., High-power 266 nm ultraviolet generation in yttrium aluminum borate. *Opt. Lett.* 2011, 36 (14), 2653–2655.
8. Krogh-Moe, J., The Crystal Structure of Lithium Diborate, $Li_2O.2B_2O_3$. *Acta Crystallogr.* 1962, 15, 190–193.
9. Kwon, T. Y.; Ju, J. J.; Cha, J. W.; Kim, J. N.; Yun, S. I., Characteristics of Critically Phase-Matched Second-Harmonic Generation of a $Li_2B_4O_7$ Crystal Grown by the Czochralski Method. *Mater. Lett.* 1994, 20, 211–215.
10. Ryu, G.; Yoon, C. S.; Han, T. P. J.; Gallagher, H. G., Growth and characterisation of $CsLiB_6O_{10}$ (CLBO) crystals. *J. Cryst. Growth* 1998, 191 (3), 492–500.
11. Masashi, Y.; Tomosumi, K.; Kouki, M.; Yusuke, M.; Hidetsugu, Y.; Masahiro, N.; Takatomo, S., Bulk Laser Damage in $CsLiB_6O_{10}$ Crystal and Its Dependence on Crystal Structure. *Jpn. J Appl. Phys.* 1999, 38 (2A), L129.
12. Chen, C. T.; Wu, B. C.; Jiang, A. D.; You, G. M., A New ultra-violet SHG crystal β-BaB_2O_4. *Sci. Sin.* 1985, 18, 235.
13. Rytz, D.; Gross, A.; Vernay, S.; Wesemann, V., $YAl_3(BO_{34})$: A Novel NLO Crystal for Frequency Conversion to UV Wavelengths. *Proc. SPIE* 2008, 6998, 699814.
14. Yu, X.; Yue, Y.; Yao, J.; Hu, Z.-G., $YAl_3(BO_3)_4$: Crystal growth and characterization. *J. Cryst. Growth* 2010, 312 (20), 3029–3033.

15. Furusawa, S.; Chikagawa, O.; Tange, S.; Ishidate, T.; Orihara, H.; Ishibashi, Y.; Miwa, K., 2nd harmonic generation in $Li_2B_4O_7$. *J. Phys. Soc. Jpn.* 1991, 60, 2691–2693.

16. Chen, C.; Luo, S.; Wang, X.; Wang, G.; Wen, X.; Wu, H.; Zhang, X.; Xu, Z., Deep UV nonlinear optical crystal: $RbBe_2(BO_3)F_2$. *J. Opt. Soc. Am. B* 2009, 26 (8), 1519–1525.

17. Zhang, X.; Wang, L.; Wang, X.; Wang, G.; Zhu, Y.; Chen, C., High-power sixth-harmonic generation of an Nd:YAG laser with $KBe_2BO_3F_2$ prism-coupled devices. *Opt. Commun.* 2012, 285 (21–22), 4519–4522.

18. Chen, C. T.; Wang, G. L.; Wang, X. Y.; Xu, Z. Y., Deep-UV nonlinear optical crystal $KBe_2BO_3F_2$ — discovery, growth, optical properties and applications. *Appl. Phys. B* 2009, 97, 9–25.

19. Nye, J. F., *Physical Properties of Crystals: Their Represntation by Tensors and Matrices.* Oxford Science Publications: Oxford, 1957.

20. Roberts, D. A., Simplified characterization of uniaxial and biaxial nonlinear optical crystals: a plea for standardization of nomenclature and conventions. *IEEE J. Quantum Electron.* 1992, 28 (10), 2057–2074.

21. Maker, P. D.; Terhune, R. W.; Nisenoff, M.; Savage, C. M., Effects of Dispersion and Focusing on the Production of Optical Harmonics. *Phys. Rev. Lett.* 1962, 8 (1), 21–22.

22. Jerphagnon, J.; Kurtz, S. K., Maker Fringes: A Detailed Comparison of Theory and Experiment for Isotropic and Uniaxial Crystals. *J. Appl. Phys.* 1970, 41 (4), 1667–1681.

23. Zhang, W.; Yu, H.; Wu, H.; Halasyamani, P. S., Phase-Matching in Nonlinear Optical Compounds: A Materials Perspective. *Chem. Mater.* 2017, 29, 2655–2668.

24. Moorthy, S.; Kumar, F.; Balakumar, S.; Subramanian, C.; Ramasamy, P., Top seeded solution growth of $KTiOPO_4$ (KTP) single crystals and their characterisation. *Mater. Sci. Eng. B* 1999, 60 (2), 88–94.

25. Pan, S.; Smit, J. P.; Lanier, C. H.; Marvel, M. R.; Marks, L. D.; Poeppelmeier, K. R., Optical Floating Zone Growth of β-BaB_2O_4 from a $LiBa_2B_5O_{10}$-Based Solvent. *Cryst. Growth Des.* 2007, 7, 1561–1564.

26. Yao, S.; Wang, J.; Liu, H.; Hu, X.; Zhang, H.; Cheng, X.; Ling, Z., Growth, optical and thermal properties of near-stoichiometric $LiNbO_3$ single crystal. *J. Alloys Compd.* 2008, 455 (1–2), 501–505.

27. Anandha Babu, G.; Subramaniyan Raja, R.; Karunagaran, N.; Perumal Ramasamy, R.; Ramasamy, P.; Ganesamoorthy, S.; Gupta, P. K., Growth improvement of $AgGaSe_2$ single crystal using the vertical Bridgman technique with steady ampoule rotation and its characterization. *J. Cryst. Growth* 2012, 338 (1), 42–46.

28. Wang, S.; Gao, Z.; Zhang, X.; Zhang, X.; Li, C.; Dong, C.; Lu, Q.; Zhao, M.; Tao, X., Crystal Growth and Effects of Annealing on Optical and Electrical Properties of Mid-Infrared Single Crystal $LiInS_2$. *Cryst. Growth & Des.* 2014, 14 (11), 5957–5961.

29. Zhang, W.; Halasyamani, P. S., Top-seeded solution crystal growth of noncentrosymmetric and polar Zn_2TeMoO_7 (ZTM). *J. Solid State Chem.* 2016, 236, 32–38.

30. Zhang, J.; Zheng, H.; Ren, Y.; Mitchell, J. F., High-Pressure Floating-Zone Growth of Perovskite Nickelate $LaNiO_3$ Single Crystals. *Crystal Growth & Des.* 2017, 17 (5), 2730–2735.

31. Wahlstrom, E. E., *Optical Crystallography.* John Wiley and Sons: New York and London, 1960.

32. Bloss, F. D., *An Introduction to the Methods of Optical Crystallography.* Holt, Rinehart and Winston, Inc.: New York, 1961.

33. Smith, G. F. H., On the Method of Minimum Deviation for the Determination of Refractive Indices. *Mineral. Mag.* 1906, 14, 191–193.

34. Eimerl, D.; Davis, L.; Velsko, S.; Graham, E. K., Optical, mechanical and thermal properties of barium borate. *J. Appl. Phys.* 1987, 62, 1968.

35. Chen, C. T.; Wu, Y. C.; Jiang, A. D.; Wu, B. C.; You, G. M.; Li, R. K.; Lin, S. J., New nonlinear-optical crystal: LiB_3O_5. *J. Opt. Soc. Am. B* 1989, 6, 616–621.

36. Wu, Y. C.; Sasaki, T.; Nakai, N.; Yokotani, A.; Tang, H. G.; Chen, C. T., CsB_3O_5: A new nonlinear optical crystal. *Appl. Phys. Lett.* 1993, 62, 2614–2615.

37. Leviton, D. B.; Madison, T. J.; Petrone Iii, P., Simple refractometers for index measurements by minimum-deviation method from far ultraviolet to near infrared. *Proc. SPIE* 1998, 3425, 148–159.

38. Yu, H.; Cantwell, J.; Wu, H.; Zhang, W.; Peoppelmeier, K. R.; Halasyamani, P. S., Top-Seeded Solution Crystal Growth, Morphology, Optical and Thermal Properties of $Ba_3(ZnB_5O_{10})PO_4$. *Cryst. Growth Des.* 2016, 16, 3976–3982.

39. Born, M.; Wolf, E., *Principles of Optics.* Pergamon Press: Oxford, 1975.

40. Armstrong, J. A.; Bloembergen, N.; Ducuing, J.; Pershan, P. S., Interactions between Light Waves in a Nonlinear Dielectric. *Phys. Rev.* 1962, 127, 1918–1939.

41. Fejer, M. M.; Magel, G. A.; Jundt, D. H.; Byer, R. I., Quasi-phase-matched second harmonic generation: tuning and tolerances. *IEEE J. Quant. Elec.* 1992, 4, 570–575.

42. Bahabad, A.; Murnane, M. M.; Kapteyn, H. C., Quasi-phase-matching of momentum and energy in nonlinear optical processes. *Nat. Photon.* 2010, 4, 570–575.

43. Zoubir, A.; Eichenholz, J.; Fujiwara, E.; Grojo, D.; Baleine, E.; Rapaport, A.; Bass, M.; Chai, B.; Richardson, M., Non-critical phase-matched second harmonic generation in $Gd_xY_{1-x}COB$. *Appl. Phys.* B 2003, 77 (4), 437–440.

44. Lambrecht, W. R. L.; Jiang, X., Noncritically phase-matched second-harmonic-generation chalcopyrites based on $CdSiAs_2$ and $CdSiP_2$. *Phys. Rev. B: Condens. Matter Mater. Phys.* 2004, 70, 045204.

45. Kleinman, D. A., Theory of Second Harmonic Generation. *Phys. Rev.* 1962, 128, 1761–1775.

46. Ashkin, A.; Boyd, G. D.; Klei nman, D. A., Phase-matched Second-Harmonic Generation in KDP without double refraction. *Appl. Phys. Lett.* 1965, 6, 179–180.

47. Hobden, M. V., Phase-Matched Second-Harminc Generation in Biaxial Crystals. *J. Appl. Phys.* 1967, 38, 4365–4372.

48. Yao, J. Q.; Fahlen, T. S., Calculations of optimum phase match parameteres for the biaxial crystal $KTiOPO_4$. *J. Appl. Phys.* 1984, 55, 65–68.

49. Wu, H.; Yu, H.; Yang, Z.; Han, J.; Wu, K.; Pan, S., Linear optical and thermo-physical properties of polar $K_3B_6O_{10}Cl$ crystal. *J. Materiomics* 2015, 1, 221–228.

50. Wu, H.; Pan, S.; Poeppelmeier, K. R.; Li, H.; Jia, D.; Chen, Z.; Fan, X.; Yang, Y.; Rondinelli, J. M.; Luo, H., $K_3B_6O_{10}Cl$: A New Structure Analogous to Perovskite with a Large Second Harmonic Generation Response and Deep UV Absorption Edge. *J. Am. Chem. Soc.* 2011, 133, 7786–7790.

51. Yu, H.; Zhang, W.; Young, J.; Rondinelli, J. M.; Halasyamani, P. S., Design and Synthesis of the Beryllium-Free Deep-Ultraviolet Nonlinear Optical Material. *Adv. Mater.* 2015, 27, 7380–7385.

52. Wu, H.; Yu, H.; Zhang, W.; Cantwell, J.; Poeppelmeier, K. R.; Pan, S.; Halasyamani, P. S., Top-Seeded Solution Crystal Growth, Linear and Nonlinear Optical Properties of $Ba_4B_{11}O_{20}F$ (BBOF). *Cryst. Growth Des.* 2017, 17 (3), 1404–1410.

53. Keszler, D. A., Borates for optical frequency conversion. *Curr. Opin. in Solid State and Mater. Sci.* 1996, 1, 204–208.

54. Keszler, D. A., Synthesis, crystal chemistry, and optical properties of metal borates. *Curr. Opin. in Solid State and Mater. Sci.* 1999, 4, 155–162.

55. Chen, C. T.; Yu, W. C.; Lin, R. K., The Relationship between the Structural Type of Anionic Group and SHG effect in Boron-Oxygen Compounds. *Chin. Phys. Lett.* 1985, 2, 389.

56. Chen, C. T., *Development of new Nonlinear Optical Crystals in the Borati Series*. Eds. Marder, S. R.; Sohn, J. E.; Stucky, G. D., Materials for Nonlinear Optics. Vol. 455; Oxford University Press: London, 1991, 360–379.

57. Chen, C. T., *Development of new Nonlinear Optical Crystals in the Borati Series*. Eds. Letoknov, V. S.; Shank, C. V.; Shen, Y. R.; Walther, H., Laser Science and Technology — An international handbook Vol. 15. Harwood Academic: Reading, 1993.

58. Schulze, G. E. R., Z. *Phys. Chem.* 1934, 24, 215.

59. Tien, T. Y.; Hummel, F. A., *J. Am. Ceram. Soc.* 1961, 44, 393.

60. Li, Z.; Lin, Z.; Wu, Y.; Fu, P.; Wang, Z.; Chen, C., Crystal Growth, Optical Properties Measurement, and Theoretical Calculation of BPO4. *Chem. Mater.* 2004, 16 (15), 2906–2908.

61. Zhang, X.; Wang, L.; Zhang, S.; Wang, G.; Zhou, S.; Zhu, Y.; Wu, Y.; Chen, C., Optical properties of the vacuum ultraviolet nonlinear optical crystal — BPO_4. *J. Opt. Soc. Am.* B 2011, 28, 2236–2239.

62. Li, Z.; Wu, Y.; Fu, P.; Pan, S.; Chen, C., Flux growth of BPO4 crystals. *J. Cryst. Growth* 2004, 270 (3–4), 486–490.

63. Zhang, S.; Zhang, E.; Fu, P.; Wu, Y., Growth of BPO_4 single crystals from Li2O–MoO3 flux. ***J. Cryst. Growth*** 2009, 311, 2433–2436.

64. Mazzeti, C.; Carli, F. D., Borates of lithium, cadium, lead, and manganese. *Gazz. Chim. Ital.* 1926, 56, 19–29.

65. Rollet, A. P.; Bouaziz, R., The binary system lithium oxide-boric anhydride. *Comp. Rend.* 1955, 240, 2417–2419.

66. Sastry, B. S. R.; Hummel, F. A., Studies in lithium oxide system: I. $Li_2O.B_2O_3$–B_2O_3. *J. Am. Ceram. Soc.* 1958, 41, 7–17.

67. Konig, H.; Hoppe, R., On borates of the alkaline metals. II. On the knowledge of LiB_3O_5. *Z. Anorg. Allg. Chem.* 1978, 439, 71–79.

68. Ihara, M.; Yuge, M.; GKrogh-Moe, J., Crystal structure of lithium triborate. Yogyo Koyokaishi 1980, 88, 179–184.

69. Shepelev, Y. F.; Bubnova, R. S.; Filatov, S. K.; Sennova, N., A.; Pilneva, N. A., LiB_3O_5 Crystal Structure at 20, 227, and 377C. *J. Solid State Chem.* 2005, 178, 2987–2997.

70. Wu, Y. C.; Jiang, A. D.; Lu, S. F.; Chen, C. T.; Shen, Y. S., Crystal growth and structure of $Li_2O.3B_2O_3$. *J. Synth. Cryst.* 1990, 19, 33–38.

71. Zhao, S. Q.; Huang, C. E.; Zhang, H. W., Crystal growth and properties of lithium triborate. *J. Cryst. Growth* 1990, 99, 805–810.

72. Ukachi, T.; Lane, R. J.; Bosenberg, W. R.; Tang, C. L., Phase-matched second-harmonic generation and growth of a LiB_3O_5 crystal. *J. Opt. Soc. Am.* B 1992, 9, 1128–1133.

73. Bruck, E.; Raymakers, R. J.; Route, R. K.; Feigeison, R. S., Surface stability of lithium triborate crystals grown from excess B_2O_3 solutions. *J. Cryst. Growth* 1993, 128, 933–937.

74. Markgraf, S. A.; Furukawa, Y.; Sato, M., Top-seeded solution growth of LiB_3O_5. *J. Cryst. Growth* 1994, 144, 343–348.

75. Shumov, D. P.; Nikolov, V. S.; Nenov, A. T., Growth of LiB_3O_5 single crystals in the $Li_2O–B_2O_3$ system. *J. Cryst. Growth* 1994, 144, 218–222.

76. Guretskii, S. A.; Ges, A. P.; Zhigunov, D. I.; Ignatenki, A. A.; Kalanda, N. A.; Kurnevich, L. A.; Luginets, A. M.; Milovanov, A. S.; Molchan, P. V., Growth of lithium triborate single crystals from molten salt solution under various temperature gradients. *J. Cryst. Growth* 1995, 156, 410–412.

77. Shumov, D. P.; Nenov, A. T.; Nihtianova, D. D., Inclusions in LiB_3O_5 crystals obtained by the top-seeded solution growth method in the $Li_2O–B_2O_3$ system. Part 1. *J. Cryst. Growth* 1996, 169, 519–526.

78. Nihtianova, D. D.; Shumov, D. P.; Macicek, J. J.; Nenov, A. T., Inclusions in the LiB_3O_5 crystals, obtained by the top-seeded solution growth method in the $Li_2O–B_2O_3$ system. Part 2. *J. Cryst. Growth* 1996, 169, 527–533.

79. Zhong, W. Z.; Tang, D. Y., Growth units and morphology of lithium triborate (LBO) crystals. *J. Cryst. Growth* 1996, 166, 91–98.

80. Hu, X. B.; Jiang, S. S.; Huang, X. R., The formation mechanisms of dislocations and negative crystals in LiB_3O_5 single crystals. *J. Cryst. Growth* 1996, 163, 266–271.

81. Kima, H. G.; Kang, J. K.; Lee, S. H.; Chung, S. J., Growth of Lithium Triborate Crystals by the TSSG Technique. *J. Cryst. Growth* 1998, 187, 455–462.

82. Hao, Z. W.; Ma, X. M., Growth of high quality nonlinear optical crystal LBO by flux method. *J. Synth. Cryst.* 2002, 31, 124–126.

83. Kim, J. W.; Yoon, C. S.; Gallagher, H. G., The effect of NaCl melt-additive on the growth and morphology of LiB_3O_5 (LBO) crystals. *J. Cryst. Growth* 2001, 222, 760–766.

84. Liu, H. B.; Shen, G. Q.; Wang, X. Q.; Wei, J. Z.; Shen, D. Z., Viscosity and IR investigations in the $Li_2O–B_2O_3$ system. *Prog. Cryst. Growth Charact.* 2000, 40, 235–241.

85. Pylneva, N. A.; Kononova, N. G.; Yurkin, A. M.; Bazarova, G. G.; Danilov, V. I., Growth and non-linear optical properties of lithium triborate crystals. *J. Cryst. Growth* 1999, 198/199, 546–550.

86. Pylneva, N. A.; Kosyakov, V.; Yurkin, A. M.; Bazarova, G. G.; Atuchin, V.; Kolesnikov, A.; Trukhanov, E.; Ziling, C., Real structure of LiB_3O_5 (LBO) crystals grown in $Li_2O–B_2O_3$-MoO_3 system. *Cryst. Res. Technol.* 2001, 36, 1377–1384.

87. Kosyakov, V. I.; Pylneva, N. A.; Bazarova, G. G.; Yurkin, A. M., Topology of liquidus surface in B_2O_3–$Li_2O.B_2O_3$–Li_2O. MoO_3–MoO_3 system: implications to the growth of lithium triborate single crystals. *Mater. Res. Bull.* 2001, 36 (573–584).

88. Parfeniuk, C.; Samaresekera, I. V.; Weinberg, F., Growth and lithium triborate crystals I. Mathematical model. *J. Cryst. Growth* 1996, 158, 514–522.

89. Parfeniuk, C.; Samaresekera, I. V.; Weinberg, F.; Edel, J.; Fjeldsted, K.; Lent, B., Growth of lithium triborate crystals II. Experimental results. *J. Cryst. Growth* 1996, 158, 523–533.

90. Kokh, A. E.; Vlezko, V. A.; Kokh, K. A., Control over te symmetry of the heat field in the station for growing LBO crystals by the Kyropoulos method. *Instrum. Exp. Tech.* 2009, 52, 747–751.

91. Kokh, A. E.; Kononova, N. G.; Mennerat, G.; Villeval, P.; Durst, S.; Lupinski, D.; Vlezko, V.; Kokh, K., Growth of high quality large size LBO crystals for high energy second harmonic generation. *J. Cryst. Growth* 2010, 312, 1774–1778.

92. Chen, C.; Wang, Y.; Wu, B.; Wu, K.; Zeng, W.; Yu, L., Design and synthesis of an ultraviolet-transparent nonlinear optical crystal $Sr_2Be_2B_2O_7$. *Nature* (London) 1995, 373, 322–324.

93. Chen, C.; Wang, Y.; Xia, Y. N.; Wu, B.; Tang, D.; Wu, K.; Wenrong, Z.; Yu, L.; Mei, L., New development of nonlinear optical crystals for the ultraviolet region with molecular engineering approach. *J. Appl. Phys.* 1995, 77, 2268–2272.

94. Kolis, J. W.; McMillen, C. D.; Franco, T., Hydrothermal Synthesis of the Deep-UV NLO Material $Sr_2Be_2B_2O_7$. *MRS Online Proc. Libr.* 2004, 848, FF1.1.

95. Ballman, A. A., A New Series of Synthetic Borates Isostrucrural with the Carbonate Mineral Huntite. *Am. Mineral.* 1962, 47, 1380–1383.

96. Filimonov, A. A.; Leonyuk, N. I.; Meissner, L. B.; Timchenko, T. I.; Rez, I. S., Nonlinear Optical Properties of an Isomorphic Family of Crystals with a Yttrium-Aluminum Borate (YAB) Structure. *Krist. Tech.* 1974, 9, 63–66.

97. Yu, J.; Liu, L.; Zhai, N.; Zhang, X.; Wang, G.; Wang, X.; Chen, C., Crystal growth and optical properties of $YAl_3(BO_3)_4$ for UV applications. *J. Cryst. Growth* 2012, 341, 61–65.

98. Ilas, S.; Loiseau, P.; Aka, G.; Taira, T., 240kW peak power at 266 nm in Nonlinear $YAl_3(BO_3)_4$ Single Crystal. *Opt. Express* 2014, 22, 30325–30332.

99. Krogh-Moe, J., Some new compounds in the system cesium oxide-boron oxide. *Arkiv. Kemi.* 1958, 12, 247–249.

100. Kato, K., Tunable UV Generation to 0.185um in CsB_3O_5. *IEEE J. Quant. Elec.* 1995, 31, 169–171.

101. Fu, P. Z.; Wang, J. X.; Hu, Z. G.; Wu, Y. C.; Yin, S. T.; Xu, Z. Y., Growth and properties of ultraviolet nonlinear optical cesium triborate. *J. Synth. Cryst.* 1999, 28, 215–218.

102. Kagebayashi, Y.; Mori, Y.; Sasaki, T., Crystal growth of cesium triborate CsB_2O_5 by Kryopoulos technique. *Bull. Mater. Sci.* 1999, 22, 971–973.

103. Komatsu, R.; Sugawara, T.; Sassa, K.; Sarukura, N.; Liu, Z.; Izumida, S.; Segawa, Y.; Uda, S.; Fukuda, T.; Yamanouchi, K., Growth and ultraviolet application of $Li_2B_4O_7$ crystasls: Genration of the fouth and fifth harmonics of $Nd:Y_3Al_5O_{12}$ lasers. *Appl. Phys. Lett.* 1997, 70, 3492–3494.

104. Sugawara, T.; Komatsu, R.; Uda, S., Linear and Nonlinear optical properties of lithium tetraborate. *Solid State Commun.* 1998, 107, 233–237.

105. Petrov, V.; Rotermund, F.; Noack, F.; Komatsu, R.; Sugawara, T.; Uda, S., Vacuum ultraviolet applicaion of $Li_2B_4O_7$ crystals: Generation of 100fs pulses down to 170 nm. *J. Appl. Phys.* 1998, 84, 5887–5982.

106. Mori, Y.; Kuroda, I.; Nakajima, S.; Sasaki, T.; Nakai, S., New nonlinear optical crystal: Cesium lithium borate. *Appl. Phys. Lett.* 1995, 67, 1818–1820.

107. Shoji, I.; Nakamura, H.; Ito, R.; Kondo, T.; Yoshimura, M.; Mori, Y.; Sasaki, T., Absolute measurement of second-harmonic nonlinear optical coefficients of $CsLiB_6O_{10}$ for visible-to-ultraviolet second-harmonic wavelengths. *J. Opt. Soc. Am. B* 2001, 18, 302–307.

108. Lin, Z.; Jiang, X.; Kang, L.; Gong, P.; Luo, S.; Lee, M.-H., Frist-principles materials applications and design of nonlinear optical crystals. *J. Phys. D: Appl. Phys.* 2014, 47, 253001 (1–19).

109. Liebertz, J.; Stahr, S., Zur tieftemperaturphae von BaB_2O_4. *Z. Kristallogr.* 1983, 165, 91.

110. Froehlich, R., Crystal structure of the low-temperature form of BaB_2O_4. *Z. Kristallogr.* 1984, 168, 109.

111. Chen, C. T.; Wu, B. C.; Jiang, A. D.; You, G. M., A new type of ultraviolet SHG crystal: β-BaB_2O_4. *Sci. Sin. B* 1985, 28, 235.

112. Tsvetkov, E. G.; Khranenko, G. G.; Solntsev, V. P., General approaches to design of a reproducible technique for the growth of large crystals of barium meteborate (BBO) for industrial application. *J. Cryst. Growth* 2005, 275, e2123–e2128.

113. Roth, M.; Perlov, D., Growth of barium borate crystals from sodium fluoride solutions. *J. Cryst. Growth* 1996, 169, 734–740.

114. Kim, H. G.; Kang, J. K.; Lee, S. H.; Chung, S. J., Growth of lithium triborate crystals by the TSSG technique. *J. Cryst. Growth* 1998, 187, 455–462.

115. Tang, D. Y.; Lin, S. T.; Dai, G. Q.; Lin, Q.; Zeng, W. R.; Zhao, Q. L.; Huang, Y. S., Growth of β-BaB_2O_4 crystals from molten salts by the top-seeded pulling technique. *J. Synth. Cryst.* 1990, 19, 21–27.

116. Bosenberg, W. R.; Lane, R. J.; Tang, C. L., Growth of large, high-quality beta-barium metaborate crystals. *J. Cryst. Growth* 1991, 108, 394–398.

117. Ltoh, K.; Marumo, F.; Kuwano, Y., β-barium borate single crystal growth by a direct Czochralski method. *J. Cryst. Growth* 1990, 106, 728–731.

118. Kouta, H.; Kuwano, Y.; Ito, K.; Marumo, F., β-BaB_2O_4 single crystal growth by Czochralski method II. *J. Cryst. Growth* 1991, 114, 676–682.

119. Kouta, H.; Imoto, S.; Kuwano, Y., β-BaB_2O_4 single crystal growth by Czochralski method using α-BaB_2O_4 and β-BaB_2O_4 single crystal as starting material. *J. Cryst. Growth* 1993, 128, 938–944.

120. Kouta, H.; Kuwano, Y., β-BaB_2O_4 phase-matching-direction growth by the Czochralski method. *J. Cryst. Growth* 1996, 166, 497–501.

121. Bordui, P. F.; Calvert, G. D.; Blachman, R., Immersion-seeded solution growth of large barium borate crystals from sodium chloride solution. *J. Cryst. Growth* 1993, 29, 371–374.

122. Eckardt, R.; Masuda, H.; Fan, Y. X.; Byer, R. L., Absolute and relativ nonlinear optical coefficients of KDP, KD*P, BaB_2O_4, $LiIO_3$, $MgO:LiNbO_3$ and KTP measured by phase-matched second harmonic generation. *IEEE J. Quant. Elec.* 1990, 26, 992–1001.

123. Wang, Y.; Pan, S., Recent development of metal borate halides: Crystal chemistry and application in second-order NLO materials. *Coord. Chem. Rev.* 2016, 323, 15–35.

124. Solov'eva, L. P.; Bakakin, V. V., Crystal structure of potassium boratofluoroberyllate $KBe_2BO_3F_2$. *Sov. Phys. Crystallogr.* 1970, 15, 802–805.

125. Baydina, I. A.; Bakakin, V. V.; Bacanova, L. P.; Pal'chik, N. A., X-ray structural study of borato-fluoroberyllates with the composition $MBe_2BO_3F_2$ (M = Na, K, Rb, Cs). *Zh. Strukt. Khim.* 1975, 16, 963–965.

126. Wu, B.; Tang, D.; Ye, N.; Chen, C., Linear and nonlinear optical properties of the $KBe_2BO_3F_2$ (KBBF) crystal. *Opt. Mater.* 1996, 5, 105–109.

127. McMillen, C. D.; Kolis, J. W., Hydrothermal crystal growth of $ABe_2BO_3F_2$ (A = K, Rb, Cs, Tl) NLO crystals. *J. Cryst. Growth* 2008, 310 (7–9), 2033–2038.

128. Ye, N.; Tang, D. Y., Hydrothermal growth of $KBe_2(BO_3)F_2$ crystals. *J. Cryst. Growth* 2006, 293, 233–235.

129. Zou, G.; Ye, N.; Huang, H.; Lin, X., Alkaline-Alkaline earth Fluoride Carbonate Crystals $ABCO_3F$ (A = K, Rb, Cs; B = Ca, Sr, Ba) as Nonlinear Optical Materials. *J. Am. Chem. Soc.* 2011, 133, 20001–20007.

130. Tran, T. T.; He, J.; Rondinelli, J. M.; Halasyamani, P. S., $RbMgCO_3F$: A New Deep-Ultraviolet Nonlinear Optical Material. *J. Am. Chem. Soc.* 2015, 137, 10504–10507.

131. Zhang, W.; Halasyamani, P. S., Crystal growth and optical properties of a UV nonlinear optical material $KSrCO_3F$. *CrystEngComm* 2017, 19, 4742–4748.

132. Tran, T. T.; Young, J.; Rondinelli, J. M.; Halasyamani, P. S., Mixed-Metal Carbonate Fluorides as Deep-Ultraviolet Nonlinear Optical Materials. *J. Am. Chem. Soc.* 2017, 139, 1285–1295 (e-print 2016).

133. Cammarate, A.; Zhang, W.; Halasyamani, P. S.; Rondinelli, J. M., Microscopic Origins of Optical Second Harmonic Generation in Noncentrosymmetric-Nonpolar Materials. **Chem. Mater.** 2014, 26, 5773–5781.

134. Pilz, T.; Jansen, M., $Li_2B_6O_9F_2$, a New Acentric Fluorooxoborate. *Z. Anorg. Allg. Chem.* 2011, 637, 2148–2152.

135. Pilz, T.; Nuss, H.; Jansen, M., $Li_2B_3O_4F_3$, a new lithium-rich fluorooxoborate. J. Solid State Chem. 2012, 186, 104–108.

136. Andriyevsky, B.; Pilz, T.; Yeon, J.; Halasyamani, P. S.; Doll, K.; Jansen, M., DFT-based ab initio study of dielectric and optical properties of bulk $LI_2B_3O_4F_3$ and $Li_2B_6O_8F_2$. *J. Phys. Chem. Solids* 2013, 74, 616–623.

137. Liang, F.; Kang, L.; Gong, P.; Lin, Z.; Wu, Y., Rational Design of Deep-Ultraviolet Nonlinear Optical Materials in Fluorooxoborates: Toward Optimal Planar Configuration. *Chem. Mater.* 2017, 29, 7098–7102.

138. Wang, X.; Wang, Y.; Zhang, B.; Zhang, F.; Yang, Z.; Pan, S., CsB_4O_6F: A Congruent-Melting Deep-Ultraviolet Nonlinear Optical Material by Combining Superior Functional Units. *Angew. Chem. Int. Ed. Engl.* 2017, 56, 14119–14123.

139. Shi, G.; Wang, Y.; Zhang, F.; Zhang, B.; Yang, Z.; Hou, X.; Pan, S.; Poeppelmeier, K. R., Finding the Next Deep-Ultraviolet Nonlinear Optical Material: $NH_4B_4O_6F$. *J. Am. Chem. Soc.* 2017, 139, 10645–10648.

140. Wu, H.; Yu, H.; Pan, S.; Halasyamani, P. S., Deep-Ultraviolet Nonlinear-Optical Mateiral $K_3Sr_3Li_2Al_4B_6O_{20}F$: Addressing the Structural Instability Problem in $KBe_2BO_3F_2$. *Inorg. Chem.* 2017, 56, 8755–8758.

141. Yu, H.; Young, J.; Wu, H.; Zhang, W.; Rondinelli, J. M.; Halasyamani, P. S., The Next Generation of Nonlinear Optical Material: $Rb_3Ba_3Li_2Al_4B_6O_{20}F$ — Synthesis, Characterization, and Crystal Growth. *Adv. Opt. Mater.* 2017, 5 (23), 1700840.

142. Zhao, S. G.; Kang, L.; Shen, Y. R.; Wang, X.; Asghar, M. A.; Lin, Z. S.; Xu, Y. P.; Zeng, S.; Hong, M. C.; Luo, J. H., Designing a Beryllium-Free Deep-Ultraviolet Nonlinear Optical Material with a Structural Instability Problem. *J. Am. Chem. Soc.* 2017, 138, 2961–2964.

143. Tran, T. T.; Yu, H.; Rondinelli, J. M.; Poeppelmeier, K. R.; Halasyamani, P. S., Deep Ultraviolet Nonlinear Optical Materials. *Chem. Mater.* 2016, 28, 5238–5258.

144. Halasyamani, P. S.; Zhang, W., Viewpoint: Inorganic Materials for UV and Deep-UV Nonlinear Optical Applications. *Inorg. Chem.* 2017, 56, 12077–12085.

4

The Solid-State Chemistry of AUO_4 Ternary Uranium Oxides: A Review

Gabriel L. Murphy[*,†], *Zhaoming Zhang*[†] *and Brendan J. Kennedy*[*]

School of Chemistry, The University of Sydney, Sydney, NSW 2006, Australia
[†]*Australian Nuclear Science and Technology Organisation, Lucas Heights, NSW 2234, Australia*

Introduction

Uranium, though relatively scarce with terrestrial abundance of 2.7 ppm,[1] is societally important in many applications including radiometric dating,[2] military ordnance,[3,4] catalysis,[5,6] and pertinently forms the keystone to the nuclear fuel cycle and the current geopolitical climate resulting from the nuclear age.[7-9] Uranium is most commonly encountered as $U(IV)O_2$, which is the most widely utilized composition in nuclear fuels and is also the iteration most commonly encountered in nature, where it exists as the mineral uraninite.[10-12] The interaction of uraninite, and its partially oxidized forms, with other elements precipitates a myriad of minerals with various physical properties, exhibiting a range of coordination environments, and oxidation states of uranium.[13,14] When subjected to prolonged nuclear fission in nuclear reactors, UO_2 nuclear fuels spawn a plethora of fission daughters which are able to recombine with the fuel matrix to form a number of fuel-derived secondary phases.[15-18] These display a range of chemistries, which can be further convoluted as the fuel is partitioned during fuel reprocessing. During reprocessing the spent fuel can become oxidized, leading to the presence of additional uranium oxidation states, and the formation of new phases. Understanding the physico-chemical properties of these environmentally hazardous and physiologically dangerous phases is considered pertinent. As a part of international plans for the disposal of spent nuclear fuels, which would ultimately be subjected to deep geological disposal, it is imperative that these phases are properly identified, to ensure they are immobilized within appropriate waste forms so as to prevent their migration into the environment.[12,15,19,20]

Surprisingly, despite the awareness of uranium compounds extending back to their discovery in 1789 by Martin Heinrich Klaproth and the subsequent development of the nuclear energy industry, a paucity of information remains regarding the structure, and properties of even some of the most simple uranium oxides. This can be attributed to the challenges in both handling radioactive materials and in their analyses.

Markedly apparent in structural elucidation methods using X-rays, the absorption of X-rays by uranium, and its strong X-ray scattering power can lead to relatively poor precision regarding the structure of these materials, particularly regarding the lighter elements such as oxygen.[21] Consequently, the precise coordination environment of uranium in many complex oxides is relatively poorly defined.

Attempts to theoretically model materials containing uranium (and other actinides) are extremely challenging, and their study can still be considered something of a developing science. This is a consequence of the large number of electrons in uranium cations, which requires the use of effective core potentials, the need to take relativistic effects into account, and the impact of correlation effects arising from the diffusivity of the valence orbitals that are often mixed between the s, p, d, and f orbitals. Traditional calculation methods require significant alterations to correctly account for these factors. These complexities place limitations upon the validity of calculations of nuclear fuels or waste forms containing uranium. The continuously improving performance of neutron and synchrotron X-ray scattering facilities has led to a number of precise structural models of both uranium- and actinide-containing systems becoming available.[22,23] These provide a means to verify the accuracy of theoretical studies and enable the physico-chemical behaviors of new phases to be more correctly predicted. The powerful combination of precise structural determination and theoretical modeling enables the solid-state chemistry, which underpins the formation and properties of actinide-containing compounds (particularly those that arise from spent nuclear fuel), to be progressively unraveled, revealing a rich vein of chemistry and crystallography that is often unique to the actinides but with potential to guide further developments in inorganic and physical sciences.

The ternary uranium oxides of the form AUO_4 are a class of materials relevant to spent nuclear fuel partitioning, waste forms, and uranium oxide concentrates.[24,25] A number of examples occur naturally as minerals.[26–28] Recently, it has been demonstrated that a number of such oxides display unusual oxygen transport phenomena, apparent ionic conductivities, and the ability to undergo unique and unexpected structural phase transformations.[29–31] Colloquially known as the monouranates, the AUO_4 oxides are distinct to the crystallographically similar tetragonal wulfenite, $AMoO_4$, and scheelite, AWO_4, structures. The AUO_4 oxides, containing either pentavalent or hexavalent uranium cations with distorted tetragonal or hexagonal bipyramid coordination environments, often contain the covalent collinear oxo uranyl species. Knowledge of monouranate compounds is relatively sparse, with the structures described with varying degrees of accuracy and precision. This chapter will focus on examining the current literature of the monouranates in relation to their structure–property relationships before discussing recent advances, made by the present authors and others, unraveling some of the unique properties of the monouranates.

Uranium Solid-State and Structural Chemistry

The solid-state chemistry of the monouranates is essentially defined by the chemistry of uranium. Thus, it is pertinent to briefly discuss the core aspects of uranium oxide chemistry in relation to the monouranates. Uranium has the ground state electron configuration $[Rn]5f^36d^17s^2$, and in the solid state it is regularly encountered in either the tetravalent ($[Rn]5f^2$), pentavalent ($[Rn]5f^1$), or hexavalent ($[Rn]5f^0$) valence state. The uranyl group, UO_2^{2+}, is synonymous with uranium oxides, and it is often found in other transuranic compounds; consequently, it is generally described as the actinyl group AnO_2^{2+} (An = actinide element). The uranyl group consists of a central uranium atom bonded to two oxygen atoms forming short, strong, covalent, and collinear bonds, $[O=U=O]^{2+}$. Although similar functional groups are known in general inorganic materials, for instance the vanadyl group, the contrasting redox chemistry and ability to access $5f$ electrons makes the uranyl/actinyl groups chemically distinct.[32–36] This is apparent in literature with several examples of unique chemical phenomena and properties being uncovered in uranium materials, where the origin can be traced to the influence of the uranyl group.[34,35,37–39] The most significant orbital contributions in the uranyl linkage is the mixing of the uranium $6d$, $5f$, $6p$, and $6s$ orbitals with the oxygen $2p$ and

Uranium
Oxygen

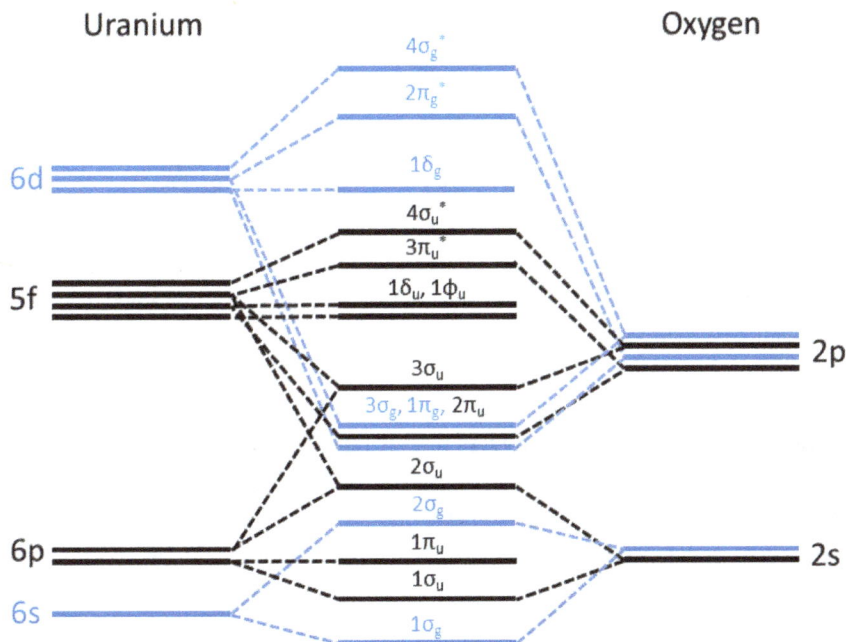

Figure 1. Molecular orbital energy-level diagram for the uranyl linkage (adapted from Figure 1 in Ref. 35).

$2s$ orbitals. A molecular orbital energy-level diagram for UO_2^{2+} is presented in Figure 1. The two uranium primary valence shells, $5f$ and $6d$, enable σ and π bonds primarily to oxygen $2p$ orbitals, forming $3\sigma_g$, $3\sigma_u$, $1\pi_g$, and $2\pi_u$ (the highest occupied molecular orbitals, HOMOs) and $4\sigma_u^*$, $3\pi_u^*$, $1\delta_u$, and $1\phi_u$ (the lowest unoccupied molecular orbitals, LUMOs). The uranium $6p$ and $6s$ shells must also be considered, because they have been shown to contribute to core electron behavior and to spatially extend beyond the $5f$ shell.[35,36,40] Figure 1 omits the uranium $7s$ and $7p$ orbitals; however, these have been shown to contribute to bonding in several cases.[35,36,40] Readers are encouraged to view references[35,36,38,40] for a more extensive discussion of the molecular orbitals and bonding of uranium. However, it should be apparent to the reader from this brief overview that understanding the bonding of uranium, both in uranyl and non-uranyl systems, is far from trivial.

The uranyl group is commonly encountered as the $U(VI)O_2^{2+}$ moiety, though it can exist as $U(V)O_2^+$. The latter is relatively unstable since it results in the addition of a single $5f$ electron into the $5f$ LUMO orbitals, which leads to the lengthening of the uranyl bond. The uranyl group is generally not expected to be encountered in U(IV) oxides. The binary U(IV) oxide UO_2 adopts the regular fluorite structure.[35] UO_2 is far more ionic than UO_2^{2+}, where the introduction of two electrons into the $5f$ shell reduces the covalency and uranyl character.[35,40] It must be stressed that although the uranyl group has been assigned to many structures, precise chemical details of its effects and interactions are not well understood in the monouranates. The current understanding of the uranyl group, its chemistry, and related theoretical studies are largely derived from liquid, gas phase, and organometallic uranium compound investigations. Such compounds are clearly distinct to the condensed oxide phases, although they provide a good foundation to the understanding of the observed structural and bonding properties found in the monouranates.

In the monouranates and other uranium oxides, the uranyl group is often surrounded by four to six additional oxygen atoms forming UO_6, UO_7, and UO_8 polyhedra with tetragonal, pentagonal, and hexagonal bipyramidal coordination environments, respectively. These motifs are illustrated in Figure 2. Of the three polyhedral arrangements, the UO_6 and UO_8 polyhedra are most frequently encountered in monouranates. UO_7 polyhedra are relatively uncommon but can be found in other complex oxides such as $Na_2[(UO_2)(MoO_4)_4]$.[41]

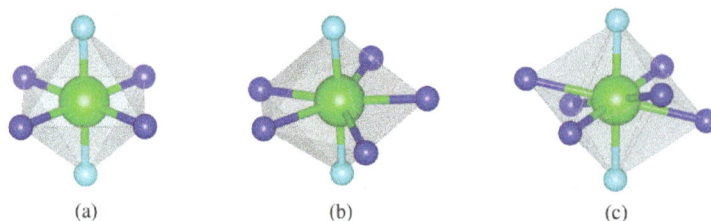

Figure 2. Uranium–oxygen coordination polyhedra motifs regularly encountered in the solid state: (a) UO_6 seen in β-$SrUO_4$,[165] (b) UO_7 seen in $Na_2[(UO_2)(MoO_4)_4]$,[41] and (c) UO_8 seen in $CaUO_4$.[165] Uranium atoms are shown in green, uranyl oxygens in aqua, and non-uranyl coordinating oxygens in blue.

The uranyl group is most often a terminal oxo ligand and it rarely bonds to other uranium centers. When such a rare bonding motif does occur, it is known as a cation–cation interaction (CCI).[42] CCIs are observed much more frequently in $An(V)O_2^+$ than in $An(VI)O_2^{2+}$ systems.[42–44] For example, they were seen in approximately 50% of Np oxides containing the NpO_2^+ group but only in around 2% of oxides containing the UO_2^{2+} group.[43,45,46] More CCIs are observed in UO_2^+ than in UO_2^{2+}, as it is more electrostatically favorable for a $[O=U=O]^+$–UO_2^+ interaction to occur than a $[O=U=O]^{2+}$–UO_2^{2+} interaction. Since hexavalent uranium is most commonly present in the monouranates, CCIs are not regularly encountered in them. Considering the geometry of uranyl-containing UO_n polyhedra ($n = 6$–8), most monouranates form either 1D chain or 2D layered structures where the terminal uranyl group is directed at the A-site cations between the 1D chain or 2D layers containing the uranium atoms. By extension, 3D framework structures generally preclude the presence of the uranyl group, or if they do they possess CCIs. Details of AUO_4 monouranate structural types will be discussed in the subsequent sections.

$R\bar{3}m$ *rhombohedral and Pbcm and Cmmm orthorhombic monouranates*

Knowledge of the monouranates first emerged from the Manhattan project. In 1944, Zachariasen isolated single crystals of $CaUO_4$ and $SrUO_4$ from molten salts of $CaCl_2$ or $SrCl_2$ and U_3O_8 or UO_3, and his work was published in 1948.[47] Due to the crystal quality, single-crystal X-ray diffraction could not be performed, rather the structures were investigated using powder X-ray diffraction methods. The two oxides were found to be isostructural, forming a rhombohedral structure in space group $R\bar{3}m$. As shown in Figure 3, the structure consists of edge-sharing UO_8 polyhedra creating 2D layers with collinear axial oxygens, geometrically consistent with the uranyl group, directed toward the Ca^{2+}/Sr^{2+} cations (which are also eight-fold coordinated to oxygens). Later studies[48–51] identified a second polymorph of $SrUO_4$ that was determined to be orthorhombic in space group $Pbcm$ and denoted as β-$SrUO_4$, whereas the rhombohedral form was denoted as the α polymorph. The structure of β-$SrUO_4$ consists of edge-sharing UO_6 polyhedra forming infinite 2D layers that are separated by Sr^{2+} cations (6-coordinated). Similar to α-$SrUO_4$, collinear axial oxo oxygens exist in β-$SrUO_4$, geometrically consistent with the uranyl group, and are directed toward the Sr^{2+} cations. The structure of β-$SrUO_4$ is illustrated in Figure 4. The uranium coordination motifs for both polymorphs can be found in Figure 2.

Neutron powder diffraction studies of $CaUO_4$ confirmed the original rhombohedral structural model.[49,51,52] However, the structural refinements for $CaUO_4$ and α-$SrUO_4$ produced relatively long uranyl bond lengths of above 1.95 Å. Such values are considered uncharacteristic of the uranyl bond.[13,21,53] Subsequent infrared and Raman spectroscopic studies by Liegeosi-Duyckaerts[54] confirmed the presence of the uranyl group in $CaUO_4$ and in a solid solution of $Ca_{1-y}Sr_yUO_4$ with y up to 0.85. Recent computational, spectroscopic, and structural investigations have further demonstrated the presence of the uranyl group in $CaUO_4$[55–59] and α-$SrUO_4$.[60] Some non-reproducibility in IR frequencies was noted in the investigation by

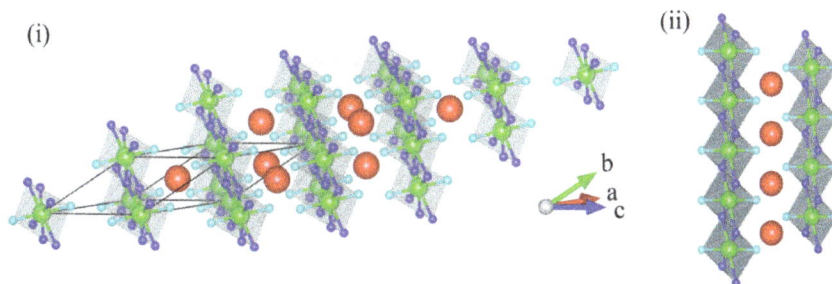

Figure 3. (i) Structural representation of the rhombohedral structure in space group $R\bar{3}m$ for CaUO₄ (see Ref. 165) (isostructural to α-SrUO₄ and α-CdUO₄) where $a = 6.26005(7)$ Å, $\alpha = 36.0375(2)°$, and unit cell volume = $75.9496(16)$ Å³. The red, green, blue, and aqua spheres represent Ca, U, non-uranyl oxygen, and uranyl oxygen atoms, respectively. The unit cell is illustrated. (ii) Highlights the terminal coordination of uranyl oxygens directed at the in-between layers with Ca²⁺ cations.

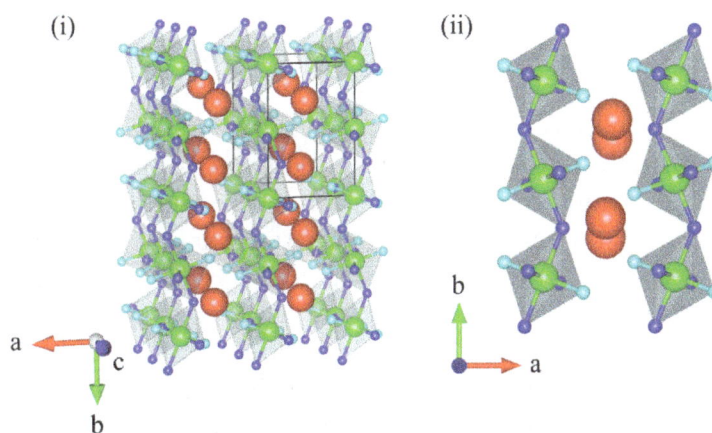

Figure 4. (i) Structural representation of the orthorhombic structure in space group *Pbcm* for β-SrUO₄ (see Ref. 165) (isostructural to BaUO₄) where $a = 5.49185(10)$ Å, $b = 7.97629(2)$ Å, $c = 8.12698(19)$ Å, and unit cell volume = $355.999(14)$ Å³. The red, green, blue, and aqua spheres represent Sr, U, non-uranyl oxygen, and uranyl oxygen atoms, respectively. (ii) Highlights the coordination motif of the terminal uranyl oxygens directed at the Sr cations; note the relative distortion compared to that observed in the rhombohedral motif of CaUO₄.

Liegeosi-Duyckaerts,[54] particularly at high Sr contents, suggestive of a miscibility gap in the solid solution. Pertinently, this study suggested the ability for both CaUO₄ and α-SrUO₄ to support oxygen vacancy defects. Inspired by this, further studies exploring how many oxygen defects could be hosted in CaUO₄ and α-SrUO₄ were initiated using either *ex situ* structural studies[61,62] or *in situ* thermogravimetric measurements.[63–65] The *ex situ* diffraction measurements provided evidence for phase separation in anion-deficient monouranates, particularly for CaUO₄.[61] Electron microscopy investigations suggested that when CaUO₄ was treated under reducing conditions, it may contain both reduced and non-reduced regions.[66,67] Furthermore, thermogravimetric measurements seemed to indicate that reduction and oxidation are not two completely complementary processes.[64,67] Based on these experimental observations, Prodan and Boswell[67] postulated that microdomains formed with variable oxygen contents in reduced CaUO₄₋ₓ, and that the nature and number of these microdomains were influenced strongly by oxygen defect–defect interactions. These interactions were argued to lead to short-range order of the oxygen vacancies which would influence its reactivity such that an oxidation thermal event would occur at a different stoichiometry from the reverse reduction event. *In situ* structural studies using laboratory X-ray powder diffraction by Pialoux and Touzelin[68,69] examined the reaction of CaCO₃ with UO₂ at high temperatures by varying the experimental condition from oxidizing to neutral. They described a $CaUO_3$ perovskite-like monoclinic phase forming in heavily reduced samples, which existed from ambient temperature to 1500°C, above which it was reported to undergo a phase transformation to an orthorhombic structure. However, they did not solve or refine the two proposed $CaUO_3$ structures, nor have there been any subsequent studies to corroborate their work. Considering its low Goldschmidt tolerance factor (t) of 0.84 calculated using the equation below, the existence of $CaUO_3$ as a perovskite is questionable.[70–72] In the following equation, r_A, r_U, and r_O correspond to ionic radius of the A-site cation, uranium and oxygen, respectively.

$$t = \frac{(r_A + r_O)}{\sqrt{2(r_U + r_O)}}$$

The structural phase transformation from rhombohedral α-$SrUO_4$ to orthorhombic β-$SrUO_4$ was originally examined by Rudorff and Pfitzer[50] who showed this occurring irreversibly above 600°C. The transformation was also examined using thermogravimetric methods by Tagawa and Fujino[73,74] who found, interestingly, mass loss upon heating α-$SrUO_4$, consistent with the loss of oxygen, before mass gain as it formed the stoichiometric β-$SrUO_4$. The two-step process associated with the transformation was also monitored with differential thermal analysis where the initial mass loss and subsequent gain steps were found to be endothermic and exothermic, respectively. The mass loss was calculated to correspond to a stoichiometry of $SrUO_{3.7}$. Tagawa and Fujino further argued, based on estimations of the partial molar free energies of oxygen for the two monouranates, that α-$SrUO_4$ is metastable and only exists on account of a high activation energy barrier that seemingly exists between itself and β-$SrUO_4$. Interestingly, it was observed that whereas the rhombohedral polymorph can exist with a range of oxygen stoichiometries, α-$SrUO_{4-x}$, orthorhombic β-$SrUO_4$ appears not to support oxygen vacancies. That Zachariasen synthesized α-$SrUO_4$ at temperatures higher than the transformation temperature might be attributed to the use of the molten flux method which may limit the amount of oxygen present and thus inhibit the formation of the orthorhombic β-$SrUO_4$.[47] It is also likely that the structure Zachariasen reported was substoichiometric and best described as α-$SrUO_{4-x}$. Flux growth methods have recently become very topical, where the ability to generate crystals of suitable quality for single-crystal X-ray diffraction studies has revealed several novel and intricate uranium and actinide phases.[39,44,56,75–87] A second phase transformation in $SrUO_4$ was reported to occur above 1200°C where β-$SrUO_4$ loses oxygen, forming a rhombohedral structure described as γ-$SrUO_{4-x}$.[73,88,89] The authors argued that this transformation is the result of the inability of the β-$SrUO_4$ structure to accommodate oxygen vacancies. From their description, however, this structural change might be better attributed to decomposition rather than a structural transformation.

Tagawa and Fujino's investigations[73,74] revealed several unusual aspects in the α–β transformation in $SrUO_4$; however, it remained unclear why orthorhombic β-$SrUO_4$ cannot support oxygen defects whereas rhombohedral α-$SrUO_4$ and $CaUO_4$ can. Further, Tagawa and Fujino speculated that the generation of a sub stoichiometric structure requires uranium to be reduced to lower valence states. Takahashi et al.,[61] from structural refinements against their laboratory X-ray powder diffraction data, suggested that the vacancies formed preferentially on the equatorial $O(2)$ site as opposed to the $O(1)$ or uranyl oxygen site. Pialoux and Touzelin,[90] using in situ high-temperature laboratory X-ray powder diffraction, investigated the reaction between strontium oxide and uranium dioxide where they controlled the partial pressure of oxygen from highly oxidizing to essentially inert. They identified both the rhombohedral and orthorhombic forms of $SrUO_4$ and determined that the former exists with variable compositions in terms of oxygen content, whereas the latter polymorph is stoichiometric. They further identified a perovskite-like $SrUO_3$ phase, which they concluded was a result of extensive reduction of $SrUO_4$. The Goldschmidt tolerance factor for $SrUO_3$ is quite low (0.88), similar to that for $CaUO_3$, suggesting that the formation of a perovskite structure is unlikely to be favorable.

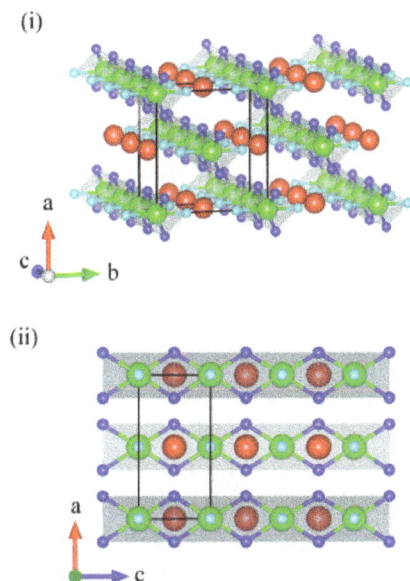

Figure 5. (i) Structural representation of the orthorhombic structure in space group *Cmmm* for β-CdUO$_4$ (see Ref. 92) where a = 7.023(4) Å, b = 6.849(3) Å, c = 3.514(2) Å, and unit cell volume = 169.03(3) Å^3. The red, green, blue, and aqua spheres represent Cd, U, non-uranyl oxygen, and uranyl oxygen atoms, respectively. (ii) Highlights the rutile-like chain configuration of the UO$_6$ and CdO$_6$ polyhedra in the [001] direction.

Ippolitova *et al.*[91] reported in 1961 that when oxides of cadmium and uranium were mixed together and heated at 570°C, a rhombohedral structure formed which was seemingly isostructural to CaUO$_4$ and α-SrUO$_4$. Denoted as α-CdUO$_4$, they reported it to undergo a phase transformation above 720°C to an orthorhombic phase denoted as β-CdUO$_4$ which transformed into an oxygen-deficient rhombohedral γ-CdUO$_{4-x}$ phase above 925°C. β-CdUO$_4$ was originally described as orthorhombic in *Pbam* by Kovba *et al.*[48] in 1961, but a later study by Yamashita *et al.*[92] using laboratory X-ray powder diffraction with refinement methods was able to produce relatively precise models of both α-CdUO$_4$ and β-CdUO$_4$ structures. Yamashita *et al.* confirmed that α-CdUO$_4$ is isostructural to CaUO$_4$ and α-SrUO$_4$ in space group $R\bar{3}m$ (see Figure 3). Although β-CdUO$_4$ was found to be orthorhombic, it is not isostructural to β-SrUO$_4$, instead the structure is described as a distorted rutile-like structure in space group *Cmmm*. This structure consists of edge-sharing UO$_6$ polyhedra forming infinite chains along the [001] direction where the uranyl groups are directed at the 6-coordinated Cd^{2+} cations that form edge-sharing chains in the [001] direction also, as displayed in Figure 5. Tagawa and Fujino[93] examined the α-CdUO$_4$ to β-CdUO$_4$ phase transformation and the transformation from β-CdUO$_4$ to the oxygen-deficient γ-CdUO$_{4-x}$ phase using thermogravimetric and differential thermal analysis measurements. They found, similar to the α-SrUO$_4$ to β-SrUO$_4$ transition, that as α-CdUO$_4$ is heated it loses mass, presumably corresponding to the loss of oxygen, followed by a reoxidation step before transforming to stoichiometric β-CdUO$_4$. The overall transition from α-CdUO$_4$ to β-CdUO$_4$ is irreversible, whereas the β to γ transition is reversible under oxidizing conditions. The reported β-CdUO$_4$ to γ-CdUO$_{4-x}$ transformation is similar to that described for the β-SrUO$_4$ to γ-SrUO$_{4-x}$ transition,[73,88,89] and involves a considerable oxygen loss from the β structure. This raises the same question of whether the observed structural change was also due to sample decomposition.

Matar and coworkers recently used *ab initio* methods, in the form of density functional theory, to examine CaUO$_4$ (see Ref. 57) and β-CdUO$_4$ (see Ref. 94) with a focus on their response to isotropic and anisotropic compression. Isotropic compression was found to amount to a relatively unexceptional bulk modulus for both monouranates, but interestingly the anisotropic compression was found to be significantly larger along the uranyl bond. This anisotropy has also been noted in computational studies of other uranium compounds,[95] although there are few experimental *in situ* studies available that have studied such anisotropic compression in monouranates and other uranyl compounds.

$BaUO_4$ was first identified by Samson and Sillen[96] in 1948, who found it to be orthorhombic in space group *Pbcm* and isostructural to β-$SrUO_4$ (see Figure 4 for structural representation). The structure was later better described by Loopstra and Rietveld in their neutron powder diffraction studies.[51] Further studies utilizing vibrational spectroscopic techniques have confirmed the presence of the uranyl group in both β-$SrUO_4$ and $BaUO_4$.[59,97,98] Reis *et al.*[99] redetermined the structure of $BaUO_4$ as orthorhombic in space group *Pbcm* using laboratory X-ray powder diffraction, solving the structure using a combination of Patterson, Fourier, and least-squares techniques. A structural investigation by Jakes and Krivy,[55] using laboratory X-ray powder diffraction, also supported the orthorhombic structural assignment for β-$SrUO_4$ and $BaUO_4$. They confirmed the model of Loopstra and Rietveld, and highlighted the peculiarity of having a uranyl species with its bond length above the typically encountered range of 1.75–1.82 Å. $BaUO_4$ has received relatively less attention than $CaUO_4$ and $SrUO_4$. It is not known to undergo any phase transformations below 1200°C under oxidizing conditions. Thermodynamic studies have shown that it exhibits unexceptional thermal behavior.[100,101] In contrast to oxygen defect formation in $CaUO_4$ and α-$SrUO_4$, $BaUO_4$ is reported to transform directly to $Ba_2U_2O_7$ when heated under mildly reducing conditions, without forming oxygen vacancies.[102–106] $Ba_2U_2O_7$ has a weberite-type Na_2MgAlF_7 structure with the two crystallographically distinct uranium cations having different distorted octahedral geometries[102] but both being pentavalent.[104,106] When exposed to highly reducing conditions, $BaUO_4$ can be reduced to a perovskite structure, $BaUO_3$.[107] $BaUO_3$ can also be prepared by reacting UO_2 and $BaCO_3$ at high temperatures.[108,109] The tolerance factor of $BaUO_3$, $t = 0.999$, suggests this would be a stable perovskite structure. The structure of $BaUO_3$ is still somewhat debated. Cordfunke *et al.*[110] and Barret *et al.*,[111] both using neutron powder diffraction, described it as orthorhombic in space group *Pbnm* (or *Pnma* in an alternative setting). They both concluded that the structure is nonstoichiometric, containing an equivalent number of uranium and barium vacancies, and is hyperstoichiometric with respect to oxygen and is best denoted as $BaUO_{3+x}$. However, Hinatsu[107] argued that the near stoichiometric structure ($BaUO_{3.03}$) is cubic in space group $Pm\bar{3}m$ and that the excess oxygen leads to an orthorhombic distortion.

In 1958, Frondel and Barnes[112] used laboratory X-ray powder diffraction to identify $PbUO_4$ from heating oxides of lead and uranium with excess lead sheets in water at 230–290°C, finding it to be isostructural to β-$SrUO_4$ and $BaUO_4$ (see Figure 4 for its structural representation). A later single-crystal X-ray diffraction study of $PbUO_4$ by Cremers *et al.*[113] provided a more accurate description of the structure confirming its orthorhombic assignment in space group *Pbcm*. Popa *et al.*[114] investigated the thermal properties of $PbUO_4$, including thermal expansivity, thermal diffusivity, and thermal conductivity from –272 to 627°C. It was found that, despite being considered as being an insulator, $PbUO_4$ exhibited an electronic contribution to its conductivity, characteristic of a semiconductor similar to that observed in $BaUO_4$.[101] Such behavior is absent in β-$SrUO_4$.[115] Other than this, $PbUO_4$ displayed unexceptional behavior and is reported to volatilize above 800°C with loss of Pb.[114]

Ibmm (Imma) and Pbcn orthorhombic monouranates

$MgUO_4$, first identified by Zachariasen in 1954, forms a distinctly different structure to the other alkaline earth metal monouranates, adopting an orthorhombic structure in space group *Imma*.[116] In this structure, described hereafter in the alternative space group setting of *Ibmm* for ease of comparison with the *Cmmm* structure, the UO_6 polyhedra edge share forming 1D chains along the [001] direction where oxo uranium oxygen bonds, consistent with the formation of the uranyl group, are directed at the Mg cations which are six-fold coordinated in a chain configuration along the [001] direction, as illustrated in Figure 6. X-ray

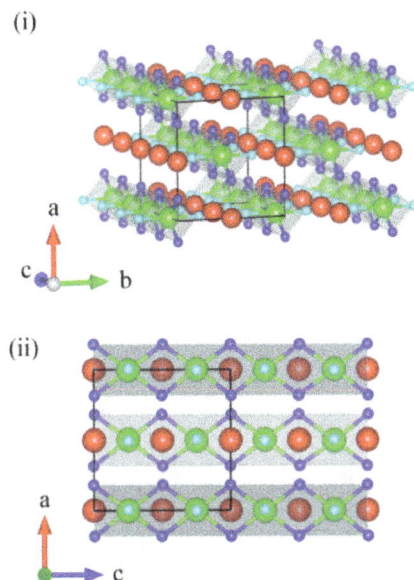

Figure 6.   (i) Structural representation of the orthorhombic structure in space group *Ibmm* for MgUO₄ (see Ref. 116) (isostructural to CoUO₄ and MnUO₄) where $a = 6.595(2)$ Å, $b = 6.520(2)$ Å, $c = 6.924(2)$ Å, and unit cell volume = 297.73(2) Å³. The red, green, blue, and aqua spheres represent Mg, U, non-uranyl oxygen, and uranyl oxygen atoms, respectively. (ii) Highlights the rutile-like chain configuration of the UO₆ and MgO₆ polyhedra in the [001] direction.

absorption spectroscopic measurements by Guo *et al.*[117] showed the oxidation state of U in MgUO₄ to be unequivocally hexavalent.

In comparison to the *Cmmm* structure, assigned to β-CdUO₄ by Yamashita *et al.*[92] and shown in Figure 5, the *Ibmm* structure has a less regular anionic sublattice resulting in the doubling of the *c*-axis.[118] The non-uranyl oxygen atoms edge share to form 1D chains in the [01] direction. Given that the subtle difference between the *Cmmm* and *Ibmm* structures lies in the oxygen sublattice and the study by Yamashita *et al.*[92] used laboratory X-ray powder diffraction, which is not very sensitive to oxygens (especially in the presence of much heavier Cd and U), it is desirable to confirm the crystal structure of β-CdUO₄ using neutron diffraction.

Both CoUO₄ and MnUO₄, first studied by Bertaut *et al.*[119] using neutron diffraction measurements, are isostructural to MgUO₄ with an orthorhombic structure in space group *Ibmm*. The magnetic structures of these two oxides were determined by Bacmann, Bertaut, and others.[119–122] Below 11K, CoUO₄ adopts an antiferromagnetic structure; Bacmann and Bertaut[121] proposed a non-collinear model for the magnetic structure of CoUO₄ based on a 2D irreducible representation with the wave vector $k = [½ \, 0 \, ½]$ in magnetic space group (Shubnikov group) $P2_1'/m$. Anisotropic exchange coupling was attributed to the Co²⁺ cation having a paramagnetic electron configuration $[Ar]d^7$ with no magnetic moment arising from the diamagnetic U⁶⁺ cation. The magnetic structure of MnUO₄ is different from that of CoUO₄. MnUO₄ undergoes a magnetic transition near 44K forming an antiferromagnetic structure similar to that of pyrolusite (MnO₂).[123–125] The magnetic structure consists of Mn²⁺ spins aligning antiferromagnetically along the [010] direction being described by the wave vector $k = [½ \, 1 \, ½]$ with the same magnetic and nuclear unit cells. Magnetization measurements by Guillot and Pauthene[126] revealed that CoUO₄ undergoes a metamagnetic transition to a ferromagnetic state that saturates above 130 kOe, whereas they did not observe saturation for MnUO₄.

The monouranates CrUO₄ and FeUO₄, first identified in the 1960s, adopt the orthorhombic structure in space group *Pbcn*.[127–132] This structure can be considered as approximately 2D layered, where UO₆ polyhedra corner share in the [100] direction forming a layered configuration, and the CrO₆ or FeO₆ octahedra

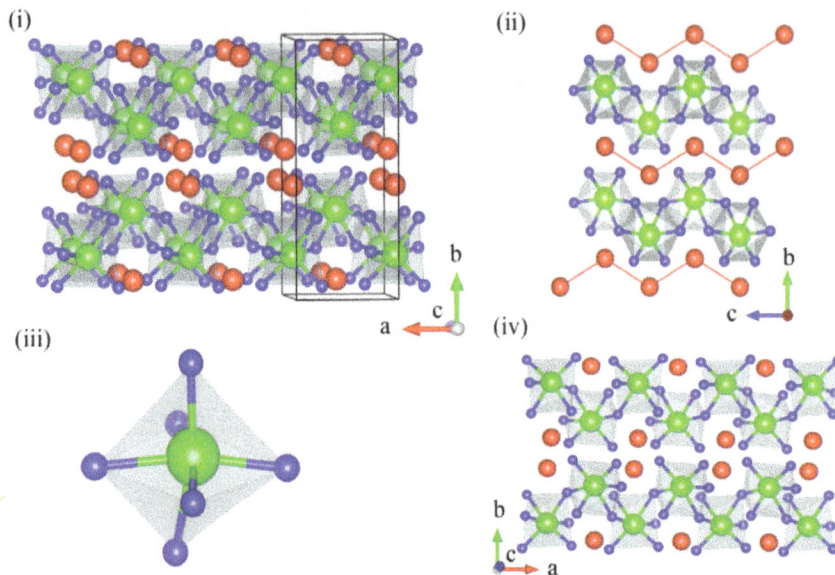

Figure 7.   (i) Structural representation of the orthorhombic structure in space group *Pbcn* for $FeUO_4$ (see Ref. 130) (isostructural to $CrUO_4$) where $a = 4.8844(2)$ Å, $b = 11.9328(5)$ Å, $c = 5.1070(2)$ Å, and unit cell volume $= 297.66(2)$ Å^3. The red, green, and blue spheres represent Fe, U, and non-uranyl oxygen atoms, respectively. (ii) Highlights the (zig-zag) chain configuration along the *c*-axis for both UO_6 and FeO_6 polyhedra (red line illustrates the zig-zag connectivity of Fe). (iii) The coordination environment of the UO_6 polyhedra, particularly the nonlinear trans oxo O–U–O configuration of the U–O bonds, suggestive of the absence of uranyl. (iv) The corner sharing motif of the UO_6 polyhedra in the [100] direction.

edge share, forming corrugated chains in the [001] direction in-between the UO_6 layers, as illustrated in Figure 7. From X-ray absorption spectroscopic measurements,[117] supported by neutron diffraction[128,129] and magnetic measurements,[120] the oxidation states of the uranium and transition metal cations were determined to be pentavalent and trivalent, respectively. Guo *et al.*[117] argued, based on *ab initio* calculations, that the strong preference of iron and chromium to be trivalent drives the reduction of uranium even when the materials are prepared under oxidizing conditions. Single-crystal X-ray diffraction studies of $FeUO_4$ by Read *et al.*[130] highlighted the peculiarity of the UO_6 polyhedra, in which the uranyl group is not observed as there are no apparent trans oxo groups, rather all the U–O bonds are of a similar length.

$CrUO_4$ is antiferromagnetic below 44.5K, and its magnetic structure was established using neutron diffraction.[129,133] The structure contains, two magnetic sublattices from the ordered magnetic Cr^{3+} and U^{5+} cations respectively, and was described in the magnetic space group *Pbcn'*. $CrUO_4$ and $NdCrTiO_5$ were the early examples of magnetoelectric materials having two magnetic sublattices.[133] $FeUO_4$ undergoes an antiferromagnetic transition below 55K forming an antiferromagnetic structure that can be described using a non-collinear model.[127] The overall magnetic moment is determined by the antiferromagnetically aligned d^5 Fe^{3+} cations (4 μ_B) as well as the antiferromagnetically aligned f^1 U^{5+} cation (2 μ_B).[127,128,131,132] This non-collinear structure can be described using the magnetic space group *Pbcn*. Below 42K, $FeUO_4$ undergoes a second magnetic ordering transition, forming a ferromagnetic structure with a magnetic moment of 2 μ_B from the ferromagnetically aligned Fe^{3+} cations, whereas the spins of the U^{5+} cation remain antiferromagnetically coupled. The resulting collinear magnetic structure can be refined in the magnetic space group *Pb'c'n*. $FeUO_4$ is a rare example of a ferromagnetic insulator.[134]

The structures of $NiUO_4$ and $ZnUO_4$ have been reported by Young[135] and Hoekstra *et al.*,[136] although the synthesis of these required the use of high-pressure–high-temperature methods. Two polymorphs of $NiUO_4$ have been described; the first was reported to be isostructural with $MgUO_4$ and the other with

$CrUO_4$, based on indexing their laboratory powder X-ray diffraction patterns without refining the structures. $ZnUO_4$, also prepared using high-pressure–high-temperature methods by Hoekstra *et al.*,[136] was reported to be isostructural to $CrUO_4$. Hoekstra *et al.* could not discern which of the $NiUO_4$ polymorphs formed at higher pressure, and no further investigations of these have been reported, prompting some authors to question their existence.[130]

Monoclinic CuUO_4 monouranate

Brisi[137] reported the synthesis of $CuUO_4$ in 1963, and its structure was solved by Siegel and Hoekstra[138] using single-crystal X-ray diffraction in 1968. $CuUO_4$ forms a distinctly different structure from the other orthorhombic transition metal monouranates as described above.[138] $CuUO_4$ has a monoclinic structure in space group $P2_1/n$, see Figure 8. In this structure, the UO_6 polyhedra, containing the uranyl group, corner share forming 2D sheets running parallel to the (010) plane. That the structure differs from the other transition metal type monouranate structures in *Ibmm* and *Pbcn* is due to the presence of the Jahn–Teller active (d^9) Cu^{2+} cation. The Jahn–Teller effect lifts the degeneracy of the e_g orbitals, leading to a distortion of CuO_6 from octahedral to tetragonal (see Figure 9), with two of the Cu–O bonds significantly lengthened

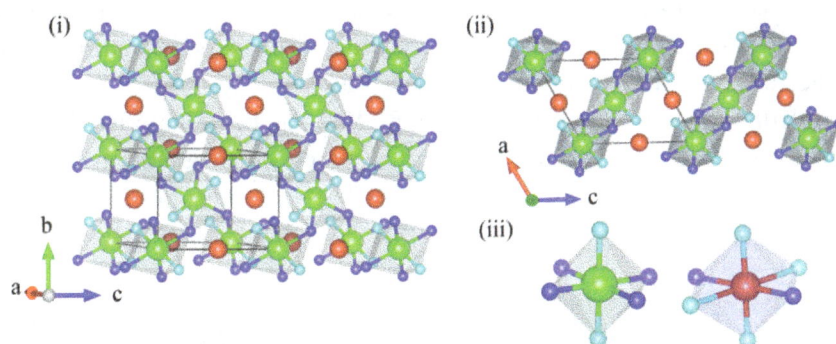

Figure 8.   (i) Structural representation of the monoclinic structure in space group $P2_1/n$ for $CuUO_4$ (see Ref. 138) where $a = 5.475(6)$ Å, $b = 4.957(6)$ Å, $c = 6.569(6)$ Å, $\beta = 118.87(15)°$, and unit cell volume = $156.12(9)$ Å^3. The red, green, aqua, and blue spheres represent Cu, U, uranyl oxygen, and non-uranyl oxygen atoms, respectively. (ii) Highlights the layered configuration of UO_6 and CuO_6 polyhedra running along the [010] direction. (iii) Highlights the distorted UO_6 polyhedra environment and also the apparently Jahn–Teller distorted CuO_6 coordination environment.

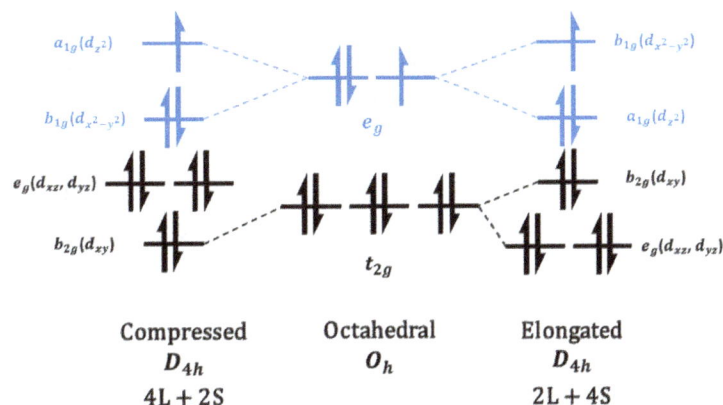

Figure 9.   The prototypical Jahn–Teller distortions experienced by Cu(II) in an octahedral coordination environment. For $CuUO_4$, it is crystallographically consistent with a 2L + 4S distortion (adapted from Ref. 172).

whereas the other four are shortened. In both the *Ibmm* and *Pbcn* transition metal monouranate structures, the transition metal coordination environments have four long and two short *A*–O bonds.[122,129,130,132,139]

Anomalous monouranates

UO_2 forms a cubic fluorite structure where the tetravalent uranium atoms are coordinated by eight oxygen anions.[22,23,140,141] UO_2 can form fluorite structured solid solutions with lanthanide, $U_{1-x}Ln_xO_{2-\delta}$, and actinide cations, $U_{1-x}An_xO_{2\pm\delta}$.[142–146] Here, uranium is encountered predominantly in its tetravalent state, with the *Ln* or *An* doping only resulting in a slight deviation from U^{4+}. However, there are exceptions, e.g. Popa *et al.*[147] recently showed that when stoichiometric quantities of UO_2 and Bi_2O_3 are mixed and heated under an argon atmosphere, they form a fluorite structure in space group $Fm\overline{3}m$ with the composition $Bi_{0.5}U_{0.5}O_2$ or $BiUO_4$ based on Rietveld refinement analysis against X-ray powder diffraction data. They also determined, using X-ray absorption spectroscopic measurements coupled with Raman spectroscopy, that the uranium cation has a pentavalent valence state (ionic radius 0.76 Å for 6-coordinate, the value for 8-coordinate is not tabulated[148]) while the bismuth cation is trivalent (ionic radius 1.17 Å for 8-coordinate, or 1.03 in 6-coordinate[148]). The structure of $BiUO_4$ is presented in Figure 10, with the fluorite UO_2 for comparison. Both the uranium and bismuth cations are disordered on the fluorite 4*a* site and have eight-fold coordination environments to oxygen. Popa *et al.*[147] described the trivalent Bi cation as being poorly bonded to oxygen compared to the smaller pentavalent uranium cation. The refined atomic displacement parameter of the oxygen atoms is unusually large, indicating considerable disorder, which was also supported by Raman spectroscopic measurements. The pentavalent uranium oxidation state in $BiUO_4$ has been supported by magnetic studies by Miyake *et al.*,[149,150] who also examined $ScUO_4$ and YUO_4. Based on magnetic measurements, Miyake *et al.*

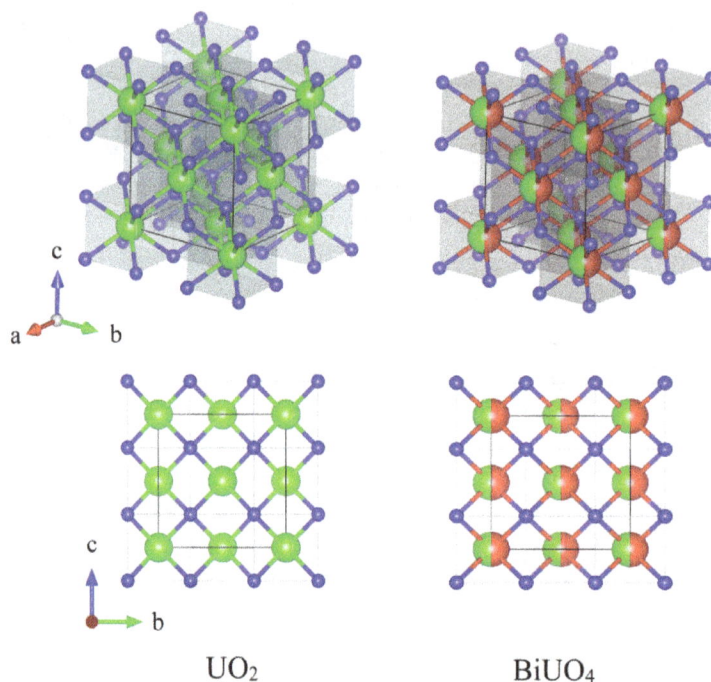

$$UO_2 \qquad\qquad BiUO_4$$

Figure 10.   Structural representations of the cubic fluorite structure in space group $Fm\overline{3}m$: (left) UO_2 (see Ref. 173) with $a = 5.468(1)$ Å and unit cell volume = 163.5(1) Å^3, and (right) $BiUO_4$[147] with $a = 5.478(1)$ Å and unit cell volume = 163.39(1) Å^3. The red, green, and blue spheres represent Bi, U, and oxygen atoms, respectively.

argued for the presence of pentavalent uranium which is consistent with trivalent Sc^{3+} ($[Ar]d^0$) and Y^{3+} ($[Kr]d^0$). No further studies of these oxides appear to have been conducted.

Vorlanite is a naturally occurring mineral with the formula $CaUO_4$. It has a fluorite structure in space group $Fm\overline{3}m$ and is isostructural with uraninite (UO_2), but contains hexavalent rather than tetravalent uranium.[26–28,151,152] Vorlanite is postulated to be a pseudo-morphic replacement of rhombohedral $CaUO_4$ (protovorlanite) in space group $R\overline{3}m$, as a consequence of radiation damage caused by the decay of uranium.[26] Although no radiation damage studies have been conducted on rhombohedral $CaUO_4$, this behavior is consistent with irradiation studies of pyrochlore materials.[153–156]

Recent progress in monouranate solid-state chemistry and crystallography

Understanding structure–property relationships is salient toward the design and manufacture of novel materials which possess advanced functional properties. As a strategy, this has been shown to be highly successful, particularly in the case of energy storage area, utilizing perovskite, pyrochlore, and Brownmillerite structured materials.[157–164] In the case of candidate nuclear waste form materials, understanding such relationships is pertinent as it may guide their storage conditions or disposal method. It is apparent from the above discussion of the AUO_4 monouranates that the majority of investigations have focused on uncovering new compositions and structural types rather than attempting to systematically understand why certain structures form and the consequence for their resulting chemistries. However, recent investigations by Read *et al.*,[130] Guo *et al.*,[117] and Murphy *et al.*[29–31,165] have shed some light on the structure–property relationships of the monouranates and unraveled some of the more remarkable properties that they may exhibit. These investigations will be discussed below.

Read *et al.*[130] examined $MnUO_4$ and $FeUO_4$ (and, but not discussed here, NiU_2O_6) produced from flux growth methods and studied them using single-crystal X-ray diffraction. They compared their structural solutions against existing literature of other monouranates, proposing that the structural preference for each monouranate, e.g. *Ibmm* orthorhombic, *Pbcn* orthorhombic, $R\overline{3}m$ rhombohedral, etc. is dictated by the size of the A-site cation. The rhombohedral $R\overline{3}m$ structure of $CaUO_4$ and others is preferred for the larger A-site cations, but as the size is reduced toward Cd^{2+} (6-coordinate ionic radius = 0.95 Å), Mn^{2+} (0.83 Å), Co^{2+} (0.745 Å), and Mg^{2+} (0.72 Å),[148] the structure reverts to a rutile-like structure in the orthorhombic *Ibmm* or *Cmmm* space group. They argued that Fe and Cr are more likely to exist as trivalent cations when the samples are prepared in air and that the ionic radii of Fe^{3+} and Cr^{3+} are too small for such lattices, resulting in the formation of the orthorhombic *Pbcn* structure. They discounted the formation of $NiUO_4$, suggesting that Ni^{2+} cation is too small for any of the known monouranate structures. This variation in structure is more dramatic than the changes induced by modulating the size of the A-site cation in ABO_3 perovskites.[166,167]

Guo *et al.*[117] used a combined experimental and theoretical approach to examine the ability of pentavalent uranium to form within $CrUO_4$ and $FeUO_4$ but not in $MgUO_4$ using X-ray powder diffraction, X-ray absorption spectroscopy, and *ab initio* calculation methods. They argued that the thermodynamic penalty for forming the less stable pentavalent (relative to hexavalent) uranium cation is benign compared to the preferred formation of trivalent chromium and iron. When the enthalpies of formation of these oxides are examined and compared with values given in the literature (both experimental and theoretical), they concluded that $CrUO_4$ and $FeUO_4$ have the least negative enthalpy of formation whereas the alkaline-earth-containing monouranates $BaUO_4$ and β-$SrUO_4$ have the most negative values. Based on the data available for AUO_4 monouranates with A = Cr, Fe, Mg, Ca, Sr, and Ba, a negative linear relationship was proposed between the thermodynamic stability of AUO_4 and the ionic radius of the A-site cation. This trend is consistent with the uranyl group contributing substantially to the structural integrity. That the uranyl moiety is absent in $CrUO_4$ and $FeUO_4$ would likely contribute to their least negative enthalpy of formation. $BaUO_4$ and β-$SrUO_4$, which have the shortest uranyl bond length of the monouranates, are expected to be more stable. However, this does not explicitly explain the ability for some monouranates to retain oxygen vacancies while others do not.

Murphy $et\ al.$[165] used neutron and synchrotron X-ray powder diffraction in conjunction with X-ray absorption spectroscopy to further understand the structural trends and relationships between $CaUO_4$, α-$SrUO_4$, β-$SrUO_4$, and $BaUO_4$. This investigation suggested that both the relative size of the A-site cations and their bonding requirements play a significant role in determining which structural type, rhombohedral or orthorhombic, forms and whether they can support oxygen vacancies. In the rhombohedral variants of $CaUO_4$ and α-$SrUO_4$, the A-site cation is inherently overbonded from bond valence sum (BVS) calculations, when the materials are stoichiometric (i.e. containing no oxygen vacancies). The formation of oxygen vacancies would reduce the BVS of both cations, leading to the formation of reduced uranium cations. The tendency for oxygen defects to form in these rhombohedral variants has been noted in several previous studies.[54,73,74,90] This overbonding is more pronounced in α-$SrUO_4$ where the A-site cation has a calculated BVS of 2.55 compared to 2.25 in $CaUO_4$, see Table 1 for details. This would imply that α-$SrUO_4$ is more inclined to support oxygen vacancies than $CaUO_4$ under ambient conditions, as experimentally observed.[165] In contrast, no oxygen vacancies were observed in orthorhombic β-$SrUO_4$ and $BaUO_4$,[73,102,110] which is consistent with their BVS values of 1.92 and 2.11 for the Sr and Ba cations, respectively (Table 1). A solid solution of $Sr_{1-x}Ca_xUO_4$ was prepared under oxidizing conditions. It was found to be single phase with an orthorhombic structure for $x = 0$ and 0.1, and single phase with a rhombohedral structure from $x = 0.7$ to 1.0.[165] In the rhombohedral phase region (i.e. Ca-rich end), it was found that the replacement of the smaller Ca^{2+} cation by the larger Sr^{2+} cation reduces the effective overbonding at both the Ca^{2+} and Sr^{2+} sites, stabilizing the bonding requirements of the A-site; see Table 2 for details. Conversely, the substitution of 0.1 f.u. of Ca^{2+} cations for Sr^{2+} in the orthorhombic structured β-$SrUO_4$ causes the Sr^{2+} to become overbonded with the BVS increasing from 1.93 to 2.36. Under these conditions, the formation of oxygen vacancies would not be favorable. This investigation strongly suggests that the ability of the rhombohedral monouranates to host oxygen vacancies (and potentially for the material to function as an oxygen ion conductor) is a consequence of the balance between reducing the A-site cations' overbonding and the uranium cations' ability to accommodate this through reduction. Consistent with previous studies,[73,74,90] the oxygen loss and gain phenomena were also apparent in our study[165] of the transformation from the rhombohedral α-$SrUO_4$

Table 1. Bond valence sums (BVS) obtained from neutron and synchrotron X-ray powder diffraction data sets refined jointly using the Rietveld method.[165]

Composition	$CaUO_4$	α-$SrUO_4$	β-$SrUO_4$	$BaUO_4$
Space group	$R\bar{3}m$	$R\bar{3}m$	$Pbcm$	$Pbcm$
A-site BVS	2.25	2.55	1.93	2.11
U BVS	6.15	6.00	5.81	5.71

Table 2. Bond valence sums (BVS) obtained from analyzing the synchrotron X-ray powder diffraction data using the Rietveld method for the series $Sr_{1-x}Ca_xUO_4$ synthesized in air with $x = 1$, 0.9, 0.8, 0.7, 0.1, and 0 (the single phase regions).[165]

Composition	$CaUO_4$	$Sr_{0.1}Ca_{0.9}UO_4$	$Sr_{0.2}Ca_{0.8}UO_4$	$Sr_{0.3}Ca_{0.7}UO_4$	$Sr_{0.9}Ca_{0.1}UO_4$	$SrUO_4$
Space group	$R\bar{3}m$	$R\bar{3}m$	$R\bar{3}m$	$R\bar{3}m$	$Pbcm$	$Pbcm$
Ca BVS	2.25	2.14	2.09	2.04	1.57	1.15
Sr BVS	3.36	3.21	3.14	3.07	2.36	1.93
U BVS	6.15	6.17	6.06	6.06	6.20	5.81

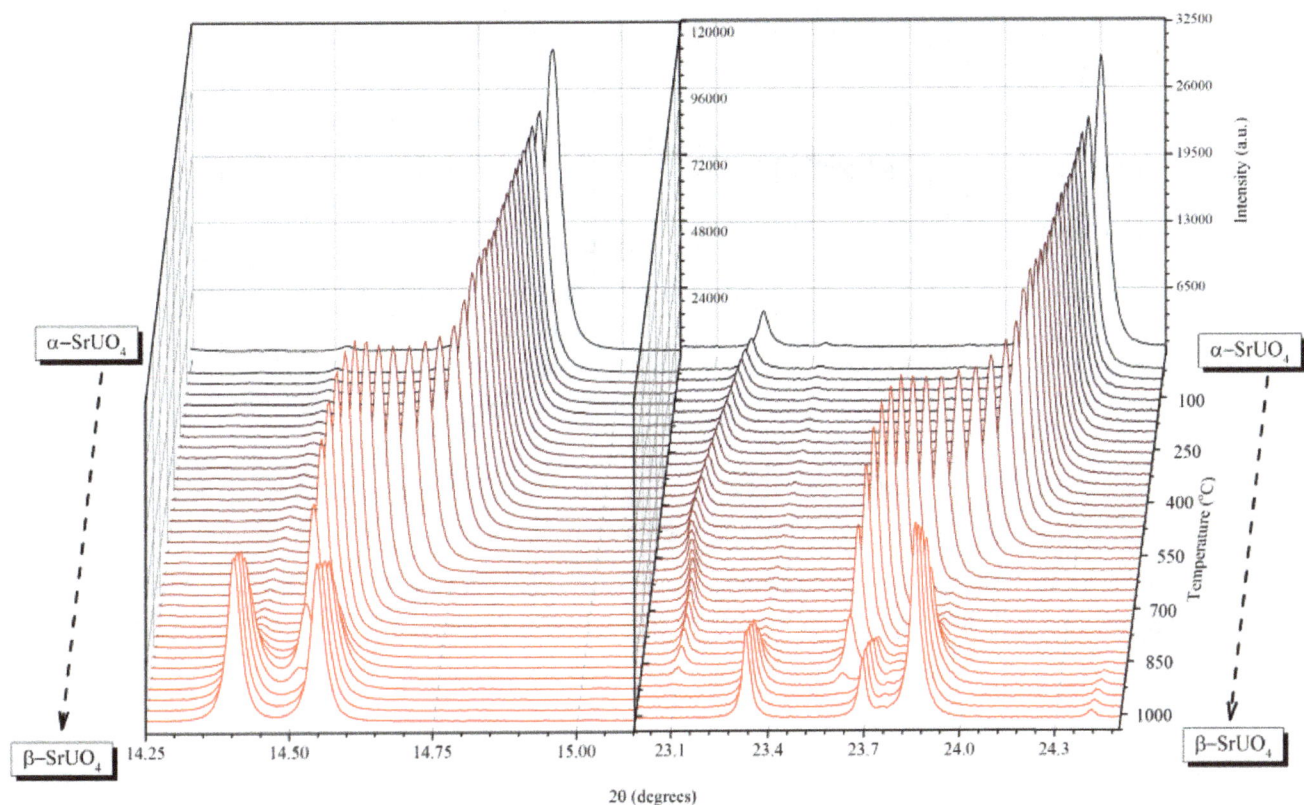

Figure 11. Temperature dependence of the synchrotron X-ray diffraction profile for α-SrUO₄ heated in air from RT to 1000°C at 14.2 ≤ 2θ ≤ 15.5 and 23 ≤ 2θ ≤ 25°.[29] Note the accelerated shift of the two main reflections around 14.8 and 24.3°C (peak positions at RT) as a result of the oxygen loss and uranium reduction from around 500°C.

to the β orthorhombic polymorph. Using thermogravimetric analysis with variable heating rates, it was further shown that the transformation has a considerable kinetic component.

Our second investigation focused on using *in situ* experimental methodologies, in addition to *ab initio* calculations, to examine the α → β transformation in SrUO₄.[29] The thermal behavior of α-SrUO₄ was examined by heating a sample in a sealed quartz capillary to 1000°C using synchrotron X-ray powder diffraction. As shown in Figure 11, when heated above 700°C, reflections corresponding to the α-SrUO₄ phase begin losing intensity, whereas new broad reflections are observed to emerge. This new phase could be refined against an orthorhombic structure in space group *Pbcm*, using the Rietveld method, consistent with β-SrUO₄. The discontinuity in the diffraction profile is indicative of a first-order reconstructive transformation. Careful examination of the diffraction patterns further shows a peculiar acceleration in the rate of reflections shifting to lower angles and, by extension thermal expansion, with temperature. This is apparent in the refined unit cell volume as a function of temperature, which is attributed to the loss of oxygen from the O(2) non-uranyl site. Figure 12 shows the O(2) site occupancy with increasing temperature, compared with the results of thermogravimetric analysis carried out under oxidizing conditions. That faster thermal expansion was observed over the same temperature range in which thermogravimetric analysis indicated oxygen loss suggests that the uranium cations are being reduced, consistent with previous postulates.[73,74,93] Refinement of the O(1) uranyl oxygen site occupancy (not shown) and O(2) non-uranyl oxygen site occupancy indicates that the loss occurred exclusively at the O(2) site, which is also reflected in the atomic displacement parameters.[29]

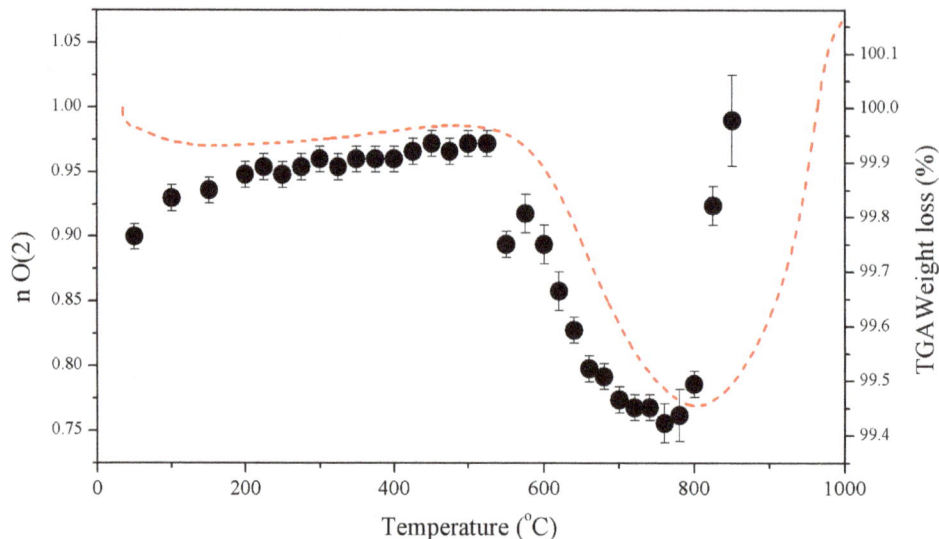

Figure 12.   Temperature dependence of the non-uranyl oxygen O(2) site occupancy for α-SrUO$_4$ obtained from Rietveld refinements against *in situ* synchrotron X-ray diffraction data.[29] Thermogravimetric analysis measurements, collected from α-SrUO$_4$ in air, are shown as the dashed red line.

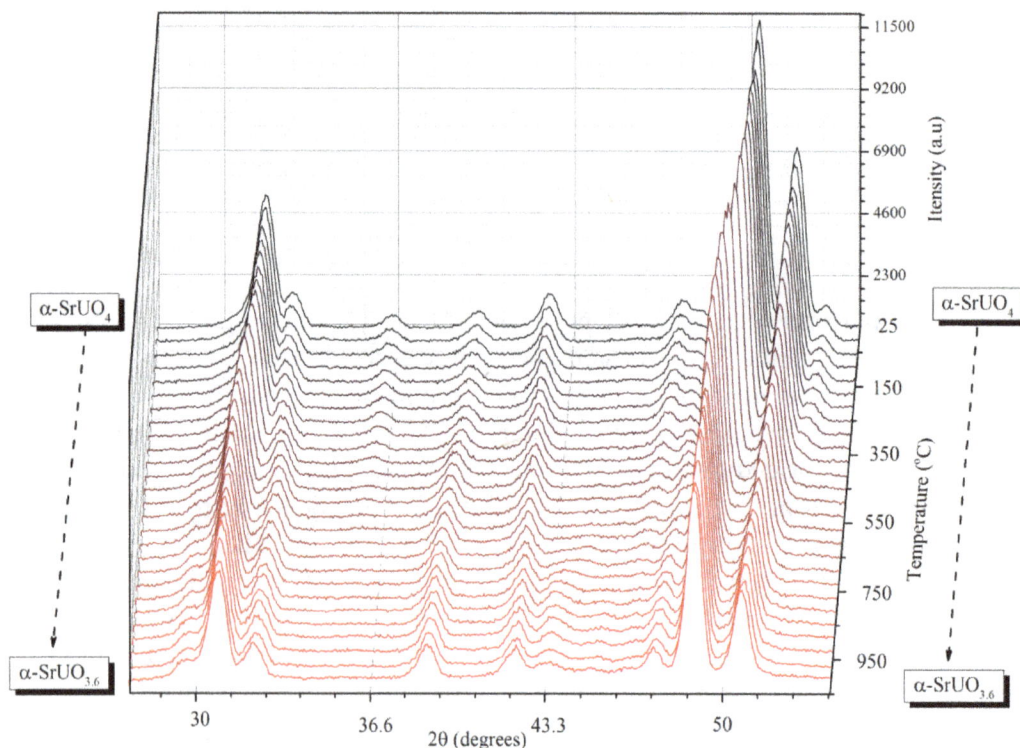

Figure 13.   Temperature dependence of a portion of the neutron powder diffraction profile for α-SrUO$_4$ heated under vacuum from RT to 1000°C.[29]

Neutron powder diffraction patterns were collected *in situ* from α-SrUO$_4$ when it was heated from RT to 1000°C inside a vacuum furnace. The diffraction profiles are presented in Figure 13. Interestingly, the α to β transformation was not observed under vacuum conditions. Analysis using the Rietveld method (results shown in Figure 14) revealed similar behavior to that of the synchrotron X-ray diffraction experiment (carried out in

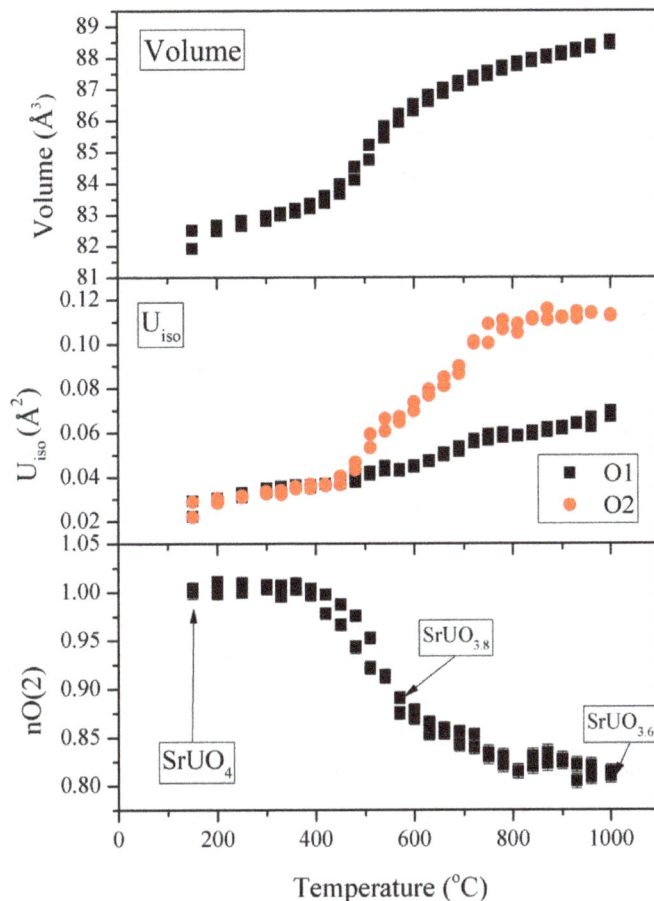

Figure 14. From top to bottom: Temperature dependence of the refined unit cell volume, isotropic displacement parameters of the uranyl O(1) and non-uranyl oxygen O(2), site occupancy of the non-uranyl oxygen O(2) for α-SrUO$_4$ heated from 150°C to 1000°C under vacuum obtained from Rietveld refinements of *in situ* neutron powder diffraction data.[29]

air) for temperatures up to around 750°C. The key difference is the absence of either the reoxidation step, observed in the synchrotron X-ray diffraction study and previous studies,[73,74,168] or the formation of the β polymorph in vacuum. Evidently, β-SrUO$_4$ cannot form without a source of additional oxygen. An implication of this is that the decreased occupancy of the non-uranyl O(2) site is not a consequence of the oxygen moving into interstitial lattice sites, but it is instead lost from the sample.[29] The reduction of uranium leading up to the transformation was examined using X-ray absorption near edge structure (XANES) measurements at the U L$_3$-edge *in situ* by heating α-SrUO$_4$ to 600°C under flowing argon. A systematic shift of the U L$_3$-edge position toward lower photon energy was observed, indicative of uranium undergoing reduction from its hexavalent state.[29]

In order to understand and elucidate the origin of the preference for oxygen vacancy formation in one structural type (rhombohedral) but not the other (orthorhombic) polymorph of SrUO$_4$, *ab initio* calculations were undertaken using a DFT (density functional theory) plane-wave package.[29] The strongly correlated character of the *f* orbitals was accounted for by the DFT+U method with the Hubbard U parameter derived from first principles. The PBE and PBEsol exchange-correlation functionals and the ultrasoft pseudo-potentials were used to reproduce the effect of core electrons.[169] The energy for oxygen vacancy formation in each polymorph was calculated as a function of oxygen content.[29] These calculations indicate that β-SrUO$_4$ is only thermodynamically favored when stoichiometric, whereas α-SrUO$_4$ is favored when substoichiometric.[29] This result, coupled with *in situ* neutron and synchrotron X-ray diffraction measurements, explains the origin of the oxygen loss and gain mechanism associated with the $\alpha \rightarrow \beta$ transformation and why a source of oxygen is critical for the transformation from α- to β-$SrUO_4$.

The phase transformation that occurs in α-$SrUO_4$ has not been observed in $CaUO_4$. As described earlier, it is strongly postulated that a key component in inducing the $\alpha \rightarrow \beta$ transformation in $SrUO_4$ is the generation of a sufficient amount of oxygen vacancies to reduce the activation energy barrier. *In situ* neutron powder diffraction measurements on $CaUO_4$ heated from RT to 1000°C under vacuum confirmed the ability for $CaUO_4$ to host oxygen vacancies and to be reduced to $CaUO_{3.7}$.[30] Without a source of oxygen, the hypothetical transformation to an orthorhombic structure, however, could not proceed. The ability for $CaUO_4$ to generate oxygen vacancies, and potentially to undergo an α to β type transformation, was also studied by *in situ* synchrotron X-ray powder diffraction under oxidizing conditions to 1000°C.[30] It was found that although no transformation was observed, $CaUO_4$ undergoes a pronounced acceleration in its rate of thermal expansion above 700°C, as shown in Figure 15, consistent with the formation of oxygen vacancies and partial reduction of the uranium cations.

Prodan and Boswell argued that oxygen–defect interactions in $CaUO_4$ could impact the reduction and potential reoxidation of the crystal lattice.[67] Since isostructural $CaUO_4$ has a smaller unit cell compared to α-$SrUO_4$, as a consequence of the smaller Ca^{2+} over the larger Sr^{2+} cation, it could be presumed that oxygen–defect interactions would be more pronounced in the former due to the relatively closer proximity of the corresponding oxygen sites. Doping Sr^{2+} in $CaUO_4$ could potentially reduce the impact of the oxygen–defect interactions by allowing lattice expansion, increasing oxygen site distances, and decreasing associated defect proximity. A series of oxides of the form α-$Sr_xCa_{1-x}UO_{4-\delta}$ was prepared under reducing conditions to establish if the rhombohedral crystal lattice could be stabilized across the series by the introduction of oxygen vacancies. A miscibility gap was observed, despite the isostructural relationship between the two end members. Single-phase samples were only obtained for $0 \leq x \leq 0.4$ and $0.9 \leq x \leq 1.0$, and two rhombohedral phases both in space group $R\bar{3}m$ but with different lattice sizes coexisted at other compositions. Murphy and coworkers.[30] postulated that these phases contain variable amounts of Ca^{2+} and Sr^{2+}, which leads to the variation in lattice size, and consequently different amounts of oxygen vacancies. That one phase can support more oxygen vacancies than the other is thought to be the origin of the observed phase miscibility.

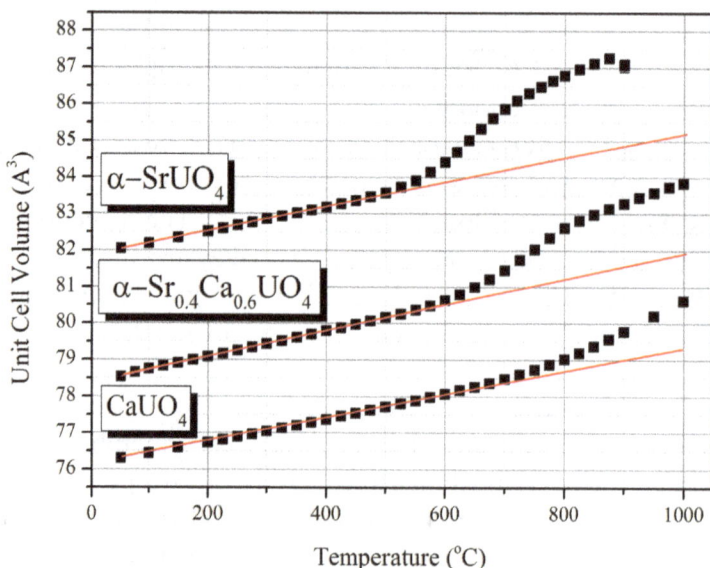

Figure 15.   Temperature dependence of the (α-phase) unit cell volume of $CaUO_4$, α-$Sr_{0.4}Ca_{0.6}UO_4$, and α-$SrUO_4$ in sealed quartz capillaries containing air heated from 25°C to 1000°C obtained from Rietveld refinements against *in situ* synchrotron X-ray powder diffraction data.[30] Note that data points are absent at $T > 900°C$ for α-$SrUO_4$ due to the phase transformation to β-$SrUO_4$.

To examine whether the doping of Sr^{2+} cations into $CaUO_4$ could enhance the spontaneous formation of oxygen vacancies at high temperatures, α-$Sr_{0.4}Ca_{0.6}UO_4$ was heated from RT to 1000°C under oxidizing conditions. Rietveld refinement of synchrotron X-ray powder diffraction data showed that the rate of change in the unit cell thermal expansion began to accelerate approximately 100°C earlier than in undoped $CaUO_4$ and approximately 80°C later than in pure α-$SrUO_4$.[30] The relative change in the unit cell size at high temperatures was greater than what would be expected from linear thermal expansion alone. This result, shown in Figure 15, is consistent with the formation of an increasing amount of oxygen vacancies with increasing Sr content: α-$SrUO_4$ > α-$Sr_{0.4}Ca_{0.6}UO_4$ > $CaUO_4$. It was argued that oxygen defects form at lower temperature and seemingly in greater quantity with increasing Sr content, as a result of the increased proximity between defects and subsequently reduced defect–defect interactions.

Neutron diffraction studies of both $CaUO_y$ and α-$SrUO_y$ indicate an apparent limiting stoichiometry of y ~ 3.6–3.7. This suggests that uranium is predominantly pentavalent; pure U(V) would correspond to $y = 3.5$. Synchrotron X-ray powder diffraction studies using highly reducing conditions, in the form of pure hydrogen gas flow, were conducted in attempts to reduce it below this limit. Serendipitously, it was found that when the substoichiometric $CaUO_{4-x}$, α-$Sr_{0.4}Ca_{0.6}UO_{4-x}$, and α-$SrUO_{4-x}$ monouranates were heated under flowing hydrogen, to above approximately 450, 350, and 200°C, respectively, they underwent first-order structural transitions involving the appearance of very weak superlattice reflections that are indicative of oxygen vacancy ordering.[30,31] Pertinent to the transformation is that a critical amount of oxygen defects must be present prior to the monouranates adopting the ordered structures. The exact amount of defects is not precisely known. The diffraction profile for α-$SrUO_{4-x}$ under these conditions is presented in Figures 16 and 17, which illustrate the symmetry lowering (evident by peak splitting) and the ordering of oxygen vacancies (based on the appearance of superlattice reflections), respectively. Remarkably, the intensity of these superlattice reflections, which corresponds to the degree of oxygen vacancy ordering, increases with increasing temperature. The ordered structures are not consistent with the proposed high-temperature phase as described by Pialoux *et al.*[90] and as such this is considered a new phase, denoted as δ-$SrUO_{4-x}$. Furthermore, it was found that cooling the δ phase results in the reverse transformation back to the oxygen disordered substoichiometric α phase.

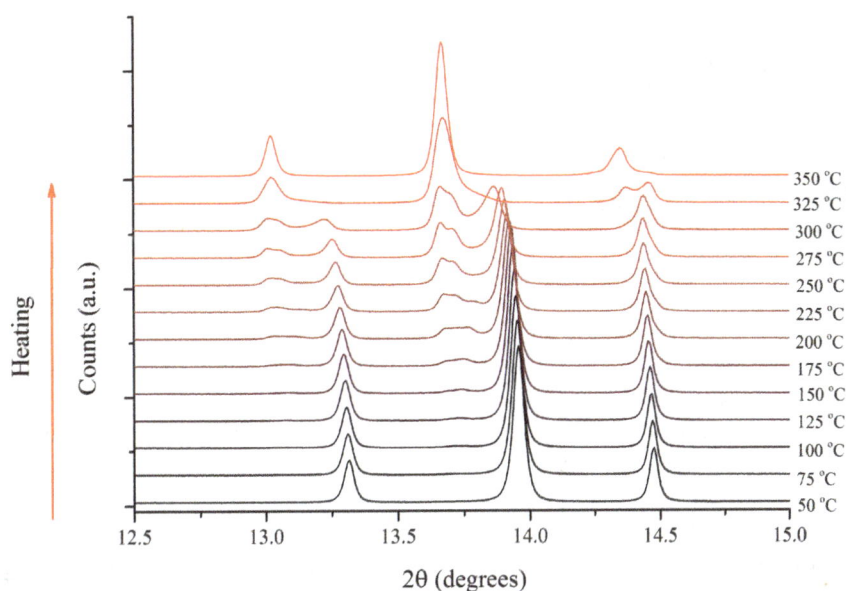

Figure 16.   Temperature dependence of the synchrotron X-ray diffraction profiles for α-$SrUO_4$ heated under flowing hydrogen from 50°C to 350°C ($12.5° \leq 2\theta \leq 15°$).[31] The discontinuity observed at $T > 150°C$ corresponds to the first-order phase transformation to disordered δ-$SrUO_{4-x}$.

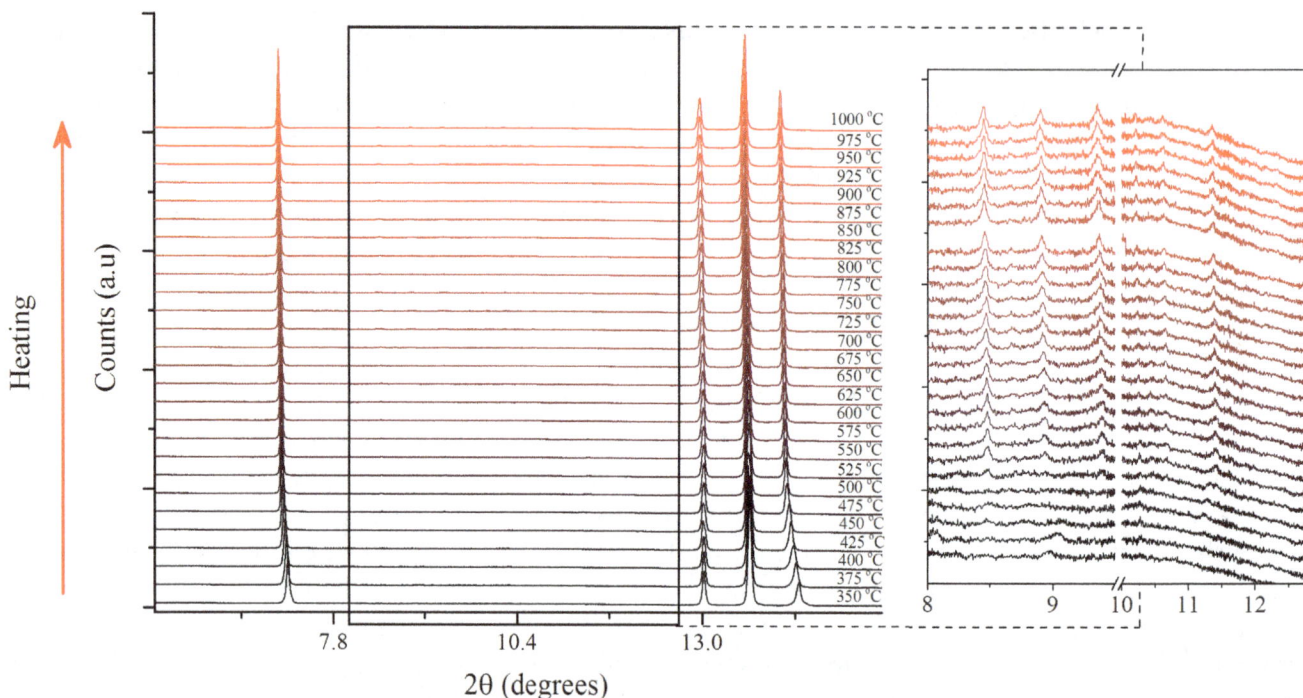

Figure 17.   Temperature dependence of the synchrotron X-ray diffraction profiles for δ-SrUO$_{4-x}$ heated under flowing hydrogen from 350°C to 1000°C (5.3° ≤ 2θ ≤ 15.5°).[31] The inset expands and highlights the weak reflections occurring between 8° < 2θ < 12.8° which are superlattice reflections corresponding to the ordering of oxygen vacancies.

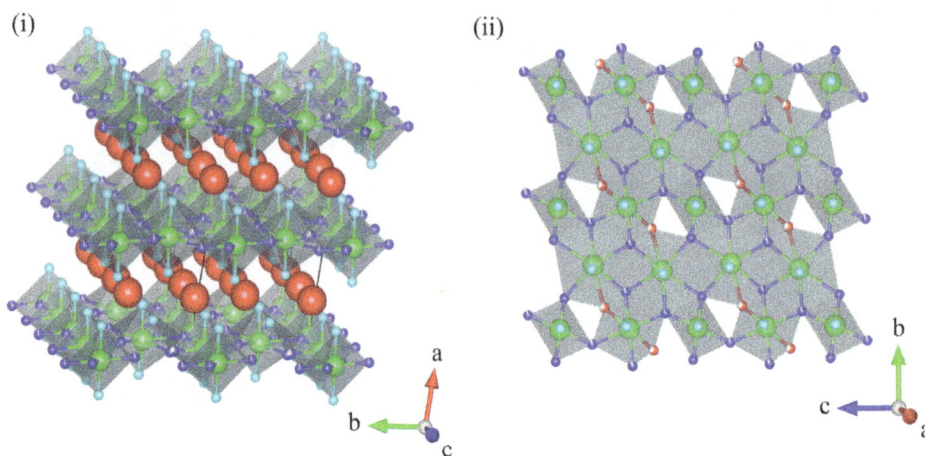

Figure 18.   (i) Structural representation of the triclinic structure in space group $P\bar{1}$ for δ-SrUO$_{4-x}$ at 1000°C[31] where a = 6.72851(3) Å, b = 6.97217(6) Å, c = 8.06596(6) Å, α = 89.8732(6)°, β = 107.3712(5)°, γ = 99.9942(8)°, and unit cell volume = 355.159(4) Å³. The red, green, and blue/aqua spheres represent Sr, U, and O atoms, respectively. (ii) Highlights the uranium–oxygen layer where the concentrated ordered defects are shown by the partially shaded red spheres representing oxygen vacancies.

The δ-SrUO$_{4-x}$ structure was solved as triclinic in space group $P\bar{1}$ using a single commensurate modulation vector. This structure is illustrated in Figure 18, which can be considered as a bridge between the rhombohedral and orthorhombic phases of SrUO$_4$: β-like UO$_6$ polyhedra edge and corner share with α-like UO$_8$ polyhedra in the δ-SrUO$_{4-x}$ structure. The analysis suggests that oxygen vacancies are associated with the UO$_8$

polyhedra and not with the UO_6 polyhedra, consistent with findings from studies of α-$SrUO_4$ and β-$SrUO_4$.[29,165] The UO_8 and UO_6 polyhedra link to form 2D sheets, where the oxo collinear oxygens are directed at the Sr^{2+} cations between the sheets. The structural configuration of these oxo oxygens is reminiscent of uranyl, although this has yet to be confirmed unequivocally. The δ-$CaUO_{4-x}$ and δ-$Sr_{0.4}Ca_{0.6}UO_{4-x}$ structures could not be solved precisely from the synchrotron X-ray diffraction data, indexing the superlattice reflections required the use of two incommensurate modulation vectors, but an average structure in space group $P\bar{1}$ could be refined. That δ-$CaUO_{4-x}$ and δ-$Sr_{0.4}Ca_{0.6}UO_{4-x}$ structurally differ from δ-$SrUO_{4-x}$ is thought to be related to the miscibility gap in the rhombohedral α-$Sr_xCa_{1-x}UO_4$ solid solution and potentially reflects a difference in short range order. Regardless, the transformations are unusual, not having been reported in any other oxide materials before. At first glance, the phase transformations seem to be perplexing, as they involve the lowering of crystallographic symmetry and ordering of oxygen vacancies at high temperatures. Thermodynamic studies into the enthalpy of formation of these phases have not been conducted, but such studies likely hold the key to understand the relative importance of entropy and enthalpy. Nevertheless, it is very intriguing for an oxide material to have the ability to induce oxygen vacancy ordering at high temperatures and to observe an order to disorder transformation with decreasing temperature, without reduction or decomposition of the sample. Such phenomena may be desirable for potential applications, e.g. in solid oxide fuels cells and ion conductors, particularly if it can be replicated in non-actinide materials.

Concluding Remarks

The above discussion demonstrates the chemical and structural complexity of the apparently simple AUO_4 monouranates. Although the history of these transcends the nuclear age, their remarkable structural flexibility and associated properties are only just emerging, reflecting the power of contemporary diffraction methods, especially when these are utilized for *in situ* studies. A unifying theme in recent investigations is the systematic approach to understand structure–property relationships under well-controlled environments (temperature and composition) rather than the traditionally applied approach of *ex situ* examination, which requires a degree of serendipity. The recent approach enables a phenomenological understanding of the solid-state chemistry and physics of monouranates that can potentially be applied to the understanding of other uranium materials, such as naturally occurring systems and those resulting directly from spent nuclear fuel or from reprocessed nuclear waste. Gaps in our knowledge and understanding of the structure–property relationships remain for the monouranates, although this is slowly being addressed. It becomes salient from the perspective of this chapter that similar structure–property investigations should be extended to the transuranic elements and compounds, although such studies will present considerable experimental challenges reflecting the high radiological activity of these. Compared to uranium, considerably less is known about the transuranics, despite the importance of these elements in high level nuclear wastes — one of the key and most challenging issues from spent nuclear fuel. In addition to the importance of understanding uranium and other actinides from the direct application perspective, the rich chemistry and properties that they possess should also be emphasized. A fascinating example[30-31] is the α to δ phase transformation in $SrUO_{4-x}$ and $CaUO_{4-x}$ with its reversible ordering of oxygen defects and accompanied symmetry lowering at high temperatures. These phenomena, which are apparently unique to these oxides, may have an impact on the stability and reactivity of these materials. Understanding this is therefore important not only from the nuclear waste form perspective but also points to a new direction in functional oxide material design where thermally induced oxygen defect ordering is desirable. Several other noteworthy examples are present in literature for different actinide systems.[23,36,39,170,171] These, and others examples, eloquently illustrate both the unique chemistry and properties of uranium and the actinides, and their potential to guide advances in general inorganic sciences.

References

1. Taylor, S. R., Abundance of chemical elements in the continental crust: A new table. *Geochim. Cosmochim. Acta* 1964, 28, 1273–1285.

2. Cayrel, R.; Hill, V.; Beers, T. C.; Barbuy, B.; Spite, M.; Spite, F.; Plez, B.; Andersen, J.; Bonifacio, P.; Francois, P.; Molaro, P.; Nordstrom, B.; Primas, F., Measurement of stellar age from uranium decay. *Nature* 2001, 409, 691–692.

3. McDiarmid, M. A.; Keogh, J. P.; Hooper, F. J.; McPhaul, K.; Squibb, K.; Kane, R.; DiPino, R.; Kabat, M.; Kaup, B.; Anderson, L.; Hoover, D.; Brown, L.; Hamilton, M.; Jacobson-Kram, D.; Burrows, B.; Walsh, M., Health effects of depleted uranium on exposed Gulf War veterans. *Environ. Res.* 2000, 82, 168–180.

4. Bleise, A.; Danesi, P. R.; Burkart, W., Properties, use and health effects of depleted uranium (DU): A general overview. *ACS Appl. Mater. Interfaces* 2003, 64, 93–112.

5. Arnold, P. L.; McMullon, M. W.; Rieb, J.; Kuhn, F. E., C–H bond activation by f-block complexes. *Angew. Chem. Int. Ed.* 2015, 54, 82–100.

6. Ephritikhine, M., The vitality of uranium molecular chemistry at the dawn of the XXIst century. *Dalton Trans.* 2006, 0, 2501–2516.

7. Hitch, C. J.; McKean, R. N., *The economics of defense in the nuclear age.* Harvard University Press: 1960.

8. Knorr, K. E., *On the uses of military power in the nuclear age.* Princeton University Press: Princeton, NJ, 2016.

9. Gustafson, J. M., *The Nuclear Delusion: Soviet–American Relations in the Atomic Age* Vol. 39. Educational Foundation for Nuclear Science, Inc. 1983, p. 35.

10. Hazen, R. M.; Ewing, R. C.; Sverjensky, D. A., Evolution of uranium and thorium minerals. *Am. Miner.* 2009, 94, 1293–1311.

11. Abram, T.; Ion, S., Generation-IV nuclear power: A review of the state of the science. *Energy Policy* 2008, 36, 4323–4330.

12. Lenzen, M., Life cycle energy and greenhouse gas emissions of nuclear energy: A review. *Energy Conv. Manag.* 2008, 49, 2178–2199.

13. Burns, P. C., U^{6+} minerals and inorganic compounds: Insights into an expanded structural hierarchy of crystal structures. *Can. Mineral.* 2005, 43, 1839–1894.

14. Burns, P. C.; Miller, M. L.; Ewing, R. C., U^{6+} minerals and inorganic phases: A comparison and hierarchy of crystal structures. *Can. Mineral.* 1996, 34, 845–880.

15. Zinkle, S. J.; Was, G. S., Materials challenges in nuclear energy. *Acta Mater.* 2013, 61, 735–758.

16. Kleykamp, H., The chemical-state of fission-products in oxide fuels at different stages of the nuclear-fuel cycle. *Nucl. Technol.* 1988, 80, 412–422.

17. Kleykamp, H., The chemical-state of the fission-products in oxide fuels. *J. Nucl. Mater.* 1985, 131, 221–246.

18. Kashparov, V. A.; Oughton, D. H.; Zvarich, S. I.; Protsak, V. P.; Levchuk, S. E., Kinetics of fuel particle weathering and Sr-90 mobility in the Chernobyl 30 km exclusion zone. *Health Phys.* 1999, 76, 251–259.

19. Cohen, B. L., High-level radioactice waste from light-water reactors. *Rev. Mod. Phys.* 1977, 49, 1–20.

20. Weber, W. J.; Navrotsky, A.; Stefanovsky, S.; Vance, E. R.; Vernaz, E., Materials science of high-level nuclear waste immobilization. *MRS Bull.* 2009, 34, 46–53.

21. Burns, P. C.; Ewing, R. C.; Hawthorne, F. C., The crystal chemistry of hexavalent uranium: Polyhedron geometries, bond-valence parameters, and polymerization of polyhedra. *Can. Mineral.* 1997, 35, 1551–1570.

22. Desgranges, L.; Baldinozzi, G.; Simeone, D.; Fischer, H. E., Structural changes in the local environment of uranium atoms in the three phases of U$_4$O$_9$. *Inorg. Chem.* 2016, 55, 7485–7491.

23. Desgranges, L.; Ma, Y.; Garcia, P.; Baldinozzi, G.; Simeone, D.; Fischer, H. E., What is the actual local crystalline structure of uranium dioxide, UO$_2$? A new perspective for the most used nuclear fuel. *Inorg. Chem.* 2017, 56, 321–326.

24. Banerjee, S.; Kutty, T. R. G., Nuclear Fuels. In *Functional Materials: Preparation, Processing and Applications* 2012, 387–466.

25. Bots, P.; Morris, K.; Hibberd, R.; Law, G. T. W.; Mosselmans, J. F. W.; Brown, A. P.; Doutch, J.; Smith, A. J.; Shaw, S., Formation of stable uranium(VI) colloidal nanoparticles in conditions relevant to radioactive waste disposal. *Langmuir* 2014, 30, 14396–14405.

26. Galuskin, E. V.; Armbruster, T.; Galuskina, I. O.; Lazic, B.; Winiarski, A.; Gazeev, V. M.; Dzierzanowski, P.; Zadov, A. E.; Pertsev, N. N.; Wrzalik, R.; Gurbanov, A. G.; Janeczek, J., Vorlanite (CaU^{6+})O$_4$ — A new mineral from the Upper Chegem Caldera, Kabardino-Balkaria, Northern Caucasus, Russia. *Am. Miner.* 2011, 96, 188–196.

27. Galuskin, E. V.; Galuskina, I. O.; Dubrovinsky, L. S.; Janeczek, J., Thermally induced transformation of vorlanite to "proto-vorlanite": Restoration of cation ordering in self-irradiated CaUO$_4$. *Am. Miner.* 2012, 97, 1002–1004.

28. Othmane, G.; Allard, T.; Menguy, N.; Morin, G.; Esteve, I.; Fayek, M.; Calas, G., Evidence for nanocrystals of vorlanite, a rare uranate mineral, in the Nopal I low-temperature uranium deposit (Sierra Pena Blanca, Mexico). *Am. Miner.* 2013, 98, 518–521.

29. Murphy, G. L.; Kennedy, B. J.; Kimpton, J. A.; Gu, Q. F.; Johannessen, B.; Beridze, G.; Kowalski, P. M.; Bosbach, D.; Avdeev, M.; Zhang, Z. M., Nonstoichiometry in strontium uranium oxide: Understanding the rhombohedral — orthorhombic transition in SrUO$_4$. *Inorg. Chem.* 2016, 55, 9329–9334.

30. Murphy, G.L.; Wang, C.; Beridze, G.; Zhang, Z.; Avdeev, M.; Muransky, O.; Gu, Q.: Kowalski, P.M.; Kennedy, B.J.; Controlling oxygen defect formation and the effect on reversible symmetry lowering and disorder-order phase transformations in non-stoichiometric ternary uranium oxides. 2018.

31. Murphy, G. L.; Wang, C.; Beridze, G.; Zhang, Z.; Kimpton, J. A.; Avdeev, M.; Kowalski, P. M.; Kennedy, B. J., An unexpected crystallographic phase transformation in non-stoichiometric SrUO$_{4-x}$: Reversible oxygen defect ordering and symmetry lowering with increasing temperature. *Inorg. Chem.* 2018, 57, 5948–5958.

32. Selbin, J.; Holmes, L. H.; McGlynn, S. P., Eletronic structure, spectra and magnetic properties of oxycations IV. Ligation effects on the infra-red spectrum of the vanadyl ion. *J. Inorg. Nucl. Chem.* 1963, 25, 1359–1369.

33. Ballhausen, C. J.; Gray, H. B., The electronic structure of the vanadyl ion. *Inorg. Chem.* 1962, 1, 111–122.

34. Denning, R. G., Electronic structure and bonding in actinyl ions. In *Complexes, Clusters and Crystal Chemistry*, Springer Berlin Heidelberg: Berlin, Heidelberg, 1992, 215–276.

35. Denning, R. G., Electronic structure and bonding in actinyl ions and their analogs. *J. Phys. Chem. A* 2007, 111, 4125–4143.

36. Vitova, T.; Pidchenko, I.; Fellhauer, D.; Bagus, P. S.; Joly, Y.; Pruessmann, T.; Bahl, S.; Gonzalez-Robles, E.; Rothe, J.; Altmaier, M.; Denecke, M. A.; Geckeis, H., The role of the 5f valence orbitals of early actinides in chemical bonding. *Nat. Commun.* 2017, 8, 1–9.

37. Chermette, H.; Rachedi, K.; Volatron, F., Trans effect and inverse trans effect in MLX$_5$ complexes (M = Mo, U; L = O, S; X = Cl, Br): A rationalization within density functional theory study. *Theochem-J. Mol. Struct.* 2006, 762, 109–121.

38. Gibson, J. K.; Haire, R. G.; Santos, M.; Marcalo, J.; de Matos, A. P., Oxidation studies of dipositive actinide ions, An^{2+} (An = Th, U, Np, Pu, Am) in the gas phase: Synthesis and characterization of the isolated uranyl, neptunyl, and plutonyl ions UO$_2{}^{2+}$(g), NpO$_2{}^{2+}$(g), and PuO$_2{}^{2+}$(g). *J. Phys. Chem. A* 2005, 109, 2768–2781.

39. Polinski, M. J.; Garner Iii, E. B.; Maurice, R.; Planas, N.; Stritzinger, J. T.; Parker, T. G.; Cross, J. N.; Green, T. D.; Alekseev, E. V.; Van Cleve, S. M.; Depmeier, W.; Gagliardi, L.; Shatruk, M.; Knappenberger, K. L.; Liu, G.; Skanthakumar, S.; Soderholm, L.; Dixon, D. A.; Albrecht-Schmitt, T. E., Unusual structure, bonding and properties in a californium borate. *Nat. Chem.* 2014, 6, 387–392.

40. Denning, R. G., Electronic-structure and bonding in actinyl ions. *Struct. Bond.* 1992, 79, 215–276.

41. Krivovichev, S. V.; Finch, R. J.; Burns, P. C., Crystal chemistry of uranyl molybdates. V. Topologically distinct uranyl dimolybdate sheets in the structures of Na$_2$(UO$_2$)(MoO$_4$)$_2$ and K$_2$(UO$_2$)$_3$O(MoO$_4$)$_2$(H$_2$O). *Can. Mineral.* 2002, 40, 193–200.

42. Krot, N. N.; Grigoriev, M. S., Cation–cation interaction in crystalline actinide compounds. *Russ. Chem. Rev.* 2004, 73, 89–100.

43. Balboni, E.; Burns, P. C., Cation–cation interactions and cation exchange in a series of isostructural framework uranyl tungstates. *J. Solid State Chem.* 2014, 213, 1–8.

44. Severance, R. C.; Smith, M. D.; zur Loye, H. C., Three-dimensional hybrid framework containing U$_2$O$_{13}$ dimers connected via cation–cation interactions. *Inorg. Chem.* 2011, 50, 7931–7933.

45. Serezhkin, V.N.; Sidorenko, G.V.; Pushkin, D.V.; Serezhkina, L.B.; Cation-Cation Interactions between Uranyl(VI) *Ions Radiochemistry* 2014, 56, 115–133

46. Forbes, T. Z.; Wallace, C.; Burns, P. C., Neptiunyl compouds: Polyhedron geomtries, bond-valence parameters, and structural hierarchy. *Can. Mineral.* 2008, 46, 1623–1645.

47. Zachariasen, W. H., Crystal chemical studies of the 5f-series of elements IV. The crystal structure of Ca(UO$_2$)O$_2$ and Sr(UO$_2$)O$_2$. *Acta Crystallogr.* 1948, 1, 281–285.

48. Kovba, T. M.; Polunina, G. P.; Ippolitova, E. A.; Simanov, Y. P.; Spitsyn, V. I., The crystalline structure of uranates .II. Uranates containing uranyl-oxygen chains. *Zhurnal Fiz. Khimii* 1961, 35, 719–722.

49. Cordfunke, E. H.; Loopstra, B. O., Preparation and properties of uranates of calcium and strontium. *J. Inorg. Nucl. Chem.* 1967, 29, 51–57.

50. Rudorff, W.; Pfitzer, F., Uber erdalkaliuranate(VI) und ihre reduktions-produkte. *Z. Naturforsch. (B)* 1954, 9, 568–569.

51. Loopstra, B. O.; Rietveld, H. M., Structure of some alkaline-earth metal uranates. *Acta Cryst. Sec. B* 1969, B 25, 787–791.

52. Brisi, C.; Appendin, M. M., Studies on calcium uranates. *Ann. Chim.* 1969, 59, 400–411.

53. Burns, P. C., The crystal chemistry of uranium. *Rev. Miner.* 1999, 38, 23–90.

54. Liegeosi-Duyckaerts, M., IR and Raman-spectrum of CaUO$_4$: New data and interpretations. *Spectroc. Acta Pt. A-Molec. Biomolec. Spectr.* 1977, 33, 709–713.

55. Jakes, D.; Krivy, I., Stoichiometric magnesium, calcium, strontium and barium monouranates. *J. Inog. Nucl. Chem.* 1974, 36, 3885–3885.

56. Read, C. M.; Bugaris, D. E.; zur Loye, H. C., Single crystal growth and structural characterization of four complex uranium oxides: $CaUO_4$, β-Ca_3UO_6, $K_4CaU_3O_{12}$, and $K_4SrU_3O_{12}$. *Solid State Sci.* 2013, 17, 40–45.

57. Matar, S. F.; Demazeau, G., Electronic band structure of $CaUO_4$ from first principles. *J. Solid State Chem.* 2009, 182, 2678–2684.

58. Allen, G. C.; Griffiths, A. J.; Vanderheijden, A. N., Electronic absorption-spectroscopy of alkaline earth metal uranate compounds. *Transit. Met. Chem.* 1981, 6, 355–360.

59. Allen, G. C.; Griffiths, A. J., Vibrational spectroscopy of alkaline-earth metal uranate compounds. *J. Chem. Soc.-Dalton Trans.* 1979, 0, 315–319.

60. Allen, G. C.; Griffiths, A. J., Vibrational spectroscopy of strontium uranate(VI) compounds. *J. Chem. Soc.-Dalton Trans.* 1977, 0, 1144–1148.

61. Takahashi, K.; Fujino, T.; Morss, L. R., Crystal-chemical and thermodynamic study on $CaUO_{4-x}$, $(Ca_{0.5}Sr_{0.5})UO_{4-x}$ and α-$SrUO_{4-x}$ (x ~ 0.5). *J. Solid State Chem.* 1993, 105, 234–246.

62. Fujino, T.; Morss, L. R., Oxygen nonstoichiometry in calcium and strontium uranates ($CaUO_{4-x}$ and $SrUO_{4-x}$): Lattice parameters and enthalpies of formation. *Inorg. Chim. Acta* 1984, 94, 123–123.

63. Anderson, J. S.; Barraclough, C. G., Heat of dissociation of non-stoichiometric calcium uranate. *Trans.Faraday Soc.* 1963, 59, 1572–1579.

64. Tagawa, H.; Fujino, T.; Yamashita, T., Formation and some chemical properities of alkaline-earth metal monouranates. *J. Inorg. Nucl. Chem.* 1979, 41, 1729–1735.

65. Holc, J.; Kolar, D., The thermal-stability of calcium uranates in a hydrogen atmosphere. *J. Solid State Chem.* 1983, 47, 98–102.

66. Krasevec, V.; Prodan, A.; Holc, J., *8th European Congress on Electron Microscopy*, Budapest, Hungary, 1984, pp. 1075–1076.

67. Prodan, A.; Boswell, F. W., The defect structure of reduced $CaUO_4$. *Acta Crystallogr. Sect. B-Struct. Commun.* 1986, 42, 141–146.

68. Pialoux, A.; Touzelin, B., The new perovskite type compound $CaUO_3$ *in situ* crystallographic study at high temperature. *Comptes Rendus Acad. Sci. Ser. II-B* 1996, 323, 687–693.

69. Pialoux, A.; Touzelin, B., High-temperature X-ray diffraction analysis of the U–Ca–O system. *J. Nucl. Mater.* 1998, 255, 14–25.

70. Goudochnikov, P.; Bell, A. J., Correlations between transition temperature, tolerance factor and cohesive energy in 2+: 4+perovskites. *J. Phys.-Condes. Matter* 2007, 19, 176201.

71. Liu, X. C.; Hong, R. Z.; Tian, C. S., Tolerance factor and the stability discussion of ABO_3 type ilmenite. *J. Mater. Sci.-Mater. Electron.* 2009, 20, 323–327.

72. Bhalla, A. S.; Guo, R. Y.; Roy, R., The perovskite structure — a review of its role in ceramic science and technology. *Mater. Res. Innov.* 2000, 4, 3–26.

73. Tagawa, H.; Fujino, T., Phase transformation of strontium monouranate(VI). *J. Inorg. Nucl. Chem.* 1978, 40, 2033–2036.

74. Tagawa, H.; Fujino, T., Anomalous behaviour in alpha to beta transition of $SrUO_4$. *Inorg. Nucl. Chem. Lett.* 1977, 13, 489–493.

75. Bugaris, D. E.; zur Loye, H. C., Materials discovery by flux crystal growth: Quaternary and higher order oxides. *Angew. Chem. Int. Ed.* 2012, 51, 3780–3811.

76. Hao, Y.; Murphy, G. L.; Bosbach, D.; Modolo, G.; Albrecht-Schmitt, T. E.; Alekseev, E. V., Porous uranyl borophosphates with unique three-dimensional open-framework structures. *Inorg. Chem.* 2017, 56, 9311–9320.

77. Hao, Y. C.; Klepov, V. V.; Murphy, G. L.; Modolo, G.; Bosbach, D.; Albrecht-Schmitt, T. E.; Kennedy, B. J.; Wang, S.; Alekseev, E. V., Influence of synthetic conditions on chemistry and structural properties of alkaline earth uranyl borates. *Cryst. Growth Des.* 2016, 16, 5923–5931.

78. Wang, S.; Alekseev, E. V.; Depmeier, W.; Albrecht-Schmitt, T. E., Surprising coordination for plutonium in the first plutonium(III) borate. *Inorg. Chem.* 2011, 50, 2079–2081.

79. Wang, S. A.; Alekseev, E. V.; Depmeier, W.; Albrecht-Schmitt, T. E., New neptunium(V) borates that exhibit the alexandrite effect. *Inorg. Chem.* 2012, 51, 7–9.

80. Wang, S. A.; Alekseev, E. V.; Depmeier, W.; Albrecht-Schmitt, T. E., Further insights into intermediate- and mixed-valency in neptunium oxoanion compounds: Structure and absorption spectroscopy of $K_2(NpO_2)_3B_{10}O_{16}(OH)_2(NO_3)_2$. *Chem. Commun.* 2010, 46, 3955–3957.

81. Wang, S. A.; Alekseev, E. V.; Depmeier, W.; Albrecht-Schmitt, T. E., Recent progress in actinide borate chemistry. *Chem. Commun.* 2011, 47, 10874–10885.

82. Wang, S. A.; Alekseev, E. V.; Ling, J.; Skanthakumar, S.; Soderholm, L.; Depmeier, W.; Albrecht-Schmitt, T. E., Neptunium diverges sharply from uranium and plutonium in crystalline borate matrixes: Insights into the complex behavior of the early actinides relevant to nuclear waste storage. *Angew. Chem.Int. Ed.* 2010, 49, 1263–1266.

83. Wang, S. A.; Alekseev, E. V.; Stritzinger, J. T.; Depmeier, W.; Albrecht-Schmitt, T. E., How are centrosymmetric and non-centrosymmetric structures achieved in uranyl borates? *Inorg. Chem.* 2010, 49, 2948–2953.

84. Roof, I. P.; Smith, M. D.; zur Loye, H. C., Crystal growth of K_2UO_4 and Na_4UO_5 using hydroxide fluxes. *J. Cryst. Growth* 2010, 312, 1240–1243.

85. Roof, I. P.; Smith, M. D.; zur Loye, H. C., High Temperature Flux Crystal Growth of Uranium-Containing Perovskites: Sr_3UO_6 and Ba_2MUO_6 (M = Cu, Ni, Zn). *J. Chem. Crystallogr.* 2010, 40, 491–495.

86. Roof, I. P.; Smith, M. D.; zur Loye, H. C., Crystal growth of uranium-containing complex oxides $Ba_2Na_{0.83}U_{1.17}O_6$, $BaK_4U_3O_{12}$ and $Na_3Ca_{1.5}UO_6$. *Solid State Sci.* 2010, 12, 1941–1947.

87. Yeon, J.; Smith, M. D.; Sefat, A. S.; zur Loye, H. C., Crystal growth, structural characterization, and magnetic properties of new uranium(iv) containing mixed metal oxalates: $Na_2U_2M(C_2O_4)_6(H_2O)_4$ (M = Mn^{2+}, Fe^{2+}, Co^{2+}, and Zn^{2+}). *Inorg. Chem.* 2013, 52, 2199–2207.

88. Fujino, T.; Masaki, N.; Tagawa, H., Crystal-structures of alpha-$SrUO_4$ and gamma-$SrUO_4$. *Z. Kristall.* 1977, 145, 299–309.

89. Brisi, C.; Montorsi, M.; Acquaron, G., Research on strontium uranates. *Rev. Int. Hautes Temp. Refract.* 1971, 8, 37–42.

90. Pialoux, A.; Touzelin, B., Étude du système U-Sr-O par diffraction X à haute température. *Can. J. Chem.* 1999, 77, 1384–1393.

91. Ippolitova, E. A.; Kovba, L. M.; Simanov, Y. P.; Bereznikova, I. A., Chemistry of the uranates of some group II elements. *Zhurnal Fiz. Khimii* 1961, 35, 145.

92. Yamashita, T.; Fujino, T.; Masaki, N.; Tagawa, H., The crystal-structures of α-$CdUO_4$ and β-$CdUO_4$. *J. Solid State Chem.* 1981, 37, 133–139.

93. Tagawa, H.; Fujino, T., Anomalous change of the oxidation-state of uranium in the phase-transformation of cadmium mon-ouranate. *Inorg. Nucl. Chem. Lett.* 1980, 16, 91–96.

94. Matar, S. F., Electronic structure and lattice anisotropy of $Cd(UO_2)O_2$. *Chem. Phys. Lett.* 2009, 476, 213–217.

95. Osman, H. H.; Pertierra, P.; Salvado, M. A.; Izquierdo-Ruiz, F.; Recio, J. M., Structure and bonding in crystalline cesium uranyl tetrachloride under pressure. *Phys. Chem. Chem. Phys.* 2016, 18, 18398–18405.

96. Samson, S.; Sillen, L. G., Die kristallstruktur des bariumuranates — Nichtexistenz der UO_4-gruppe. *Ark. Kemi. Mineral. Geol.* 1948, 25, 1–16.

97. Appel, H.; Bickel, M.; Melchior, S.; Kanellakopulos, B.; Keller, C., Structural and magnetic-properities of $BaUO_4$ and $BaNpO_4$. *J. Less-Common Met.* 1990, 162, 323–334.

98. Bickel, M.; Kanellakopulos, B., Structure, vibrational-spectrum and magnetism of $SrUO_4$ — comment. *J. Less-Common Met.* 1990, 163, L19–L20.

99. Reis, A. H.; Hoekstra, H. R.; Gebert, E.; Peterson, S. W., Redetermination of crystal-structure of barium uranate. *J. Inorg. Nucl. Chem.* 1976, 38, 1481–1485.

100. Tanaka, K.; Tokushima, K.; Kurosaki, K.; Ohishi, Y.; Muta, H.; Yamanaka, S., Thermophysical properties of $BaUO_4$. *J. Nucl. Mater.* 2013, 443, 218–221.

101. Popa, K.; Colineau, E.; Wastin, F.; Konings, R. J. M., The heat capacity of $BaUO_4$. *J. Chem. Thermodyn.* 2007, 39, 104–107.

102. Cordfunke, E. H. P.; Ijdo, D. J. W., $Ba_2U_2O_7$: Crystal-structure and phase-relationships. *J. Phys. Chem. Solids* 1988, 49, 551–554.

103. Cordfunke, E. H. P.; Ouweltjes, W., Standard enthalpies of formation of uranium-compounds XIV. BaU_2O_7 and $Ba_2U_2O_7$. *J. Chem. Thermodyn.* 1988, 20, 235–238.

104. Liu, J. H.; Van den Berghe, S.; Konstantinovic, M. J., XPS spectra of the U^{5+} compounds KUO_3, $NaUO_3$ and $Ba_2U_2O_7$. *J. Solid State Chem.* 2009, 182, 1105–1108.

105. Nakamura, A.; Doi, Y.; Hinatsu, Y., Magnetic properties of barium uranate $Ba_2U_2O_7$. *J. Solid State Chem.* 2011, 184, 531–535.

106. Nakamura, A.; Kenji, Y., $Ba_2U_2O_7$: A canted ferromagnet with double-perovskite related orthorhombic structure. *Physica B Condens. Matter* 2006, 378–380, 548–549.

107. Hinatsu, Y., Magnetic studies on $BaUO_3$, $BaPrO_3$ and $BaTbO_3$. *J. Alloy. Compd.* 1993, 193, 113–115.

108. Matsuda, T.; Yamanaka, S.; Kurosaki, K.; Uno, M.; Kobayashi, S., Heat capacity measurement of $BaUO_3$. *J. Alloy. Compd.* 2001, 322, 77–81.

109. Yamanaka, S.; Kurosaki, K.; Matsuda, T.; Uno, M., Thermophysical properties of $BaUO_3$. *J. Nucl. Mater.* 2001, 294, 99–103.

110. Cordfunke, E. H. P.; Booij, A. S.; SmitGroen, V.; vanVlaanderen, P.; Ijdo, D. J. W., Structural and thermodynamic characterization of the perovskite-related $Ba_{1+y}UO_{3+x}$ and $(Ba, Sr)_{1+y}UO_{3+x}$ phases. *J. Solid State Chem.* 1997, 131, 341–349.

111. Barrett, S. A.; Jacobson, A. J.; Tofield, B. C.; Fender, B. E. F., The preperation and structure of barium uranium oxide $BaUO_{3+x}$. *Acta Crystallogr. Sect. B-Struct. Commun.* 1982, 38, 2775–2781.

112. Frondel, C.; Barnes, I., Structural relations of UO_2, isometric $PbUO_4$ and orthorhombic $PbUO_4$. *Acta Crystallogr.* 1958, 11, 562–563.

113. Cremers, T. L.; Eller, P. G.; Larson, E. M.; Rosenzweig, A., Single-crystal structure of lead uranate (VI). *Acta Crystallogr. Sect. C-Cryst. Struct. Commun.* 1986, 42, 1684–1685.

114. Popa, K.; Benes, O.; Staicu, D.; Griveau, J. C.; Colineau, E.; Raison, P. E.; Vigier, J. F.; Pagliosa, G.; Sierig, M.; Valu, O. S.; Somers, J.; Konings, R. J. M., Thermal properties of $PbUO_4$ and Pb_3UO_6. *J. Nucl. Mater.* 2016, 479, 189–194.

115. Konings, R. J. M.; Popa, K.; Colineau, E.; Wastin, F., The low-temperature heat capacity of $CaUO_4$ and $SrUO_4$. *J. Chem. Thermodyn.* 2008, 40, 220–224.

116. Zachariasen, W. H., Crystal chemical studies of the 5f-series of elements XXI. The crystal structure of magnesium orthouranate. *Acta Crystallogr.* 1954, 7, 788–791.

117. Guo, X. F.; Tiferet, E.; Qi, L.; Solomon, J. M.; Lanzirotti, A.; Newville, M.; Engelhard, M. H.; Kukkadapu, R. K.; Wu, D.; Ilton, E. S.; Asta, M.; Sutton, S. R.; Xu, H. W.; Navrotsky, A., U(V) in metal uranates: A combined experimental and theoretical study of $MgUO_4$, $CrUO_4$, and $FeUO_4$. *Dalton Trans.* 2016, 45, 4622–4632.

118. Baur, W. H.; Joswig, W.; Pieper, G.; Kassner, D., $CoReO_4$, A new rutile-type derivative with ordering of 2 cations. *J. Solid State Chem.* 1992, 99, 207–211.

119. Bertaut, E. F.; Delapalme, A.; Forrat, F.; Pauthenet, R., Etudes des uranates de cobalt et de manganese. *J. Phys. Radium* 1962, 23, 477–485.

120. Miyake, C.; Kondo, T.; Takamiya, T.; Yoneda, Y., A mangetic study of U–M–O ternary mixed oxides (M = Mn, Fe, Co, Ni, Cu). *J. Alloy. Compd.* 1993, 193, 116–118.

121. Bacmann, M.; Bertaut, E. F., A non colinear model for magnetic structure of $UCoO_4$. *J. Phys.* 1969, 30, 949.

122. Bacmann, M.; Bertaut, E. F., Parameters atomiques et structure magnetique de $MnUO_4$. *J. Phys.* 1966, 27, 726.

123. Ghodbane, O.; Pascal, J. L.; Favier, F., Microstructural effects on charge-storage properties in MnO_2-based electrochemical Supercapacitors. *ACS Appl. Mater. Interfaces* 2009, 1, 1130–1139.

124. Kohler, T.; Armbruster, T.; Libowitzky, E., Hydrogen bonding and Jahn–Teller distortion in groutite, alpha-MnOOH, and manganite, gamma-MnOOH, and their relations to the manganese dioxides ramsdellite and pyrolusite. *J. Solid State Chem.* 1997, 133, 486–500.

125. Chabre, Y.; Pannetier, J., Structural and electrochemical properities of the proton gamma-MnO_2 system. *Prog. Solid State Chem.* 1995, 23, 1–130.

126. Guillot, M.; Pauthene.R., Magnetic properties of several substances in strong pulsed fields. *J. Appl. Phys.* 1965, 36, 1003–1004.

127. Bacmann, M.; Bertaut, E. F., Structure du nouveau compose $UFeO_4$. *Bull. Soc. fr. Mineral. Crist.* 1967, 90, 257.

128. Bacmann, M.; Bertaut, E. F.; Blaise, A., Neutron diffraction and magnetic measurement study of ferromagnetic $UFeO_4$. *C. R. Hebd. Seances Acad. Sci. B* 1968, 266, 45.

129. Bacmann, M.; Bertaut, E. F.; Bassi, G., Parametres atomiques et structure magnetique de $UCrO_4$. *Bull. Soc. fr. Mineral. Crist.* 1965, 88, 214.

130. Read, C. M.; Smith, M. D.; zur Loye, H. C., Single crystal growth and structural characterization of ternary transition-metal uranium oxides: $MnUO_4$, $FeUO_4$, and NiU_2O_6. *Solid State Sci.* 2014, 37, 136–143.

131. Bacmann, M.; Bertaut, E. F.; Blaise, A.; Chevalie.R; Roult, G., Magnetic structures and properties of $UFeO_4$. *J. Appl. Phys.* 1969, 40, 1131.

132. Bacmann, M.; Chevalie.R; Bertaut, E. F.; Roult, G.; Belakhov. M., Mossbauer effect and neutron diffraction of $UFeO_4$. *C. R. Hebd. Seances Acad. Sci. Ser. B* 1968, 267, 518.

133. Greenblatt, M.; Hornreich, R. M.; Sharon, B., Magnetoelectric compounds with 2 sets of magnetic sublattices: $UCrO_4$ and $NdCrTiO_5$. *J. Solid State Chem.* 1974, 10, 371–376.

134. Althammer, M.; Meyer, S.; Nakayama, H.; Schreier, M.; Altmannshofer, S.; Weiler, M.; Huebl, H.; Geprags, S.; Opel, M.; Gross, R.; Meier, D.; Klewe, C.; Kuschel, T.; Schmalhorst, J.-M.; Reiss, G.; Shen, L.; Gupta, A.; Chen, Y.-T.; Bauer, G. E. W.; Saitoh, E.; Goennenwein, S. T. B., Quantitative study of the spin Hall magnetoresistance in ferromagnetic insulator/normal metal hybrids. *Phys. Rev. B* 2013, 87, 224401.

135. Young, A. P., Nickel orthouronate: High-pressure synthesis. *Science* 1966, 153, 1380–1381.

136. Hoekstra, H. R.; Marshall, R. H., Some uranium-transition element double oxides. *Adv. Chem. Ser.* 1967, 71, 211–227.

137. Brisi, C., The uranates of the MU_3O_{10} type. *Ann. Chim.* 1963, 53, 325–332.

138. Siegel, S.; Hoekstra, H. R., Crystal structure of copper uranium tetroxide. *Acta Crystallogr. Sec. B* 1968, B 24, 967–970.

139. Bacmann, M.; Bertaut, E. F., A non colinear model of magnetic structure of $UCoO_4$. *J. Phys.* 1969, 30, 949–953.

140. Andersson, D. A.; Baldinozzi, G.; Desgranges, L.; Conradson, D. R.; Conradson, S. D., Density functional theory calculations of UO_2 oxidation: Evolution of UO_{2+x}, U_4O_{9-y}, U_3O_7, and U_3O_8. *Inorg. Chem.* 2013, 52, 2769–2778.

141. Conradson, S. D.; Manara, D.; Wastin, F.; Clark, D. L.; Lander, G. H.; Morales, L. A.; Rebizant, J.; Rondinella, V. V., Local structure and charge distribution in the UO_2–U_4O_9 system. *Inorg. Chem.* 2004, 43, 6922–6935.

142. Hubert, S.; Purans, J.; Heisbourg, G.; Moisy, P.; Dacheux, N., Local structure of actinide dioxide solid solutions $Th_{1-x}U_xO_2$ and $Th_{1-x}Pu_xO_2$. *Inorg. Chem.* 2006, 45, 3887–3894.

143. Bohler, R.; Quaini, A.; Capriotti, L.; Cakir, P.; Benes, O.; Boboridis, K.; Guiot, A.; Luzzi, L.; Konings, R. J. M.; Manara, D., The solidification behaviour of the UO_2–ThO_2 system in a laser heating study. *J. Alloy. Compd.* 2014, 616, 5–13.

144. Aronson, S.; Clayton, J. C., Themorydnamic properities of nonstoichiometric urania-thoria solids solutions. *J. Chem. Phy.* 1960, 32, 749–754.

145. Dash, S.; Parida, S. C.; Singh, Z.; Sen, B. K.; Venugopal, V., Thermodynamic investigations of ThO_2–UO_2 solid solutions. *J. Nucl. Mater.* 2009, 393, 267–281.

146. Anderson, J. S.; Edgington, D. N.; Roberts, L. E. J.; Wait, E., The oxides of uranium part IV. The system UO_2–ThO_2–O. *J. Chem. Soc.* 1954, 0, 3324–3331.

147. Popa, K.; Prieur, D.; Manara, D.; Naji, M.; Vigier, J. F.; Martin, P. M.; Blanco, O. D.; Scheinost, A. C.; Prussmann, T.; Vitova, T.; Raison, P. E.; Somers, J.; Konings, R. J. M., Further insights into the chemistry of the Bi–U–O system. *Dalton Trans.* 2016, 45, 7847–7855.

148. Shannon, R., Revised effective ionic radii and systematic studies of interatomic distances in halides and chalcogenides. *Acta Crystallogr. Sec. A* 1976, 32, 751–767.

149. Miyake, C.; Isobe, T.; Yoneda, Y.; Imoto, S., Magnetic properties of pentavalent uranium ternary oxides with fluorite structure: $ScUO_4$, YUO_4, CaU_2O_6 and CdU_2O_6. *Inorg. Chim. Acta* 1987, 140, 137–140.

150. Miyake, C.; Kawasaki, O.; Gotoh, K.; Nakatani, A., Magnetic properties of pentavalent uranium ternary oxides with fluorite structure .2. *J. Alloy. Compd.* 1993, 200, 187–190.

151. Galuskin, E. V.; Kusz, J.; Armbruster, T.; Galuskina, I. O.; Marzec, K.; Vapnik, Y.; Murashko, M., Vorlanite, $(CaU^{6+})O_4$, from Jabel Harmun, Palestinian Autonomy, Israel. *Am. Miner.* 2013, 98, 1938–1942.

152. Khoury, H. N.; Sokol, E. V.; Clark, I. D., Calcium uranium oxide minerals from central Jordan: Assemblages, chemistry, and alteration products. *Can. Mineral.* 2015, 53, 61–82.

153. Weber, W. J.; Ewing, R. C.; Catlow, C. R. A.; de la Rubia, T. D.; Hobbs, L. W.; Kinoshita, C.; Matzke, H.; Motta, A. T.; Nastasi, M.; Salje, E. K. H.; Vance, E. R.; Zinkle, S. J., Radiation effects in crystalline ceramics for the immobilization of high-level nuclear waste and plutonium. *J. Mater. Res.* 1998, 13, 1434–1484.

154. Sickafus, K. E.; Grimes, R. W.; Valdez, J. A.; Cleave, A.; Tang, M.; Ishimaru, M.; Corish, S. M.; Stanek, C. R.; Uberuaga, B. P., Radiation-induced amorphization resistance and radiation tolerance in structurally related oxides. *Nat. Mater.* 2007, 6, 217–223.

155. Sickafus, K. E.; Minervini, L.; Grimes, R. W.; Valdez, J. A.; Ishimaru, M.; Li, F.; McClellan, K. J.; Hartmann, T., Radiation tolerance of complex oxides. *Science* 2000, 289, 748–751.

156. Wang, S. X.; Wang, L. M.; Ewing, R. C.; Was, G. S.; Lumpkin, G. R., Ion irradiation-induced phase transformation of pyrochlore and zirconolite. *Nucl. Instr. Meth. Phys. Res. Sec. B* 1999, 148, 704–709.

157. Steele, B. C. H., Oxygen ion conductors and their technological applications. *Mater. Sci. Eng. B-Solid State Mater. Adv. Technol.* 1992, 13, 79–87.

158. Steele, B. C. H.; Heinzel, A., Materials for fuel-cell technologies. *Nature* 2001, 414, 345–352.

159. Motohashi, T.; Hirano, Y.; Masubuchi, Y.; Oshima, K.; Setoyama, T.; Kikkawa, S., Oxygen storage capability of Brownmillerite-type $Ca_2AlMnO_{5+\delta}$ and its application to oxygen enrichment. *Chem. Mat.* 2013, 25, 372–377.

160. Burschka, J.; Pellet, N.; Moon, S. J.; Humphry-Baker, R.; Gao, P.; Nazeeruddin, M. K.; Gratzel, M., Sequential deposition as a route to high-performance perovskite-sensitized solar cells. *Nature* 2013, 499, 316–319.

161. Kreuer, K. D.; Paddison, S. J.; Spohr, E.; Schuster, M., Transport in proton conductors for fuel-cell applications: Simulations, elementary reactions, and phenomenology. *Chem. Rev.* 2004, 104, 4637–4678.

162. Wu, Z. S.; Ren, W. C.; Wen, L.; Gao, L. B.; Zhao, J. P.; Chen, Z. P.; Zhou, G. M.; Li, F.; Cheng, H. M., Graphene anchored with Co_3O_4 nanoparticles as anode of lithium ion batteries with enhanced reversible capacity and cyclic performance. *ACS Nano* 2010, 4, 3187–3194.

163. Jiang, J.; Li, Y. Y.; Liu, J. P.; Huang, X. T.; Yuan, C. Z.; Lou, X. W., Recent advances in metal oxide-based electrode architecture design for electrochemical energy storage. *Adv. Mater.* 2012, 24, 5166–5180.

164. Liu, C.; Li, F.; Ma, L. P.; Cheng, H. M., Advanced materials for energy storage. *Adv. Mater.* 2010, 22, E28–62.

165. Murphy, G.; Kennedy, B. J.; Johannessen, B.; Kimpton, J. A.; Avdeev, M.; Griffith, C. S.; Thorogood, G. J.; Zhang, Z. M., Structural studies of the rhombohedral and orthorhombic monouranates: $CaUO_4$, alpha-$SrUO_4$, beta-$SrUO_4$ and $BaUO_4$. *J. Solid State Chem.* 2016, 237, 86–92.

166. Ranson, P.; Ouillon, R.; Pinan-Lucarre, J. P.; Pruzan, P.; Mishra, S. K.; Ranjan, R.; Pandey, D., The various phases of the system $Sr_{1-x}Ca_xTiO_3$ — A Raman scattering study. *J. Raman Spectrosc.* 2005, 36, 898–911.

167. King, G.; Woodward, P. M., Cation ordering in perovskites. *J. Mater. Chem.* 2010, 20, 5785–5796.

168. Fujino, T.; Tagawa, H.; Adachi, T.; Hashitani, H., Gravimetric method for determination of oxygen in uranium-oxides and ternary uranium-oxides by additon of alkaline-earth compounds. *Anal. Chim. Acta* 1978, 98, 373–383.

169. Perdew, J. P.; Ruzsinszky, A.; Csonka, G. I.; Vydrov, O. A.; Scuseria, G. E.; Constantin, L. A.; Zhou, X. L.; Burke, K., Restoring the density-gradient expansion for exchange in solids and surfaces. *Phys. Rev. Lett.* 2008, 100, 136406.

170. Wang, S. A.; Alekseev, E. V.; Juan, D. W.; Casey, W. H.; Phillips, B. L.; Depmeier, W.; Albrecht-Schmitt, T. E., NDTB-1: A Supertetrahedral Cationic Framework That Removes TcO_4^- from Solution. *Angew. Chem. Int. Ed.* 2010, 49, 1057–1060.

171. Elgazzar, S.; Rusz, J.; Amft, M.; Oppeneer, P. M.; Mydosh, J. A., Hidden order in URu_2Si_2 originates from Fermi surface gapping induced by dynamic symmetry breaking. *Nat. Mater.* 2009, 8, 337.

172. Halcrow, M. A., Jahn–Teller distortions in transition metal compounds, and their importance in functional molecular and inorganic materials. *Chem. Soc. Rev.* 2013, 42, 1784–1795.

173. Desgranges, L.; Baldinozzi, G.; Rousseau, G.; Niepce, J. C.; Calvarin, G., Neutron diffraction study of the in situ oxidation of UO_2. *Inorg. Chem.* 2009, 48, 7585–7592.

5

Transparent Conductors: Complex Coordination in Complex Metal Oxides

Karl Rickert, Steven Flynn and Kenneth R. Poeppelmeier

Department of Chemistry, Northwestern University, 2145 Sheridan Rd. Evanston, IL 60208-3113, U.S.A.

Transparency to visible light and electrical conductivity appear at first glance to be mutually exclusive — highly conductive materials such as copper are typically opaque and colored, while transparent materials such as glass are often electrically insulating. Despite this apparent mismatch, not only are transparent conductors (TCs) possible, they are in high demand for energy harvesting, energy consumption reduction, and consumer electronics. These materials are widely deployed for electricity generation in solar cells and are effective substrates in solar-driven water splitting.[1,2] They lower energy use by serving as integral components in functional glass, as both smart windows and passive low-emissivity windows, which can be deployed on an entire building or a much smaller scale, depending on the need.[3] High-end consumer electronics that have a touch screen or a liquid crystal display also require TCs as a transparent electrode.[4] In fact any device containing a photoactive (emits or receives light) layer that is part of an electrical circuit needs a 'transparent electrode' to complete the circuit without blocking the photoactive layer. Given the ubiquity of these devices, it is not surprising that the production of TCs represents a multibillion dollar industry that is expected to continue growing.[3,5]

Obviously, from these widespread applications, TCs are important, but how can they maintain both properties? The key to understanding this duality is in the electronic structures of the three classes of electronic materials: conductors, insulators, and semiconductors. The two crucial features in the electronic band structures of TCs are the valence band, or the fully occupied orbitals with the highest energy, and the conduction band, which is the unoccupied or partially occupied orbitals with the lowest energy. These concepts are similar to the highest occupied molecular orbital (HOMO) and lowest unoccupied molecular orbital (LUMO), respectively, but are extrapolated to account for the interactions of an infinite number of atoms in a periodic lattice, namely a crystalline material, instead of in an isolated molecule. As a result, 'bands' are formed from the continuous overlapping orbitals and collectively form the electronic band structure of a material. In conductors, the Fermi level crosses one or more bands, leaving them being partially occupied and enabling high electrical conductivity. This partial occupancy also enables a multitude of electron orbital transitions that can be triggered by incident light, resulting in opacity. Insulators do not have partially-filled bands at the Fermi level, but instead possess a large energy difference, known as the band gap, between the valence and the conduction bands. Thus, all bands are either full or empty and the wide band gap limits carriers from crossing into empty bands. If the band gap is greater than the highest energy of visible light (~3.1 eV), then incident light lacks the energy necessary to promote electrons into the conduction band and the material is transparent. Semiconductors are more complex than the other electronic materials, and can be classified as either intrinsic or extrinsic. An intrinsic semiconductor is similar to an insulator, except the band gap is small enough for external energy, such as heat or light, to excite electrons from the valence band into the conduction band. If the band gap is small, visible light will be able to produce this excitation and are the material will not be transparent. Intrinsic semiconductors are typically weakly conducting, and therefore insufficient for the applications desired of TCs. The difficulties in producing a TC in these classes of materials are overcome in extrinsic semiconductors. In an extrinsic semiconductor, an 'extrinsic' dopant is added to a host material (typically an intrinsic semiconductor) which can improve the conductivity by fundamentally altering the band structure, the carrier concentration, the carrier mobility, defect states, or some combination of all of these. For example, Sn^{4+} can replace In^{3+} in the intrinsic semiconductor In_2O_3 to produce tin-doped In_2O_3 (ITO or $Sn:In_2O_3$), which is a major commercial TC. Doping alters the fundamental band structure, as shown in Figure 1, and therefore changes the electronic and optical properties.[6] In this example, tin doping changes the conduction band in In_2O_3 so that it is degenerate with the Fermi level and partially occupied, producing the desired conductivity. The isolation of this band within the gap avoids the issue of interband electron transitions present in metals, making the transparency depend mainly on the band gap. In this figure, the band gap is low (~1 eV) compared to the experimentally determined value (~3.6 eV), but these band structures are calculated with the local density approximation, which can underestimate the band gap. Extrinsic semiconductors can be further separated into *n*- and *p*-type materials, depending on their dominant charge carrier. In the case of ITO, the dopant produces excess electrons, resulting in an *n*-type material. If electron holes are the main charge carriers, then the material is *p*-type. Thus, extrinsic semiconductors are the major materials of interest for developing TCs as they can possess both a high conductivity and transparency to visible light. However, several alternatives also exist. The earliest TCs were noble metals machined to an incredible thinness that allowed a limited transmission of light through the material while maintaining the metallic conductivity. More recently, thin

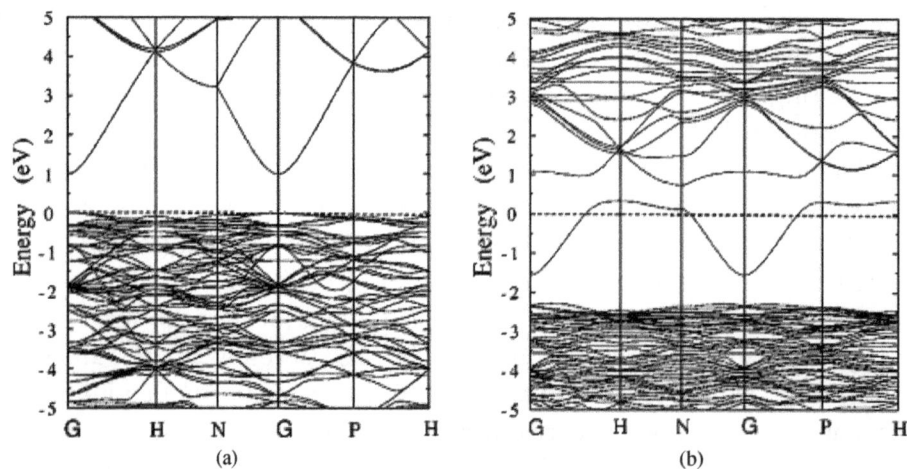

Figure 1.   Electronic band structures for (a) the intrinsic semiconductor In_2O_3 and (b) the extrinsic, *n*-type semiconductor $Sn:In_2O_3$. From Ref. [6].

materials have again been considered for TCs in the form of graphene and carbon nanostructures which are conductive and again gain transparency from being extremely thin.

The aforementioned highly specific electronic structure is required for a material to display both transparency and conductivity, which presents an odd contrast to the simple crystalline structure design principle for TCs: a high density of corner- and/or edge-sharing polyhedra must be present (for *n*-type TCs). This requirement originates from a need for an intact electron pathway without which the material would not be electrically conductive. In terms of the electronic band structure, this pathway is present as a highly dispersed, *s*-type band defining the bottom of the conduction band. This type of band is necessary for *n*-type TC oxides and is illustrated in Figure 1(b), where the band that satisfies this criterion is degenerate with the Fermi level. Despite this seemingly simple principle, the structures of TCs become more complex the more in-depth one delves. The intact pathway guideline holds true on the unit cell scale, but when employing a local atomic site perspective, a whole host of details must be considered. The concentration of dopant in extrinsic semiconductors is typically insufficient to wholly occupy a distinct crystallographic position. The resulting partially occupied sites bring differences in coordination into play. Different coordination environments will have unique local energy landscapes, which can impact the site preference for a dopant, the efficacy of a dopant on specific sites, and, if filled, alter the structure itself. Even sites with the same coordination number can maintain different coordination environments (e.g. octahedral vs trigonal prismatic) which will also exhibit different energy environments. Despite the complexity of TC oxides, there are several large structural families that dominate *n*- and *p*-type materials. The majority of the presented discussion will focus on *n*-type TCs because they are more numerous, have been known for longer, and have been extensively studied. TCs with a *p*-type nature, which were only recently discovered (comparatively), will be discussed to a lesser extent. It is worth noting, however, that in both *n*- and *p*-type TCs the property of greatest interest is the electrical conductivity and most efforts have focused on increasing this conductivity up to the Bellingham limit (the theoretical maximum conductivity of a transparent material).[7] As a result of the importance of TCs, the field has been thoroughly and excellently reviewed a number of times, and these reviews are recommended for a more detailed look at TCs.[1,8–10] TCs have also been the subject of technical books and book chapters, of which the *Handbook of Transparent Conductors* is particularly recommended as providing a thorough, in-depth coverage of the subject.[11] Other books and reports that cover TCs tend to have a more specific focus or a more simplified global perspective, depending on the objective and intended audience.[12–14]

The materials that are discussed here are all metal oxides. As mentioned earlier, thin sheets of pure metals or carbon nanostructures have demonstrated TC properties, but oxides represent a clear majority of the materials currently in use and form the focus of this chapter. Several practical factors have led to the dominance of oxide materials. For example, oxides generally demonstrate stability with regard to device operating conditions while being chemically inert with regard to neighboring layers. Both of these attributes contribute to a higher longevity of electronic devices. Compared to the thin metal layers that were originally used, oxide ceramics have a much greater durability and can be deposited in thicker layers over larger areas without suffering performance disadvantages, leading to more resilient devices. Another inherent advantage of oxides is that many of them have wide (>2 eV) band gaps.[15] Although close to the desired band gap (≥ 3.1 eV) for TCs, these are typically a little too low. Upon doping, however, these band gaps can effectively increase through a phenomenon known as the Burstein–Moss effect, as illustrated in Figure 2.[16,17] Despite some absorption resulting from intraband transitions, this effect produces an experimentally measured band gap that is greater than the fundamental band gap, meaning that materials with fundamental band gaps less than 3.1 eV still have potential as TCs. This effect is limited, thus the wide band gaps of metal oxides provide a needed starting point close enough to the desired band gap so that doping can feasibly be used to achieve the 3.1 eV band gap. Provided that a dopant can be found to produce a degenerate *n*-type material, the

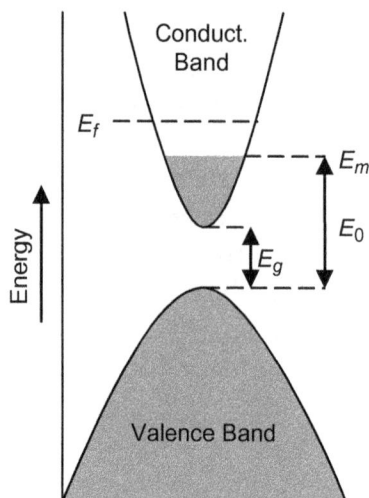

Figure 2.   Conceptual band diagram illustrating the Burstein–Moss effect. In degenerate n-type semiconductors, the fundamental band gap (E_g) is lower than the experimentally measured band gap (E_0). The n-type character of the material results in a partially filled conduction band, and the degenerate character results in a Fermi level (E_f) inside the conduction band. Thus, optically driven electronic transitions occur between the valence band and the lowest unoccupied portion of the conduction band (E_m).

Burstein–Moss effect can enable materials with low fundamental band gaps, and thus a potentially higher conductivity, to be transparent to visible light, thereby satisfying both requirements needed to make a useful TC.[3,5]

In order to be an effective TC, however, a material needs to be more than just transparent and conductive. Many of the applications are in layered devices, and a TC needs to remain chemically inert with respect to the layers it contacts and any additional environmental stimuli. For a device to be fully optimized, the transparency window needs to match the range of energies useable by the device, which may verge into the infrared or ultraviolet wavelengths. This ensures that no useful energy is lost as a result of the TC.

Transparent Conductors: n-type

The two routes to obtain an extrinsically doped n-type TC are to produce an anion deficiency or a cation excess. Although both these result in a surplus of electrons, the mechanisms to produce these off-stoichiometries are different. For example, a common method of introducing an anion deficiency is to replace a cation that has with one oxidation state with a different cation with a higher oxidation state while keeping the oxygen constant (or removing oxygen afterward through a reduction step). This maintains the cation lattice and disrupts the anion lattice. In contrast, a cation surplus can be performed by adding an interstitial cation dopant that occupies a new crystallographic site, disrupting the cation lattice and maintaining the anion lattice. Weak Lewis acid elements with filled d-orbitals have been the dopants of choice for producing extrinsically doped n-type TCs, as summarized in Figure 3. The choice for weaker Lewis acid cations is a result of the weaker metal–oxygen bond that forms, which is much more conducive to promoting anion vacancies. Elements with filled d-orbitals are preferred because they eliminate the possibility of intraorbital d–d transitions which would severely reduce the transparency of a material. These selection factors resulted in a thorough study of a small section of the periodic table, and the structural families of some of the most important cations from this section will be discussed. Strong Lewis acid cations, however, have recently been found to hold potential as TC materials and will also be considered.

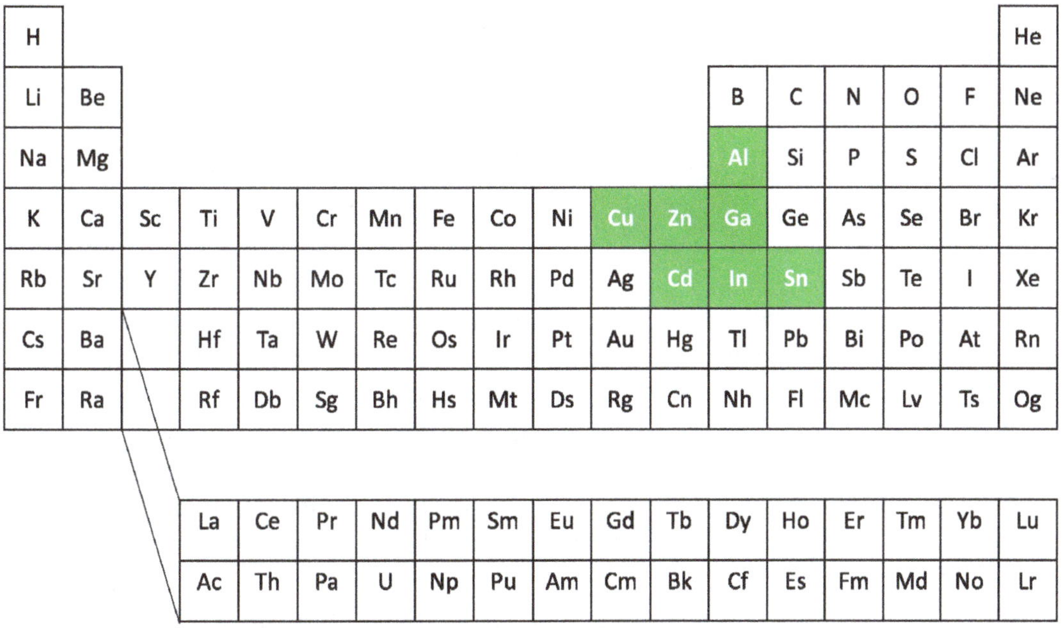

Figure 3. Elements of greatest interest as cations in TC oxide materials are in gray.

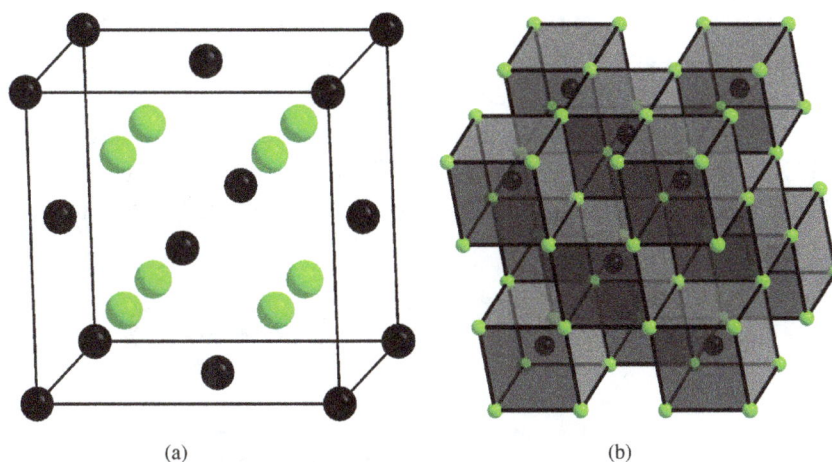

(a) (b)

Figure 4. The (a) unit cell of fluorite (CaF_2) as viewed in the (115) direction and (b) an expanded fluorite structure showing the corner- and edge-sharing lattice of coordination cubes surrounding the cation sites as viewed in the (945) direction. Calcium atoms are black and fluorine atoms are gray.

Indium-based Transparent Conductors

The first family that will be discussed is arguably the most important as well as one of the most prolific — those that have In as a majority of the cation content. Many of these materials adopt anion-deficient, fluorite-related structures, and as such the fluorite structure will be discussed first. The unit cell of CaF_2, where fluorite derives its name, is provided in Figure 4(a). There are a few key attributes of this structure that will be used here to describe when a structure is 'fluorite related': (1) the cation lattice must be cubic close packed, (2) the anions must only occupy tetrahedral holes. Furthermore, in discussing 'anion-deficient', fluorite-related structures, (a) anion vacancies can be present, but (b) there cannot be cation vacancies, and (c) distortions in both the lattices (i.e. atomic relaxations) that accommodate the anion vacancies can be present. As can be observed from Figure 4(b), the basic fluorite structure maintains an interconnected network of edge-sharing anion cubes. This forms the necessary basis for conductivity by providing an unbroken pathway for charge carriers to move through. An intuitive hypothesis with introducing anion deficiencies into this structure is that conductivity would decrease, as the connectivity would by definition decrease, but widespread, systematic deficiencies can give rise to new, entirely connected systems. One example of this is the bixbyite structure of In_2O_3 — which is an anion-deficient, fluorite-related structure.

The bixbyite structure has 25% of the fluorite anion sites vacant ($MO_{1.5}$ vs fluorite's MF_2) which reduces the average cation coordination number from 8 to 6 and, in the case of bixbyite, all of the cation sites are indeed 6-coordinate. Despite the decrease in coordination number, the cation sites still maintain a continuous network of corner- and edge-sharing polyhedra. Unlike CaF_2, however, bixbyite maintains two distinct, non-equivalent cation sites that have the same coordination number, as shown in Figure 5.[18] When compared to the perfect cubes of fluorite, each coordination environment is missing two anions, but one coordination environment (b-site, named after the Wyckoff position) is missing the anions along a body diagonal and the other (d-site) is missing the anion pair along a face diagonal. Upon relaxation of the lattice, the b-site forms an octahedron that is approximately regular, but the d-site forms a distorted octahedron. Undoped In_2O_3 thin films are conductive in their own right, but the conductivity is increased upon reduction (producing oxygen vacancies) and decreased upon oxidation (filling oxygen vacancies).[19,20] In the reduced case ($In_2O_{3-\delta}$), the anion vacancies are acting as a 'dopant', that is, they are producing additional carriers to compensate for the 'excess' In^{3+}. Furthermore, this means that some number of cations have a coordination number less than 6 in the extended lattice. When doped to form an extrinsic semiconductor, the differences in these sites becomes extremely important. $Sn:In_2O_3$ is mentioned previously both because it is an understandable example of an extrinsic semiconductor and because it is one of the most important TC materials to date. Specifically, it has one of the highest reported conductivities of commercial TCs, 10,000 S/cm, while maintaining a wide band gap and excellent transparency. The high

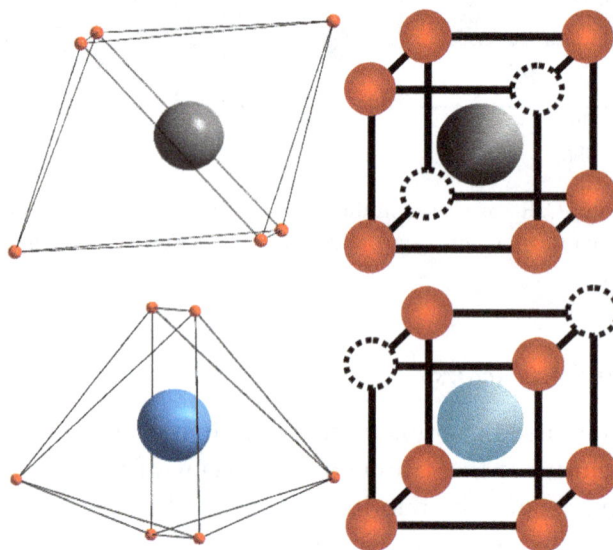

Figure 5.   The non-equivalent, 6-coordinate cation sites (anions are on vertices) in bixbyite (left) and the ideal cubic analogies in the fluorite structure (right). The b-site (top) is missing a pair of anions along a body-diagonal, whereas the d-site has the missing pair along a face-diagonal (bottom). From Ref. [18].

conductivity makes ITO the TC of choice for liquid crystal television and computer displays, solar cells, electrochromic windows, and light emitting diodes. An additional beneficial property of ITO is that it can be chemically etched, meaning that intricate shapes that are required for these applications can be readily obtained. As a result of its desirable properties, ITO has been thoroughly studied both with regard to property optimization and the underlying causes of these desirable properties. A major cause of the conductivity is the formation of Frank–Köstlin clusters that results from the introduction of Sn into In_2O_3. The greater the amount of Sn, the greater the concentration of clusters and the greater the conductivity. As a result, the solubility of Sn in the In_2O_3 lattice has been of preeminent concern. To examine the solubility mechanism in more detail, the site preference (if any) of the Sn dopant is a chief concern. Experimental and computational investigations agree that Sn dopes on the *b*-site and only the *b*-site.[6,21,22] The reducible Frank–Köstlin clusters, which are composed of Sn atoms on In sites (producing an antisite) and an interstitial O atom, are centered on and limited by the number of *b*-sites.[23] As the *b*-site represents 25% of the cation sites in the bixbyite unit cell, there is an inherent limit on the solubility of Sn. This limit assumes that all of the *b*-site can be replaced with Sn, but Sn does not fully occupy the *b*-site, instead having a solubility limit of ~7% of the cation sites. Given this site preference, seeking new In-based structures with a greater number of regular octahedra (*b*-site analogues) is an understandable next step in achieving high solubility limits of Sn. Unfortunately, such structures are difficult to target and obtain. For example, one of the only examples currently reported was recently discovered as $Zn_{0.456}In_{1.084}Ge_{0.460}O_3$ and 50% of the cation sites are *b*-site analogues.[18] Despite having the same anion content ($MO_{1.5}$) as bixbyite, $Zn_{0.456}In_{1.084}Ge_{0.460}O_3$ has four different coordination sites: 6-coordinate (*b*-site analogue), 6-coordinate (*d*-site analogue), 4-coordinate, and 8-coordinate. Although the higher *b*-site presence is desirable, the impact of the new sites remains to be determined.

Seeking new structures with a higher *b*-site presence is only one research route that has been pursued to increase the Sn solubility in In_2O_3. Another avenue that has been explored is cosubstitution. Briefly, cosubstitution consists of replacing cations with a given overall oxidation state with a combination of different cations that maintains the same overall oxidation state, but the oxidation states of the individual cations can differ. For example, a well-investigated cosubstitution route with In_2O_3 is replacing an In^{3+}/In^{3+} pair with a M^{2+}/Sn^{4+} (where M = Mg, Ca, Ni, Cu, Zn, or Cd) pair, as both pairs have overall oxidation states of 6+, but a pair of trivalent cations is replaced with a divalent and tetravalent combination. Nor is this restricted to pairs of cations, as $In^{3+}/In^{3+}/In^{3+}$ triplets can also be replaced with $M^{2+}/M^{2+}/Sb^{5+}$ (where M = Cu, Zn) triplets. In either case, the resulting $M_xIn_{2-2x}Sn_xO_3$ (or $M_{2x/3}In_{2-2x}Sb_{x/3}O_3$) is isostructural to bixbyite and forms a solid solution, where the solubility limits depend on the identity of M.[24,25] As can be observed from the aforementioned $Zn_{0.456}In_{1.084}Ge_{0.460}O_3$, this route is not necessarily as successful when Sn is replaced with a different element, despite the success of Sb. Unlike ITO, where multiple reports confirm that Sn prefers the *b*-site, the cosubstituted systems do not have a unified, clear-cut site preference. At higher concentrations (>25%), the substitutes cannot physically be present only on the *b*-site. Existing computational models, X-ray diffraction experiments, and extended X-ray absorption fine structure (EXAFS) studies suggest that both *b*- and *d*-sites are partially occupied by the substitutes both above and below the 25% level, and no overarching preference for either substitute is consistently present. The cosubstitution method, however, brings an interesting quandary to TCs: it is well documented that the Sn solubility limit can be dramatically increased (up to 25% of the cation sites), but why does the conductivity not show the same dramatic increase? Relating cosubstituted bixbyite back to ITO shows that ITO has a very obvious dopant — replacing In^{3+} with Sn^{4+} — but cosubstituted bixbyite balances the tetravalent cation with a divalent cation, preventing the tetravalent cation from acting purely as a dopant. That being said, the cosubstituted materials are still conductive, particularly with M = Zn, but this conductivity is attributed to an inherent off-stoichiometry of the two substitutes, producing an anion deficiency that acts as a dopant effect.

The defect mechanisms have been examined at length, both experimentally and computationally, to understand the changes to the structure and the related property alterations.

As a result of ITO's prominence as a TC and the strong correlation between the Sn content and the conductivity, trying to increase the Sn solubility has been a natural avenue of investigation. Sn, however, is not the only tetravalent cation that can dope In_2O_3. Indeed, with the success of ITO, massive doping studies of In_2O_3 have been undertaken to determine if there is a better dopant than Sn.[26] Of the stable, common tetravalent cations, the only potential rivals are Si, Ti, or Ge, which display similar conductivity values to that of ITO with Ge being of particular interest as its mobility is almost twice that of ITO.[27-31] The desirable performance of these specific dopants appears to be counterintuitive as they have a higher Lewis acid strength than Sn. This higher affinity for electron pair acceptance promotes a stronger dopant–oxygen bond which would hinder the creation of anion vacancies. These vacancies are either present in as-synthesized materials or introduced with a reduction step and increase the conductivity of a material.[32] For example, oxygen atoms can balance the tetravalent charge of Sn in ITO, nullifying its dopant character. Removing these atoms produces carriers that are characteristic of a highly conducting material. As such, weaker metal–oxygen bonds are preferred to make an abundance of oxygen vacancies possible, while inversely, stronger interactions hinder this. In addition, these bonds and the increased potential for interstitial anion positions would mean that new scattering centers could be introduced, which would further decrease the mobility of the carriers. Despite these concerns, the higher (relative to Sn) Lewis acid strength elements perform well as dopants in both In_2O_3 and ITO. This increased performance is attributed to a higher charge mobility. Instead of the new scattering centers, the high Lewis acid strength dopants are hypothesized to act as electron screens, weakening the role of the O $2p$ valence band as a scattering center.[28]

Furthermore, if tetravalent cations make excellent dopants in In_2O_3, pentavalent and hexavalent cations have the potential to be even better dopants, increasing the carrier concentration by two or three times per replaced ion more than what would be produced with a tetravalent dopant. A key point about these oxidation states, however, is that they are most commonly encountered among the transition metals, often with empty d-orbitals which could lead to a wide variety of electron transitions that would make a material opaque. Anionic doping offers a similar mechanism to tetravalent doping if, for example, O^{2-} is replaced with a halogen such as F^-.[33,34] This rationale has been expanded to truly mixed-anion materials, such as InOF, which can be transparent to both visible and ultraviolet light and whose conductivity can be tuned over a wide range.[35] Despite these various options and the investigations that have pursued them, ITO remains a dominant TC material. Another fluorite related, anion-deficient TC, $Ga_{3-x}In_{5+6}Sn_2O_{16}$ for $0.3 \leq x \leq 1.6$, helps illustrate why transition metals are inferior dopants. A competitive TC in its own right, $Ga_{3-x}In_{5+6}Sn_2O_{16}$ can have its Sn replaced with Ti, Zr, or Hf to form an isostructural solid solution.[36-39] When this substitution is performed, however, the conductivity decreases. The cause for this decrease is the formation of covalent necks, as depicted in Figure 6, which increase the electron scattering and decrease the electron mobility.[39] Thus, a major strength of Sn as a dopant is that Sn^{4+} has filled, core-like d-orbitals while other potential competitors from the transition metals have partially occupied d-orbitals that can interact with the anion lattice. This is a fundamental reason why TC research concentrates on the main block elements and those cations with filled valence d-orbitals (i.e. Zn^{2+}, Cd^{2+}, Cu^+, and Ag^+). This rationale, that having filled valence d-orbitals avoids harmful covalent interactions between the cation and anion lattices, explains the competitive conductivity of Ge:In_2O_3 and Si:In_2O_3, but not that of Ti:In_2O_3 (4700 S/cm vs. ITO's 4900 S/cm in non-optimized bulk powder samples).[31] Owing to the small size of Ti (an ionic radius of 0.605 Å vs. the 0.8 Å of In), in small quantities it may not distort the bixbyite lattice enough to enable the formation of these covalent necks. Regardless, the easiest way to avoid them is to stay away from elements that have the potential for this interaction. This justification may also explain why

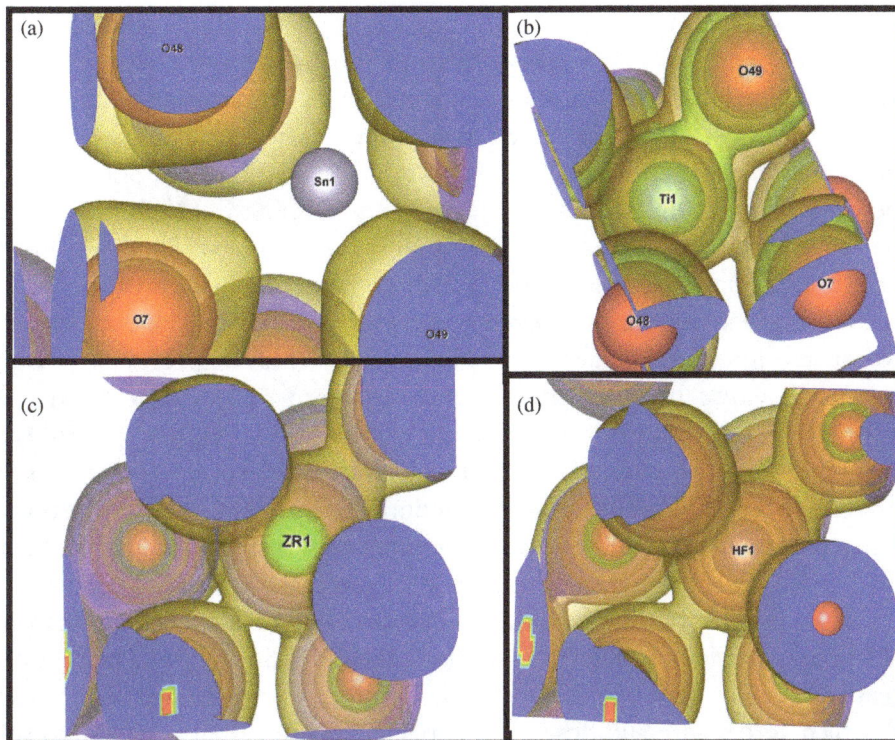

Figure 6. Valence charge density (e/Å^3) around a 7-coordinate site in Ga$_2$In$_6$Sn$_2$O$_{16}$ that is occupied by (a) Sn, (b) Ti, (c) Zr, and (d) Hf. Colors denote different levels, which are nested inside each other. Covalent 'necks' are present for Ti, Zr, and Hf, but not Sn. From Ref. [39].

attempts at replacing the In in ITO with an alternative trivalent transition metal cation (i.e. Y^{3+} or Sc^{3+}) also results in a decreased conductivity.[40] The strong covalent interaction of metals with empty *d*-orbitals is similar to that predicted for highly Lewis acidic dopants, likely suppressing oxygen vacancies in the material and decreasing the number of charge carriers.

Despite the interesting chemistry and desirable TC properties of In-based TCs, In itself is an undesirable material for widespread applications because it is uncommon, concentrated in certain geographic locations, and the refinement process necessary to extract In is costly. In addition to these supply concerns, In-based TCs have some physical properties that are undesirable. Typical moisture contents in air are sufficient to react with In$_2$O$_3$ to form InOOH or In(OH)$_3$, which decrease the efficacy of electronic devices and leads to delamination over time and lowers the lifetime of a device. These shortcomings make finding competitive TCs with reduced or no In content a major focus of the TC industry. TCs without indium have been known and deployed since the discovery of the first TCs, but they tend to lack the high conductivity present in In-based TCs.

Tin-based transparent conductors

The importance of Sn in In-based systems cannot be overstated, but SnO$_2$ also functions as an excellent TC in its own right, when doped, and is the earliest TC to be widely deployed. Indeed, TCs based on SnO$_2$ have an inherent low emissivity, facile deposition, and lower cost that make them the materials of choice for wide-area depositions, such as functional glass.[3] 'Functional glass' is a very broad term and can be separated into several categories that require a TC, namely (1) low-emission, (2) electrochromic, (3) heat dispersion,

(4) deicing, and (5) static dissipating glasses.[41] These categories are typically applied to windows, ranging from the deicing capability of automobile windshields to the heat dispersive ability of oven windows.

Similar to In_2O_3, SnO_2 displays transparency and conductivity in an undoped state, but 'undoped' is again a misnomer, as these properties are attributed to the inherent non-stoichiometry of as-deposited materials.[8] Unlike In, Sn is stable as either Sn^{2+} or Sn^{4+}, and thus the possibility of SnO forming has been considered and can be avoided by tuning deposition parameters. Doping SnO_2, however, leads to improved properties, and the two most successful dopants are Sb and F. The use of Sb as a mainstream dopant for SnO_2 departs from the trend in In_2O_3, as Sb has a greater Lewis acid strength than Sn. Although these dopants obtain the best conductivity and transparency, it is worth noting that one of the earliest dopants was Cl, which was incorporated into SnO_2 as a side product of using chloride precursors during the deposition process.[42,43] Despite not having a conductivity as high as In-based TCs, $Sb:SnO_2$ and $F:SnO_2$ are in high demand because (1) SnO_2 is much more abundant and cheaper than In_2O_3 and (2) SnO_2 is easy to deposit on extremely large surface areas via pyrolysis. Similar to bixbyite, the rutile structure of SnO_2 maintains a continuous network of edge- and corner-sharing octahedra which can function as a pathway for charge carriers and facilitates high conductivity. Unlike In-based systems, however, both cation and anion doping make SnO_2 a highly effective TC. Furthermore, the doping does have a substitutional component to it (i.e. site replacement), but a major portion of the dopant also occupies interstitial sites inside the rutile lattice.[44] The possible interstitial sites are provided in Figure 7 for $Sb:SnO_2$ and $F:SnO_2$.[45,46] The Sb interstitial sites have been determined experimentally from neutron diffraction, with a partial occupancy of 0.103(10) for a 10% doping level of Sb, which agrees with all of the Sb occupying the interstitial sites and no substitutional doping occurring. The F interstitial sites, in contrast, are first determined computationally and then compared to experimental results to confirm whether substitutional or interstitial doping occurs.[45,47] In the case of $F:SnO_2$, both substitutional and interstitial doping occur, but it has been hypothesized that the interaction between interstitial and substitutional F^- produces defect complexes that have the interstitial F^- act as an acceptor, compensating for the doping and decreasing the conductivity.[48] This decreased conductivity is observed as the concentration of F increases, suggesting the acceptor interactions are not present or significantly weaker when F is present in low concentrations. Unlike F, which has a known oxidation state of 1- (replacing the 2- of O), Sb has stable oxidation states of both 3+ and 5+, and determining which oxidation state is present has been a matter of interest for $Sb:SnO_2$. As $Sb:SnO_2$ is n-type, Sb^{5+} would be the expected

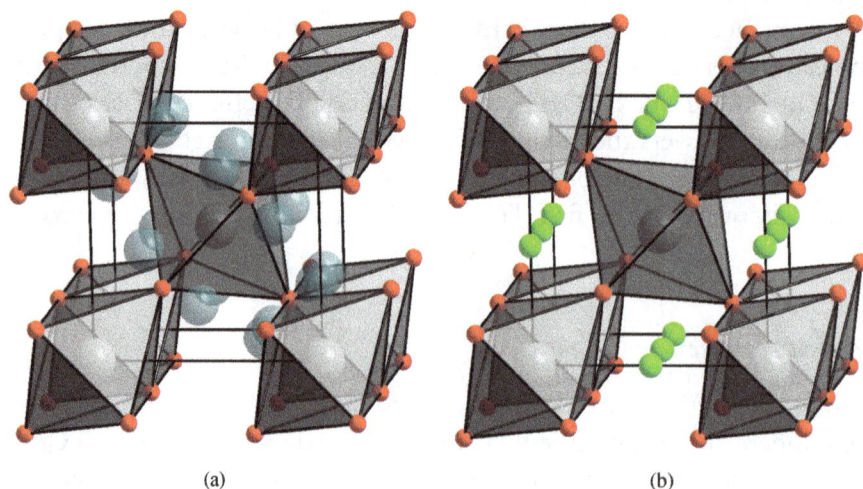

(a)   (b)

Figure 7.   The rutile structure of SnO_2 (Sn is in the center of the octahedra, O is on the vertices) with interstitial positions for (a) Sb and (b) F dopants, as viewed in the (7 9 31) direction.

oxidation state. Experimental investigations into this phenomenon have focused on different thermal treatments of various Sb:SnO$_2$ compositions, but concur in that low levels of doping are almost exclusively Sb^{5+}, but higher levels of doping result in an increasing Sb^{3+} content, which counterbalances the Sb^{5+} cation and decreases the conductivity.[49,50] Thus, the major difference between the conductivity mechanisms of Sn- and In- based TCs is that the major role is played by the dopant lattice (Sn-based) or the host lattice (In-based).

TiO$_2$ is similar to SnO$_2$, but it can adopt the anatase structure instead of the rutile structure or is a material that has been investigated for TC applications.[51] Ta^{5+} has been successfully employed as a dopant in TiO$_2$. When compared to Sb, Ta has oxidation states that are more distinct, in that they tend not to be stable under similar conditions. In many cases, Ta^{5+} is the preferred oxidation state. Sn:TiO$_2$, which is still in the anatase structure, has also been investigated and demonstrates a potential for photoenergy conversion devices.[52] Ta:TiO$_2$ has been of interest as being stable under reducing conditions and as a potential material for water splitting devices.[53] On a less applied level, Ta:TiO$_2$ is important because it breaks away from the elements with filled *d*-orbitals that have thus far characterized TC materials. It switches to the opposite end of the transition metal family, where both Ta and Ti will have completely empty *d*-orbitals when in their stable cation state.

Zinc-based transparent conductors

Zn-based TCs depart from the standard 6-coordinated cation structural building block that is employed in both In- and Sn-based TCs to form the bixbyite and rutile structures. Instead, as shown in Figure 8, 4-coordinated cation sites form the basic structural building block in the wurtzite structure. Each tetrahedral building block shares its corners with several other tetrahedra, forming an unbroken network that is conducive to electron mobility. In addition to this inherent property, the wurtzite structure has sufficient

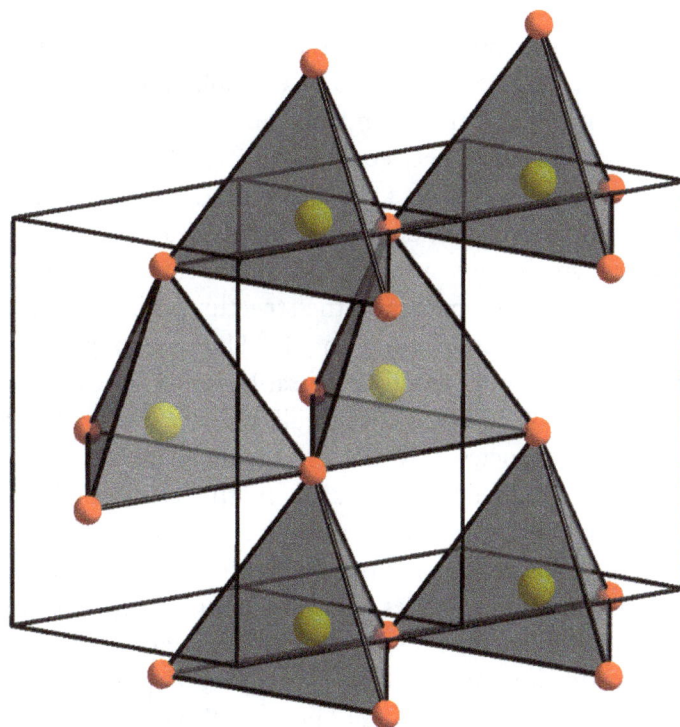

Figure 8.   A doubled (along the *x*-direction) unit cell of wurtzite ZnO as viewed in the [1 1 1] direction, showing Zn (center of tetrahedrons) and O (vertices).

space to enable both interstitial and substitutional dopants, similar to the case encountered in the rutile structure of SnO_2. As with both In_2O_3 and SnO_2, ZnO can be a TC in its own right, but its properties dramatically improve upon doping. In contrast, however, the origins of the conductivity in undoped ZnO are unclear and controversial. The wurtzite structure can accommodate Zn or O vacancies, Zn or O interstitials, Zn or O antisites, and/or a combination of these to form defect complexes. In addition to these purely structural possibilities, hydrogen incorporation onto interstitial sites from synthetic methods has also been considered as a potential cause for ZnO's properties.[54] The potentially low concentration of these defects has resulted in ZnO being examined closely by several computational investigations, each supporting one or more of the possible intrinsic defects. Despite the difficulty inherent in experimentally examining these small concentrations, it has been investigated in a variety of ways.

In addition to the difficult-to-characterize origin of ZnO's TC properties, doping studies of ZnO are slightly different from In- and Sn-based TCs. A major goal of many investigations is to dope ZnO so that it is *p*-type, complementing the successful *n*-type dopants that have already been identified. Although both In_2O_3 and SnO_2 have also been considered as potential *p*-type hosts, there is a more concerted attempt on making ZnO a *p*-type material. Despite this interest, the most successful dopant in ZnO, Al^{3+}, produces an *n*-type TC, aluminum-doped zinc oxide (AZO). AZO in particular is the TC of choice for solar cells with a photoactive layer of copper indium gallium selenide (CIGS) and is under consideration for expansion to other solar cell designs.[55-57] There is less controversy over the origins of the AZO's conductivity and transparency than there is for ZnO, but again the case is not absolutely clear cut. Unintended hydrogen doping is again a potential factor that may contribute to the conductivity.[58] Concerning the Al dopant, computational studies agree that below the solubility limit (~3%) Al substitutes onto Zn sites.[59,60] Once the solubility limit is exceeded, Al forces Zn onto interstitial sites, which results in an overall decrease in TC properties. Ga is a second successful trivalent dopant into ZnO, but has been the subject of less study when compared to AZO.[61] The most common method is to introduce the defect of interest to the investigation and subsequently analyze the material.

The interest in doping ZnO to make both *n*- and *p*-type TCs has resulted in many structures beyond the simple wurtzite structure that have been considered. For example, the $(In,Ga)_2O_3(ZnO)_n$ is of interest to combine the desired properties of In with the abundance and price of Zn.[62-64] The structure consists of mirrored ZnO wurtzite layers sandwiched between bixbyite $(In,Ga)_2O_3$ layers.[65-68] As the value of *n* increases, the wurtzite layers become thicker and the bixbyite layers further apart. Thus, there are three distinct cation networks in the structure: corner-sharing tetrahedra (wurtzite), corner- and edge-sharing octahedra (bixbyite), and corner- and edge-sharing 5-coordinate cation sites (above and below the mirror plane of wurtzite layers). Given the layered nature of the structure, none of these cation networks are fully continuous in 3D, rather they are only continuous in 2D. Unless a single crystal is used for direction-dependent conductivity measurements, this would necessarily result in lower conductivity values. This is experimentally observed, with values of < 300 S/cm in bulk powder specimens after a reduction procedure that is intended to produce oxygen vacancies.[62] As the ZnO content increases, which makes the $(In,Ga)_2O_3$ layers more dispersed, the overall conductivity decreases. In this case, combining both In_2O_3 and ZnO makes a material with the structural characteristics of both, but inferior TC properties.

As with the other TC families, the reason for the conductivity of the $(In,Ga)_2O_3(ZnO)_n$ series is not intuitive. Most of the characterization on this series has relied upon amorphous thin films, in which case, the composition is the same, but the local environments may differ widely. In these studies, the electrical properties of the films depended on the amount of O present during deposition, which suggests that oxygen vacancies are a cause of the conductivity.[63,69-71] The end point of this series (*n* = 1, with Ga = In), $InGaZnO_4$, has been studied more in depth to elucidate the mechanisms that are contributing to the electrical properties.[72] This is the ZnO-poor endpoint and results in a structure that does not have any wurtzite layers.

Instead, the bixbyite layers, containing only In, surround the mirrored planes of wurtzite that contain Ga and Zn. The end result is a structure with 5- and 6-coordinate cation sites only. Based on this and the composition, this material is not truly 'Zn-based', but is related to both ZnO and In_2O_3 nearly equally. In a combined approach of computational and experimental techniques, O vacancies were not found to be the major contributor to conductivity. Instead, a Ga presence on Zn sites, producing Zn antisites, was found to be the major defect that caused the conductivity. This antisite is likely favored as a result of the unusual coordination environments present in $InGaZnO_4$. O vacancies are energetically feasible, but they are less favored than cation vacancies, which would counterbalance the O vacancy and not contribute to conductivity.[72]

Cadmium-based transparent conductors

The first oxide material that was investigated for and demonstrated transparency and conductivity was CdO in 1907.[73] Despite being the first discovered and having some of the highest conductivity values of transparent materials, Cd-based TCs are rarely used commercially. Indeed, as shown in Figure 9, the most conductive oxide-based TC is In:CdO, and many of the other highly conductive TCs have Cd as a major component.[74] The desirable transparency and conductivity of Cd-based TCs, however, are countered by the toxicity of CdO and Cd in general.[75–78] The relatively low melting point of cadmium (321°C) increases the risk of exposure, making it particularly unsuitable for large-scale production.[79] Other cadmium-based materials, most notably cadmium (II) telluride (CdTe), are utilized in the photovoltaic industry despite this risk, but they are physically more stable and less soluble than cadmium and CdO, which lowers the risk of exposure. Although direct toxicity comparisons have never been performed, it is generally assumed that the toxicity of Cd, CdO, and CdTe are similar.[80] The higher stability of CdTe, combined with its position in solar cells, which embed it behind several layers of benign materials, lowers the risk of contact. Even with these reduced risks, there is widespread public disapproval over the use of CdTe, and research is ongoing to replace it.[81]

The health concerns of Cd-based TCs prevent their widespread commercial deployment, but this family exhibits interesting structural diversity that is at odds with the other families discussed here. In:CdO maintains the simple, rock salt structure of CdO, but Cd_2SnO_4 has the Sr_2PbO_4 structure that transitions to a spinel structure upon the introduction of In and $CdSnO_3$ displays a perovskite structure. The highest conductivity is obtained with the rock salt structure, which still maintains a continuous corner- and edge-sharing

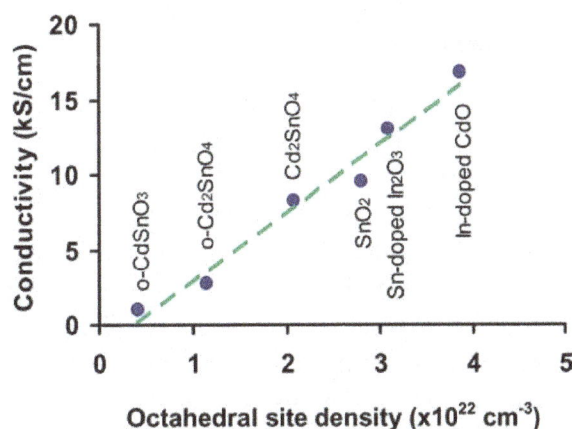

Figure 9. The highest reported conductivities of the most conductive TC materials as a function of 6-coordinate cation site density. From Ref. [74].

network of octahedra, but the remaining structures are significantly different from the In- and Sn-based systems. The second highest conductivity is exhibited by the cubic form of Cd_2SnO_4, which has a continuous network of edge- and corner-sharing octahedra but is also the only material that unambiguously has 4-coordinate sites in addition to 6-coordinate octahedral sites.[82] The other structures, orthorhombic Cd_2SnO_4 and perovskite $CdSnO_3$, also have the continuous network of 6-coordinate sites, assuming a reasonable Cd–O bond length to be < 2.7Å, but whereas all of the other TCs have this network composed of regular or distorted octahedra, these two materials have the network composed of a mixture of regular octahedra and triangular prisms.[82,83] As such, it is not unexpected that these two materials would have lower conductivities, as multiple local energy environments would be present and charge mobility would be limited by the energy environment that is least favorable.

The simple rock salt structure of CdO provides an interesting contrast to the fluorite structure of In_2O_3. In rock salt, each O atom occupies octahedral holes, whereas in fluorite each O atom occupies tetrahedral holes. Thus, the removal of a single oxygen atom in CdO alters the coordination environments of 6 cations instead of the four in fluorite. As a result, minor changes have much greater consequences that have the potential to destabilize the structure. This means that fluorite can be more resilient to change and produce a wider variety of related structures. This, combined with the toxicity of Cd, helps to explain why Cd-based materials are receiving less of a research effort than In-based materials.

In some instances, however, Cd-based and In-based materials cross paths. As mentioned in the "Indium-based Transparent Conductors" section, Cd^{2+} is used in conjunction with Sn^{4+} to cosubstitute In_2O_3, but beyond the conditions for creating these bixbyite phases lies the spinel phase space of $(CdIn_2O_4)_{1-y}(Cd_2SnO_4)_y$ ($0 \leq y \leq 0.75$).[84–87] The spinel structure has one-third of the cations in 4-coordinate site and the remainder in 6-coordinate sites. Although the 6-coordinate sites form a continuous corner- and edge-sharing network, the 4-coordinate sites are completely isolated from each other. In a classic spinel, the single species (Cd) would occupy the 4-coordinate site and the double species (In) the 6-coordinate site. In this $CdIn_2O_4$, however, the spinel has an inversion factor, with cations that are mixed on both sites.[85] Cd_2SnO_4, in contrast, is fully inverted, with all Sn on octahedral sites and has a self-doping mechanism based on the Sn-on-Cd antisite.[88] Upon combining the two systems ($y > 0$), Sn retains its exclusive occupancy of octahedral sites. Cd accompanies Sn onto the octahedral sites, decreasing the overall In content of these sites. Unusually, Cd_2SnO_4 has a higher conductivity than $CdIn_2O_4$, and upon combining the two the conductivity decreases. These spinels also have a symmetry forbidden transition from the valence band maximum to the conduction band minimum which inflates the optical band gap above the fundamental band gap.[89]

Aluminum-based transparent conductors

While Al has already been mentioned as an important dopant for Zn-based TCs and will be mentioned again as a fundamental component in p-type TCs, it is also the majority element in its own class of TC materials: mayenite. Unlike the prior families, mayenite, with a standard formula of $Ca_{12}Al_{14}O_{33}$ (also written as $12CaO \cdot 7Al_2O_3$), adopts a cage-like structure, as shown in Figure 10(a) that contains two formula units in the form of 12 cages (Figure 10(b)) per unit cell. This structure is inherently important for mayenite's potential as a TC, as only two of the cages contain an oxygen ion. Removing this free O ion thus produces an excess of electrons and converts all cages to a uniform structure. This removal converts mayenite from a wide band gap insulator to a metal-like conductor and can be combined with H dopants.[90–92] This important property has resulted in a convention of describing mayenite's formula as $[Ca_{12}Al_{14}O_{32}]^{2+}O^{2-}$ (or doubled for the two formula units), which becomes $[Ca_{12}Al_{14}O_{32}]^{2+} + 2e^-$ upon removing the free O.

The removal of O transforms $Ca_{12}Al_{14}O_{33}$ from an insulator to a metal-like conductor, but the reason behind this transition is debated. If the electron concentration is low, then computational studies suggest

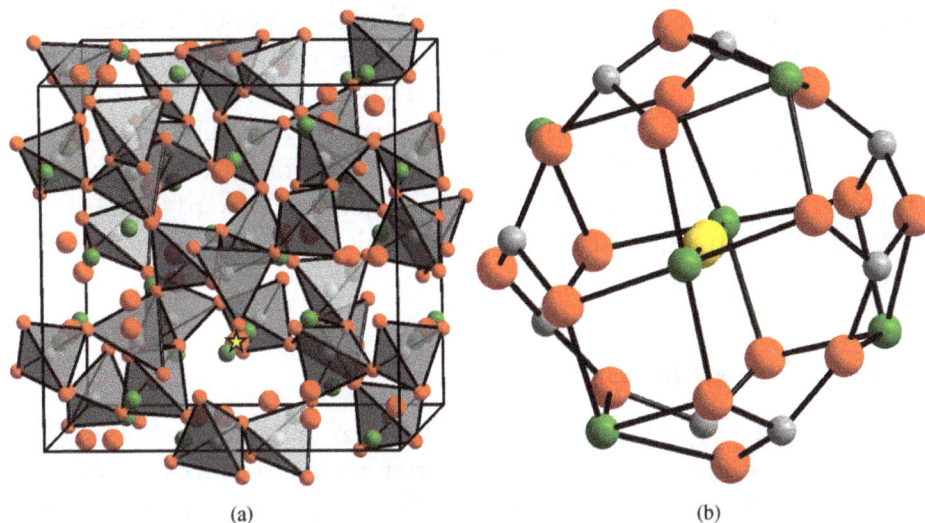

Figure 10.   (a) One unit cell of the mayenite structure for $Ca_{12}Al_{14}O_{33}$ as viewed in the [8 1 1] direction showing grey tetrahedra centered on the Al atoms. Yellow star is the center of a cage structure, which is shown in b). (b) Cage structure, with the O3 positions in yellow. Only 2 of the 24 O3 sites are occupied. Removal of these O atoms is key in mayenite's transition from an insulator to a conductor. Green atoms are Ca, grey atoms are Al, and red are O.

that the charges are localized.[93] If the electron concentration is high, then charges are believed to be delocalized and electrons are considered to have free-electron like behavior.[94] A contributing factor to this transition is that the free carriers that are present after the transition do not appear to compete with polarons. Rather they both contribute to the conductivity.[95] Such a removal is conducive to producing *n*-type material, and the lattice itself is accepting of a wide variety of dopants and substitutions. The various elements that can be inserted into the structure grant the ability to tune various properties of mayenite to a specific application. Mayenite has been well-known as a material for decades, but it is only recently that its insulator-to-conductor transition has been discovered. Despite this fact, it has been rigorously studied at a substitutional and dopant level.[96–103]

Transparent Conductors: *p*-type

As has been discussed, *n*-type TCs have been successful, containing materials that are highly conductive and transparent. There are some undesirable attributes of the current materials, such as a reliance on expensive elements such as In, chemical instability over time, and not quite achieving the Bellingham limit of maximum conductivity for a transparent material. Overall, however, these materials suffice for many of current technology's demands and are employed in a wide variety of commercial processes and products. TCs with a *p*-type character have a very different history from *n*-type TCs, as the latter have been known and experimented with for 90 years more than the former, which were first discovered in 1997.[73,104] A major reason of this dichotomy is the difference in the carrier type, as the former require mobile holes, whereas *n*-type TCs require mobile electrons. Oxide materials, so successful in *n*-type TCs, pose a challenge for *p*-type materials because of the strong localization of holes around oxygen atoms at the valence band maximum. This localization reduces the mobility of the holes to a negligible value, even if the holes are present in high concentrations. Cations with valence bands of similar energies to those of oxygen overcome this obstacle. The similarity in energies enables the formation of highly covalent bonds and maintains antibonding orbitals as the valence band maximum.[105] As in *n*-type TCs, the best cations have full *d*-orbitals to prevent transitions between orbitals that produce a colored material instead of a transparent one. The cations that meet both of these criteria the best are Cu^+, Ag^+, and Au^+, and the cost of Au explains why materials with a preponderance of the first two cations are dominating the *p*-type TC research field. As a result of the particular requirements of *p*-type conductivity and transparency, research investigations have moved beyond the *p*-block elements of *n*-type TCs and can contain d^5, d^6, or d^{10} metals as a major component despite the unoccupied *d*-orbitals.[106–111] A highly conductive *p*-type TC would provide a major step forward in the electronics industry as such a device would enable fully transparent *p*–*n* junctions. This would unlock the possibility of completely transparent electronics and invisible circuits.

Copper-based transparent conductors

Since the discovery of the first *p*-type TC, $CuAlO_2$, the field has burgeoned with interest in improving its performance and in finding new materials.[112,113] This interest has led to a focus on the delafossite structure, as it is not only the structure of $CuAlO_2$, but some of its key features are hypothesized to be conducive to *p*-type conductivity. For example, the delafossite structure, idealized as ABO_2 where A is a monovalent cation and B is a trivalent cation and shown in Figure 11, has layers of A^+ cations next to 4-coordinate O atoms. With the similar valence energies of Cu^+ and Ag^+, and the specific coordination environment of O, this crys-

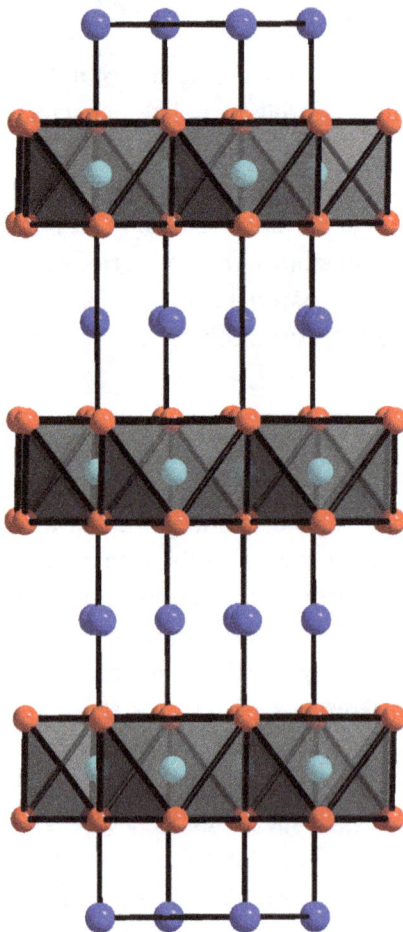

Figure 11.   Unit cell of delafossite $CuAlO_2$ as viewed along the [1 30 1] direction showing Cu (blue), Al (teal), and O (red) atoms. Gray polyhedra are regular octahedra.

tal structure supports a bonding nature in the O atoms which reduces hole localization. The A^+ layers are unique in that each A atom is only coordinated by two O atoms, resulting in a 'linear' cation family, contrasting with the octahedra and tetrahedra found in the *n*-type TCs.[74] The delafossite structure has also demonstrated a capacity to support an overabundance of oxygen ($ABO_{2+\delta}$), which can aid in making a material *p*-type, as oxygen vacancies aid in making a material *n*-type.[114] Similar to the *n*-type families, however, $CuAlO_2$ has inherent conductivity without an obvious dopant. Furthermore, the reported conductivity values vary in order of magnitude between 10^{-3} and 10^1 S/cm, suggesting that sample preparation itself strongly influences the conductivity by altering the hole concentration. This could be the result of a cation deficiency (such as Cu vacancies), an anion surplus, or unintended contaminants.[115] Comparison studies of hydrothermally prepared and high-temperature solid-state synthesized $CuAlO_2$ have demonstrated that the preparation method reproducibly impacts the conductivity and identified a defect that contributes to the conductivity. This defect is a combination of a Cu antisite and partially occupied interstitial oxygen positions. That is, an Al atom occupies a Cu site and two additional oxygen atoms surround this site.[116] This is an unusual defect mechanism, as the Cu antisite would generally be anticipated to produce an *n*-type material, but the surplus oxygen atoms alter the material to make it *p*-type. It is worth noting that in this study unintended contaminants were also considered. Cu vacancies, however, are also a possibility, and a computational approach has asserted that they are energetically plausible.[115]

As is typical with a host system, myriad dopings and substitutions have been performed on ABO_2 with A = Cu in an effort to improve its properties, and p-type character is common in these studies, such as in $CuBO_2$, with B = B, Fe, (Mg, Cr), and many more.[117–120] $Mg:CuScO_2$ and $Ca:CuYO_2$, for example, maintain unit cells that are more accomodating to oxygen interstitials than $CuAlO_2$ and demonstrate O pressure-dependent conductivity that is not present in $CuAlO_2$.[74] Although the complete ABO_2 system, with A = Cu, Ag, or Au, and B = Al, Ga, Sc, In, or Y, has been considered, Cu appears to be key to obtaining transparency and conductivity.[118] Interestingly, Cu-based delafossites have also been known to exhibit tunable semiconductor type, that is, $CuInO_2$ can display either *p*- or *n*-type character based on which dopant and deposition conditions are employed.[121] Departing from the trends of the other families discussed herein, the Cu delafossite family is successful as a sulfide, such as $Zn:CuAlS_2$.[122]

Other Cu-based TCs depart from the delafossite structure type of $CuAlO_2$, but maintain similarly layered structures. One example is $La_5Cu_6O_4S_7$, which has a structure composed of combinations of fluorite and antifluorite layers. In addition to this structure, $La_5Cu_6O_4S_7$ has a calculated band gap of 4.44 eV, which is more than ample for transparency to visible light. $La_5Cu_6O_4S_7$ has a reported metallic conductivity along one crystalline direction and predicted metallic conductivity along a second crystalline direction.[123] These measured properties bring in an unusual facet of layered structures: they can have asymmetric properties. This is because the layers can have a given property, such as high conductivity, in a direction parallel to the layers and then lose this property in a direction orthogonal to the layers. Viewing this from a connectivity perspective, corner- and edge-sharing layers of the same types form 2D electron pathways. Orthogonal to these pathways, however, each layer comes into play, and they will commonly have different local energy environments that would change the measured values.

Additional p-type transparent conductors

A major part of the research into *p*-type TCs has focused on Cu-containing materials as a result of their initial success. A second major focus mentioned earlier is finding a successful *p*-type dopant for ZnO. These are not the only options for *p*-type TCs, however. Ag-based TCs in particular have been considered as a result of Ag's similarity to Cu.[106] One such example is Ag_3VO_4, which has α-, β-, and γ-phases. Of these, the α-phase in particular was predicted to be a *p*-type TC and demonstrated a measurable conductivity.[107]

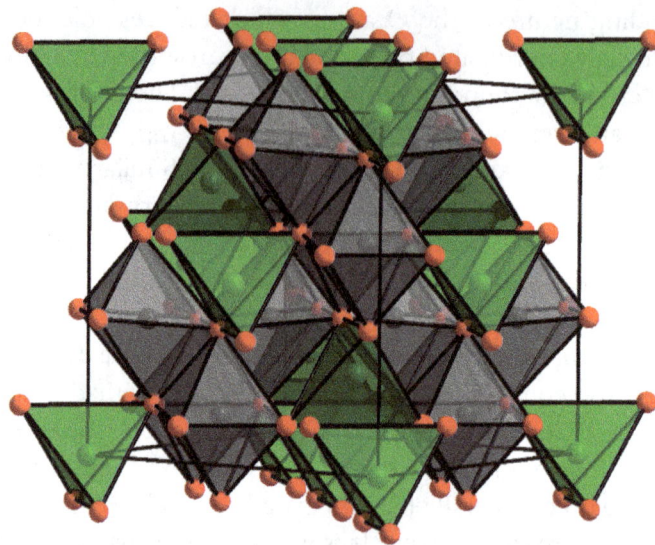

Figure 12.   Unit cell of $CrMn_2O_4$ viewed along the [19 3 28] direction. Grey octahedra centered on Cr form a continuous corner and edge-sharing network throughout the structure. Green tetrahedra centered on Mn are isolated from each other. Red spheres are O atoms.

The spinel structure, depicted in Figure 12, has already been discussed in relation to n-type TCs, but in addition to its n-type capability, the spinel structure and the inverse spinel structure are considered to have great potentials for p-type TCs.[124–126] Furthermore, the elements of interest for p-type TCs are abundant and easily procured, making them very attractive systems to work with.[127] The spinel family, of the generic form A_2BO_4, consists of a wide variety of compositions. A theory-based survey of many different spinels demonstrated that a major source of the conductivity is the presence of antisite defects. Based on the stability of the different intrinsic types of antisite defects in a given spinel, it can be classified into one of four distinct doping types related to the type of conductivity it will favor. These theory insights, exclusive to the spinel family, has allowed for the targeting of high-potential candidate p-type TCs, such as Co_2ZnO_4 and Co_2NiO_4, and the development of design principles for optimizing their properties via doping. Such doping can be intrinsic, as an inherent or induced off-stoichiometry, or extrinsic through aliovalent substitution.[104,121–124] For example, the non-equilibrium growth of Co_2ZnO_4 or Co_2NiO_4 can produce a wide range of A/B ratios, and the Zn-rich end of the Co_2ZnO_4 phase space is found to be conductive p-type materials.[128,129] Similarly, doping with Mg in both Co_2ZnO_4 and Co_2NiO_4 has been observed to improve the p-type conductivity by factors of 20 and 10^4, respectively.[124] Antisites may be a major source of the conductivity in p-type spinels, but the conduction mechanism has not been studied on such a systematic scale. Indeed, the conduction mechanism of the spinel Rh_2ZnO_4 can be attributed to either localized polaronic or band-like transport, depending on what investigative method is used.[130]

Summary

The families presented in this chapter represent some of the major systems of interest for transparent conducting metal oxide ceramics. In each case, the structures, coordination environments, and continuous cation networks have been discussed to demonstrate the importance of crystal structure with regard to properties. When relevant, the defect or defects that contribute to the desired properties have been reviewed, to the best of current reports. These defects have ranged from anion and cation vacancies and interstitials to antisites and inherent off-stoichiometry.

The purpose of this chapter is to introduce several of the most important structural and compositional families of materials that are of interest for transparent conducting applications. This list is by no means exhaustive, focusing on oxides and ceramic materials and excluding carbon nanostructures, pure metals, and materials with anions other than oxygen. Mixed anion materials such as oxide-fluorides and oxide-sulfides are only discussed briefly. Each family is only considered in their crystalline forms; the study of amorphous films is mentioned but not discussed in great depth. Even in the discussed crystalline oxides, the main systems are chosen, omitting some of the less prominent families. If information on these topics is desired, the reader is recommended to the reviews and books mentioned in the introduction of this chapter and repeated here.[1,8–14]

TCs with n-type character are the more established type of material, representing the vast bulk of commercial TCs and of reported research results. The conductivity can be optimized to approach the Bellingham limit, over 10,000 S/cm in thin film form, and transparency and conductivity both have been targets for tuning studies. These properties, superior to those of p-type materials, are partially a result of the n-type materials having been known for quite a bit longer, providing researchers the time necessary to improve original devices and materials. The structures of n-type TCs are varied, including fluorite, rutile, wurtzite, rock salt, spinel, and cage structures. This variety enables a high probability of finding a TC that matches the lattice of adjacent layers in a device, enabling the production of high-quality layered devices that resist delamination. As a result, n-type TCs can be employed in small, highly advanced devices such as handheld electronics (typically ITO), or in large area applications such as architectural glass (usually SnO_2). The n-type materials have been pursued with weak Lewis acids, because this produces a weaker metal–oxygen bond that would promote oxygen vacancies and n-type character. Strong Lewis acids, however, are showing that they may yet have a role to play in high-quality TCs as they can act as electron screens which reduces scattering and increases charge mobility. Oxygen vacancies have certainly been shown to be beneficial to n-type materials, but extrinsic dopants, antisite defects, and interstitial defects also contribute to the desired properties of these materials.

TCs with a p-type character are comparatively newer and have lower optimized conductivity values, typically <10 S/cm. The delafossite family of $CuAlO_2$ has proven to be a breakthrough, both in demonstrating the existence of conductive p-type TCs and in providing a starting point for further investigations. Although p-type TCs contain many mixed-anion varieties, departing from Cu-based materials is not easy. The spinel class of compounds perhaps offers the opportunity to move to more abundant materials that are easier to obtain. Indeed, this structure type has been shown to be quite resilient to a variety of elemental components, both as structural members and as dopants.

The structural aspects of the major TC materials have been described here, but what will be the next evolution of TCs? What steps are being pursued to reach the Bellingham limit and obtain other functionalities? A few potentially 'game changing' discoveries are eagerly sought after by industries and researchers alike. If ceramic TCs can be constructed of cheaper, more abundant materials than the existing commercial leaders while maintaining similar or superior properties, then the TC industry would be able to improve dramatically while continuing to use the existing infrastructure. With regard to the solar cell industry, increasing conversion efficiency is always desirable, but increasing device robustness and longevity is also a major goal. To advance this second objective, solar cells composed of only metal oxides are desired. Such prototype devices exist, such as one made of Cu_2O and ZnMgO, and the current focus is increasing their conversion efficiencies to be competitive with non-metal oxide devices.[131] Another high-profile research path is moving away from inorganic ceramic thin films and into organic materials. Some of the major materials of interest are graphene sheets and carbon nanostructures. If this transition should occur, some of the major benefits would be highly flexible TCs, more abundant starting material, and lower synthetic temperatures. This being said, commercially viable scale-up processes need to be developed, implemented, and broken into the market. In either case, a highly conductive and transparent p-type material is highly sought after, as such a material is required to make fully transparent *p–n* junctions. Such junctions could then make many everyday electronics fully transparent, rather than requiring an opaque contact.

References

1. Granqvist, C. G., Transparent conductors as solar energy materials: a panoramic review. *Sol. Energy Mater. Sol. Cells* 2007, 91, 1529–1598.

2. Hajibabaei, H.; Zandi, O.; Hamann, T. W., Tantalum nitride films integrated with transparent conductive oxide substrates via atomic layer deposition for photoelectrochemical water splitting. *Chem. Sci.* 2016, 7, 6760–6767.

3. Ginley, D. S.; Bright, C., Transparent conducting oxides. *MRS Bull.* 2000, 25, 15–18.

4. Lewis, B. G.; Paine, D. C., Applications and processing of transparent conducting oxides. *MRS Bull.* 2000, 25, 22–27.

5. Liu, H.; Avrutin, V.; Izyumskaya, N.; Özgür, Ü.; Morkoç, H., Transparent conducting oxides for electrode applications in light emitting and absorbing devices. *Superlattices Microstruct.* 2010, 48, 458–484.

6. Mryasov, O. N.; Freeman, A. J., Electronic band structure of indium tin oxide and criteria for transparent conducting behavior. *Phys. Rev. B* 2001, 64, 233111.

7. Bellingham, J. R.; Phillips, W. A.; Adkins, C. J., Intrinsic Performance Limits in Transparent Conducting Oxides. *J. Mater. Sci. Lett.* 1992, 11, 263–265.

8. Chopra, K. L.; Major, S.; Pandya, D. K., Transparent conductors — a status review. *Thin Solid Films* 1983, 102, 1–46.

9. Freeman, A. J.; Poeppelmeier, K. R.; Mason, T. O.; Chang, R. P. H.; Marks, T. J., Chemical and thin-film strategies for new transparent conducting oxides. *MRS Bull.* 2000, 25, 45–51.

10. Minami, T.; Miyata, T., Present status and future prospects for development of non- or reduced-indium transparent conducting oxide thin films. *Thin Solid Films* 2008, 517, 1474–1477.

11. *Handbook of Transparent Conductors.* 1 ed. Eds. Ginley, D. S.; Hosono, H.; Paine, D. C., Springer Boston, MA 2011.

12. Dawar, A. L.; Jain, A. K.; Jagadish, C.; Hartnagel, H., *Semiconducting Transparent Thin Films.* 1 ed. CRC Press. 1995.

13. Barquinha, P.; Martins, R.; Pereira, L.; Fortunato, E., *Transparent Oxide Electronics: From Materials to Devices*; 1 ed.; Wiley: 2012.

14. *Transparent Conductors and Barrier Layers for Thin Film Solar Cells: Final Technical Report*, National Renewable Energy Laboratory (NREL), 2012.

15. Zhang, H.; Chen, G.; Bahnemann, D. W., Photoelectrocatalytic materials for environmental applications. *J. Mater. Chem.* 2009, 19, 5089–5121.

16. Burstein, E., Anomalous Optical Absorption Limit in InSb. *Phys. Rev.* 1954, 93, 632–633.

17. Moss, T. S., The Interpretation of the Properties of indium Antimonide. *Proc. Phys. Soc., London, Sect. B* 1954, 67, 775–782.

18. Rickert, K.; Sedefoglu, N.; Malo, S.; Caignaert, V.; Kavak, H.; Poeppelmeier, K. R., Structural, Electrical, and Optical Properties of the Tetragonal, Fluorite-Related $Zn_{0.456}In_{1.084}Ge_{0.460}O_3$. *Chem. Mater.* 2015, 27, 5072–5079.

19. Muller, H. K., Electrical and Optical Properties of Sputtered In_2O_3 Films. I. Electrical Properties and Intrinsic Absorption. *Phys. Status Solidi* 1968, 27, 723–731.

20. Manifacier, J. C.; Szepessy, L.; Bresse, J. F.; Perotin, M.; Stuck, R., In_2O_3: (Sn) and SnO_2: (F) Films — Application to Solar-Energy Conversion. II. Electrical and Optical Properties. *Mater. Res. Bull.* 1979, 14, 163–175.

21. Nomura, K.; Ujihira, Y.; Tanaka, S.; Matsumoto, K., Characterization and Estimation of ITO (Indium-Tin-Oxide) by Mossbauer Spectrometry. *Hyperfine Interact.* 1988, 42, 1207–1210.

22. Nadaud, N.; Lequeux, N.; Nanot, M.; Jové, J.; Roisnel, T., Structural Studies of Tin-Doped Indium Oxide (ITO) and $In_4Sn_3O_{12}$. *J. Solid State Chem.* 1998, 135, 140–148.

23. Gonzalez, G. B.; Mason, T. O.; Quintana, J. P.; Warschkow, O.; Ellis, D. E.; Hwang, J. H.; Hodges, J. P.; Jorgensen, J. D., Defect structure studies of bulk and nano-indium-tin oxide. *J. Appl. Phys.* 2004, 96, 3912–3920.

24. Bizo, L.; Choisnet, J.; Retoux, R.; Raveau, B., The great potential of coupled substitutions in In_2O_3 for the generation of bixbyite-type transparent conducting oxides, $In_{2-2x}M_xSn_xO_3$. *Solid State Commun.* 2005, 136, 163–168.

25. Bizo, L.; Choisnet, J.; Raveau, B., Coupled substitutions in In_2O_3: New transparent conductors $In_{2-x}M_{2x/3}Sb_{x/3}O_3$ (M = Cu, Zn). *Mater. Res. Bull.* 2006, 41, 2232–2237.

26. Lu, Y. B.; Yang, T. L.; Ling, Z. C.; Cong, W. Y.; Zhang, P.; Li, Y. H.; Xin, Y. Q., How does the multiple constituent affect the carrier generation and charge transport in multicomponent TCOs of In-Zn-Sn oxide. *J. Mater. Chem. C* 2015, 3, 7727–7737.

27. Kanai, Y., Electrical Properties of In_2O_3 Single Crystals Doped with Metallic Donor Impurity. *Jpn. J. Appl. Phys. 1* 1984, 23, 127.

28. Wen, S. J.; Campet, G.; Portier, J.; Couturier, G.; Goodenough, J. B., Correlations between the Electronic Properties of Doped Indium Oxide Ceramics and the Nature of the Doping Element. *Mater. Sci. Eng., B* 1992, 14, 115–119.

29. Campet, G.; Han, S. D.; Wen, S. J.; Manaud, J. P.; Portier, J.; Xu, Y.; Salardenne, J., The electronic effect of $Ti4^+$, $Zr4^+$ and Ge^{4+} dopings upon the physical properties of In_2O_3 and Sn-doped In_2O_3 ceramics: application to new highly-transparent conductive electrodes. *Mater. Sci. Eng. B* 1993, 19, 285–289.

30. Maruyama, T.; Tago, T., Germanium-Doped and Silicon-Doped Indium-Oxide Thin Films Prepared by Radiofrequency Magnetron Sputtering. *Appl. Phys. Lett.* 1994, 64, 1395–1397.

31. Guilmeau, E.; Berardan, D.; Simon, C.; Maignan, A.; Raveau, B.; Ovono, D. O.; Delorme, F., Tuning the transport and thermoelectric properties of In_2O_3 bulk ceramics through doping at In-site. *J. Appl. Phys.* 2009, 106, 053715.

32. Lany, S.; Zakutayev, A.; Mason, T. O.; Wager, J. F.; Poeppelmeier, K. R.; Perkins, J. D.; Berry, J. J.; Ginley, D. S.; Zunger, A., Surface Origin of High Conductivities in Undoped In_2O_3 Thin Films. *Phys. Rev. Lett.* 2012, 108.

33. Avaritsiotis, J. N.; Howson, R. P., Fluorine Doping of In_2O_3 Films Employing Ion-Plating Techniques. *Thin Solid Films* 1981, 80, 63–66.

34. Avaritsiotis, J. N.; Howson, R. P., Composition and Conductivity of Fluorine-Doped Conducting Indium Oxide Films Prepared by Reactive Ion Plating. *Thin Solid Films* 1981, 77, 351–357.

35. Mori, T.; Kajihara, K.; Kanamura, K.; Toda, Y.; Hiramatsu, H.; Hosono, H., Indium-Based Ultraviolet-Transparent Electroconductive Oxyfluoride InOF: Ambient-Pressure Synthesis and Unique Electronic Properties in Comparison with In_2O_3. *J. Am. Chem. Soc.* 2013, 135, 13080–13088.

36. Edwards, D. D.; Mason, T. O.; Goutenoire, F.; Poeppelmeier, K. R., A new transparent conducting oxide in the Ga_2O_3–In_2O_3–SnO_2 system. *Appl. Phys. Lett.* 1997, 70, 1706–1708.

37. Edwards, D. D.; Mason, T. O.; Sinkler, W.; Marks, L. D.; Goutenoire, F.; Poeppelmeier, K. R., A structural investigation of $Ga_{3-x}In_{5+x}Sn_2O_{16}$. *J. Solid State Chem.* 1998, 140, 242–250.

38. Dolgonos, A.; Lam, K.; Poeppelmeier, K. R.; Freeman, A. J.; Mason, T. O., Electronic and optical properties of $Ga_{3-x}In_{5+x}Sn_2O_{16}$: An experimental and theoretical study. *J. Appl. Phys.* 2014, 115, 013703.

39. Rickert, K.; Huq, A.; Lapidus, S. H.; Wustrow, A.; Ellis, D. E.; Poeppelmeier, K. R., Site Dependency of the High Conductivity of $Ga_2In_6Sn_2O_{16}$: The Role of the 7-Coordinate Site. *Chem. Mater.* 2015, 27, 8084–8093.

40. Ambrosini, A.; Duarte, A.; Poeppelmeier, K. R.; Lane, M.; Kannewurf, C. R.; Mason, T. O., Electrical, optical, and structural properties of tin-doped In_2O_3–M_2O_3 solid solutions (M = Y, Sc). *J. Solid State Chem.* 2000, 153, 41–47.

41. Gordon, R. G., Criteria for choosing transparent conductors. *MRS Bull.* 2000, 25, 52–57.

42. Aboaf, J. A.; Marcotte, V. C.; Chou, N. J., Chemical Composition and Electrical Properties of Tin Oxide Films Prepared by Vapor Deposition. *J. Electrochem. Soc.* 1973, 120, 701–702.

43. Raccanelli, A.; Maddalena, A. Electrical Conductivity of Heat-Treated SnO_2 Films. *J. Am. Ceram. Soc.* 1976, 59, 425–430.

44. Gonzalez, G. B., Investigating the defect structures in transparent conducting oxides using X-ray and neutron scattering techniques. *Materials* 2012, 5, 818–850.

45. Berry, F. J.; Greaves, C., A Neutron Diffraction Investigation of the Defect Rutile Structure of Tin-Antimony Oxide. *J. Chem. Soc. Dalton Trans.* 1981, 2447–2451.

46. Canestraro, C. D.; Oliveira, M. M.; Valaski, R.; da Silva, M. V. S.; David, D. G. F.; Pepe, I.; da Silva, A. F.; Roman, L. S.; Persson, C., Strong inter-conduction-band absorption in heavily fluorine doped tin oxide. *Appl. Surf. Sci.* 2008, 255, 1874–1879.

47. Fantini, M. C. A.; Torriani, I., The Compositional and Structural Properties of Sprayed SnO_2 — F Thin Films. *Thin Solid Films* 1986, 138, 255–265.

48. Canestraro, C. D.; Roman, L. S.; Persson, C., Polarization dependence of the optical response in SnO_2 and the effects from heavily F doping. *Thin Solid Films* 2009, 517, 6301–6304.

49. Berry, F. J.; Laundy, B. J., Sb-121 Mossbauer Study of the Effects of Calcination on the Structure of Tin-Antimony Oxides. *J. Chem. Soc. Dalton Trans.* 1981, 1442–1444.

50. Kojima, M.; Kato, H.; Gatto, M., Blackening of Tin Oxide Thin Films Heavily Doped with Antimony. *Philos. Mag. B* 1993, 68, 215–222.

51. Taro, H.; Yutaka, F.; Atsuki, U.; Kinnosuke, I.; Kazuhisa, I.; Yasushi, H.; Go, K.; Yukio, Y.; Toshihiro, S.; Tetsuya, H., Ta-doped Anatase TiO_2 Epitaxial Film as Transparent Conducting Oxide. *Jpn. J. Appl. Phys.* 2005, 44, L1063.

52. Duan, Y.; Fu, N.; Liu, Q.; Fang, Y.; Zhou, X.; Zhang, J.; Lin, Y., Sb-Doped TiO_2 Photoanode for Dye-Sensitized Solar Cells. *J. Phys. Chem. C* 2012, 116, 8888–8893.

53. Ni, M.; Leung, M. K. H.; Leung, D. Y. C.; Sumathy, K., A review and recent developments in photocatalytic water-splitting using TiO_2 for hydrogen production. *Renewable Sustainable Energy Rev.* 2007, 11, 401–425.

54. Van de Walle, C. G., Hydrogen as a cause of doping in zinc oxide. *Phys. Rev. Lett.* 2000, 85, 1012–1015.

55. Contreras, M. A.; Egaas, B.; Ramanathan, K.; Hiltner, J.; Swartzlander, A.; Hasoon, F.; Noufi, R., Progress toward 20% efficiency in Cu(In,Ga)Se$_2$ polycrystalline thin-film solar cells. *Prog. Photovoltaics* 1999, 7, 311–316.

56. Chopra, K. L.; Paulson, P. D.; Dutta, V., Thin-film solar cells: An overview. *Prog. Photovoltaics* 2004, 12, 69–92.

57. Sundaramoorthy, R.; Repins, I. L.; Gennett, T.; Pern, F. J.; Albin, D.; Li, J. V.; DeHart, C.; Glynn, S.; Perkins, J. D.; Ginley, D. S.; Gessert, T., Comparison of Amorphous InZnO and Polycrystalline ZnO: Al Conductive Layers for CIGS Solar Cells. In *IEEE Photovoltaic Spec. Conf. 34th*. Philadelphia, PA, 2009, Vol. 1–3, 2225–2230.

58. Chang, H. P.; Wang, F. H.; Wu, J. Y.; Kung, C. Y.; Liu, H. W., Enhanced conductivity of aluminum doped ZnO films by hydrogen plasma treatment. *Thin Solid Films* 2010, 518, 7445–7449.

59. Lany, S.; Zunger, A., Dopability, intrinsic conductivity, and nonstoichiometry of transparent conducting oxides. *Phys. Rev. Lett.* 2007, 98, 045501.

60. Bazzani, M.; Neroni, A.; Calzolari, A.; Catellani, A., Optoelectronic properties of Al:ZnO: Critical dosage for an optimal transparent conductive oxide. *Appl. Phys. Lett.* 2011, 98, 121907.

61. Ma, Y.; Zeng, Y.; Perrodin, D.; Bourret, E.; Jiang, Y., Single-Crystal Growth of ZnO:Ga by the Traveling-Solvent Floating-Zone Method. *Cryst. Growth Des.* 2017, 17, 1008–1015.

62. Moriga, T.; Edwards, D. D.; Mason, T. O.; Palmer, G. B.; Poeppelmeier, K. R.; Schindler, J. L.; Kannewurf, C. R.; Nakabayashi, I., Phase relationships and physical properties of homologous compounds in the zinc oxide–indium oxide system. *J. Am. Ceram. Soc.* 1998, 81, 1310–1316.

63. Orita, M.; Ohta, H.; Hirano, M.; Narushima, S.; Hosono, H., Amorphous transparent conductive oxide InGaO$_3$(ZnO)$_m$ (m <= 4): a Zn4s conductor. *Philos. Mag. B* 2001, 81, 501–515.

64. Nomura, K.; Ohta, H.; Ueda, K.; Kamiya, T.; Hirano, M.; Hosono, H., Thin-film transistor fabricated in single-crystalline transparent oxide semiconductor. *Science* 2003, 300, 1269–1272.

65. Cannard, P. J.; Tilley, R. J. D., New Intergrowth Phases in the ZnO–In$_2$O$_3$ System. *J. Solid State Chem.* 1988, 73, 418–426.

66. Nakamura, M.; Kimizuka, N.; Mohri, T., The Phase Relations in the In$_2$O$_3$–Fe$_2$ZnO$_4$–ZnO System at 1350 Degree C. *J. Solid State Chem.* 1990, 86, 16–40.

67. Kimizuka, N.; Isobe, M.; Nakamura, M., Syntheses and Single-Crystal Data of Homologous Compounds, In$_2$O$_3$(ZnO)$_M$ (M=3, 4, and 5), InGaO$_3$(ZnO)$_3$, and Ga$_2$O$_3$(ZnO)$_M$ (M=7, 8, 9, and 16) in the In$_2$O$_3$–ZnGa$_2$O$_4$–ZnO System. *J. Solid State Chem.* 1995, 116, 170–178.

68. Lv, M.; Liu, G.; Xu, X., Homologous Compounds Zn$_n$In$_2$O$_{3+n}$ (n = 4, 5, and 7) Containing Laminated Functional Groups as Efficient Photocatalysts for Hydrogen Production. *ACS Appl. Mater. Interfaces* 2016, 8, 28700–28708.

69. Nomura, K.; Takagi, A.; Kamiya, T.; Ohta, H.; Hirano, M.; Hosono, H., Amorphous oxide semiconductors for high-performance flexible thin-film transistors. *Jpn. J. Appl. Phys.* 1 2006, 45, 4303–4308.

70. Kamiya, T.; Nomura, K.; Hirano, M.; Hosono, H., Electronic structure of oxygen deficient amorphous oxide semiconductor a–InGaZnO4$_{4-x}$:Optical analyses and first-principle calculations. *Phys. Status Solidic* 2008, 5, 3098–3100.

71. Noh, H.-K.; Chang, K. J.; Ryu, B.; Lee, W.-J., Electronic structure of oxygen-vacancy defects in amorphous In-Ga-Zn-O semiconductors. *Phys. Rev. B* 2011, 84, 115205.

72. Murat, A.; Adler, A. U.; Mason, T. O.; Medvedeva, J. E., Carrier Generation in Multicomponent Wide-Bandgap Oxides: InGaZnO$_4$. *J. Am. Chem. Soc.* 2013, 135, 5685–5692.

73. Badeker, K., Concerning the electricity conductibility and the thermoelectric energy of several heavy metal bonds. *Ann. Phys.* 1907, 22, 749–766.

74. Ingram, B. J.; Gonzalez, G. B.; Kammler, D. R.; Bertoni, M. I.; Mason, T. O., Chemical and structural factors governing transparent conductivity in oxides. *J. Electroceram.* 2004, 13, 167–175.

75. Friberg, L., Proteinuria and kidney injury among workmen exposed to cadmium and nickel dust — preliminary report. *J. Ind. Hyg. Toxicol.* 1948, 30, 32–36.

76. Kahan, E.; Derazne, E.; Rosenboim, J.; Ashkenazi, R.; Ribak, J., Adverse health-effects in workers exposed to cadmium. *Am. J. Ind. Med.* 1992, 21, 527–537.

77. Wang, A.; Babcock, J. R.; Edleman, N. L.; Metz, A. W.; Lane, M. A.; Asahi, R.; Dravid, V. P.; Kannewurf, C. R.; Freeman, A. J.; Marks, T. J., Indium-cadmium-oxide films having exceptional electrical conductivity and optical transparency: clues for optimizing transparent conductors. *Proc. Natl. Acad. Sci. U. S. A.* 2001, 98, 7113–7116.

78. Nordberg, G. F., Cadmium and health in the 21st Century — historical remarks and trends for the future. *Biometals* 2004, 17, 485–489.

79. Lane, R. E.; Campbell, A. C. P., Fatal Emphysema in 2 Men Making a Copper Cadmium Alloy. *Br. J. Ind. Med.* 1954, 11, 118–122.

80. Fthenakis, V. M., Overview of Potential Hazards, In *Solar Cells*. Elsevier: 2013, 533–545.

81. Major, J. D.; Treharne, R. E.; Phillips, L. J.; Durose, K., A low-cost non-toxic post-growth activation step for CdTe solar cells. *Nature* 2014, 511, 334–337.

82. Bowden, M. E.; Cardile, C. M., Structures of Orthorhombic and Cubic Dicadmium Stannate by Rietveld Refinement. *Powder Diffr.* 1990, 5, 36–40.

83. Mizoguchi, H.; Eng, H. W.; Woodward, P. M. Probing the electronic structures of ternary perovskite and pyrochlore oxides containing Sn^{4+} or Sb^{5+}. *Inorg. Chem.* 2004, 43, 1667–1680.

84. Kammler, D. R.; Mason, T. O.; Young, D. L.; Coutts, T. J.; Ko, D.; Poeppelmeier, K. R.; Williamson, D. L., Comparison of thin film and bulk forms of the transparent conducting oxide solution $Cd_{1+x}In_{2-2x}Sn_xO_4$. *J. Appl. Phys.* 2001, 90, 5979–5985.

85. Mason, T. O.; Gonzalez, G. B.; Kammler, D. R.; Mansourian-Hadavi, N.; Ingram, B. J., Defect chemistry and physical properties of transparent conducting oxides in the $CdO–In_2O_3–SnO_2$ system. *Thin Solid Films* 2002, 411, 106–114.

86. Ko, D.; Poeppelmeier, K. R.; Kammler, D. R.; Gonzalez, G. B.; Mason, T. O.; Williamson, D. L.; Young, D. L.; Coutts, T. J., Cation distribution of the transparent conductor and spinel oxide solution $Cd_{1+x}In_{2-2x}Sn_xO_4$. *J. Solid State Chem.* 2002, 163, 259–266.

87. Mason, T. O.; Kammler, D. R.; Ingram, B. J.; Gonzalez, G. B.; Young, D. L.; Coutts, T. J., Key structural and defect chemical aspects of Cd-In-Sn-O transparent conducting oxides. *Thin Solid Films* 2003, 445, 186–192.

88. Zhang, S. B.; Wei, S.-H. Self-doping of cadmium stannate in the inverse spinel structure. *Appl. Phys. Lett.* 2002, 80, 1376–1378.

89. Segev, D.; Wei, S.-H., Structure-derived electronic and optical properties of transparent conducting oxides. *Phys. Rev. B* 2005, 71, 125129.

90. Hayashi, K.; Matsuishi, S.; Kamiya, T.; Hirano, M.; Hosono, H. Light-induced conversion of an insulating refractory oxide into a persistent electronic conductor. *Nature* 2002, 419, 462–465.

91. Matsuishi, S.; Toda, Y.; Miyakawa, M.; Hayashi, K.; Kamiya, T.; Hirano, M.; Tanaka, I.; Hosono, H. High-density electron anions in a nanoporous single crystal: $[Ca_{24}Al_{28}O_{64}]^{4+}(4e-)$. *Science* 2003, 301, 626–629.

92. Kim, S.-W.; Hayashi, K.; Hirano, M.; Hosono, H.; Tanaka, I., Electron Carrier Generation in a Refractory Oxide $12CaO·7Al_2O_3$ by Heating in Reducing Atmosphere: Conversion from an Insulator to a Persistent Conductor. *J. Am. Ceram. Soc.* 2006, 89, 3294–3298.

93. Sushko, P. V.; Shluger, A. L.; Hirano, M.; Hosono, H., From Insulator to Electride: A Theoretical Model of Nanoporous Oxide $12CaO·7Al_2O_3$. *J. Am. Chem. Soc.* 2007, 129, 942–951.

94. Medvedeva, J. E.; Freeman, A. J., Hopping versus bulk conductivity in transparent oxides: $12CaO·7Al_2O_3$. *Appl. Phys. Lett.* 2004, 85, 955–957.

95. Lobo, R. P. S. M.; Bontemps, N.; Bertoni, M. I.; Mason, T. O.; Poeppelmeier, K. R.; Freeman, A. J.; Park, M. S.; Medvedeva, J. E., Optical Conductivity of Mayenite: From Insulator to Metal. *J. Phys. Chem. C* 2015, 119, 8849–8856.

96. Bertoni, M. I.; Mason, T. O.; Medvedeva, J. E.; Freeman, A. J.; Poeppelmeier, K. R.; Delley, B., Tunable conductivity and conduction mechanism in an ultraviolet light activated electronic conductor. *J. Appl. Phys.* 2005, 97,103713.

97. Ingram, B. J.; Bertoni, M. I.; Poeppelmeier, K. R.; Mason, T. O., Point defects and transport mechanisms in transparent conducting oxides of intermediate conductivity. *Thin Solid Films* 2005, 486, 86–93.

98. Hayashi, K.; Sushko, P. V.; Ramo, D. M.; Shluger, A. L.; Watauchi, S.; Tanaka, I.; Matsuishi, S.; Hirano, M.; Hosono, H., Nanoporous Crystal $12CaO·7Al_2O_3$: A Playground for Studies of Ultraviolet Optical Absorption of Negative Ions. *J. Phys. Chem. B* 2007, 111, 1946–1956.

99. Palacios, L.; Bruque, S.; Aranda, M. A. G., Structure of gallium-doped mayenite and its reduction behaviour. *Phys. Status Solidi B* 2008, 245, 666–672.

100. Hayashi, K.; Ueda, N.; Matsuishi, S.; Hirano, M.; Kamiya, T.; Hosono, H., Solid State Syntheses of $12SrO·7Al_2O_3$ and Formation of High Density Oxygen Radical Anions, O^- and O^{2-}. *Chem. Mater.* 2008, 20, 5987–5996.

101. Hayashi, K.; Muramatsu, H.; Matsuishi, S.; Kamiya, T.; Hosono, H., Humidity-Sensitive Electrical Conductivity in $Ca_{12}Al_{4-x}Si_xO_{32}Cl_{2+x}$ ($0 \leq x \leq 3.4$) Ceramics. *Electrochem. Solid State Lett.* 2009, 12, J11-J13.

102. Maurelli, S.; Ruszak, M.; Witkowski, S.; Pietrzyk, P.; Chiesa, M.; Sojka, Z. Spectroscopic CW-EPR and HYSCORE investigations of Cu^{2+} and O^{2-} species in copper doped nanoporous calcium aluminate ($12CaO \cdots 7Al_2O_3$). *Phys. Chem. Chem. Phys.* 2010, 12, 10933–10941.

103. Polfus, J. M.; Toyoura, K.; Hervoches, C. H.; Sunding, M. F.; Tanaka, I.; Haugsrud, R. Nitrogen and hydrogen defect equilibria in $Ca_{12}Al_{14}O_{33}$: a combined experimental and computational study. *J. Mater. Chem.* 2012, 22, 15828–15835.

104. Kawazoe, H.; Yasukawa, M.; Hyodo, H.; Kurita, M.; Yanagi, H.; Hosono, H., P-type electrical conduction in transparent thin films of $CuAlO_2$. *Nature* 1997, 389, 939–942.

105. Kawazoe, H.; Yanagi, H.; Ueda, K.; Hosono, H., Transparent p-type conducting oxides: Design and fabrication of p–n heterojunctions. *MRS Bull.* 2000, 25, 28–36.

106. Trimarchi, G.; Peng, H.; Im, J.; Freeman, A. J.; Cloet, V.; Raw, A.; Poeppelmeier, K. R.; Biswas, K.; Lany, S.; Zunger, A., Using design principles to systematically plan the synthesis of hole-conducting transparent oxides: Cu_3VO_4 and Ag_3VO_4 as a case study. *Phys. Rev. B* 2011, 84, 165116.

107. Cloet, V.; Raw, A.; Poeppelmeier, K. R.; Trimarchi, G.; Peng, H.; Im, J.; Freeman, A. J.; Perry, N. H.; Mason, T. O.; Zakutayev, A.; Ndione, P. F.; Ginley, D. S.; Perkins, J. D., Structural, Optical, and Transport Properties of alpha- and beta-Ag_3VO_4. *Chem. Mater.* 2012, 24, 3346–3354.

108. Peng, H. W.; Lany, S., Semiconducting transition-metal oxides based on d^5 cations: Theory for MnO and Fe_2O_3. *Phys. Rev. B* 2012, 85, 201202.

109. Im, J.; Trimarchi, G.; Peng, H.; Freeman, A. J.; Cloet, V.; Raw, A.; Poeppelmeier, K. R., $KAg_{11}(VO_4)_4$ as a candidate p-type transparent conducting oxide. *J. Chem. Phys.* 2013, 138, 194703.

110. Peng, H.; Zakutayev, A.; Lany, S.; Paudel, T. R.; d'Avezac, M.; Ndione, P. F.; Perkins, J. D.; Ginley, D. S.; Nagaraja, A. R.; Perry, N. H.; Mason, T. O.; Zunger, A., Li-Doped Cr_2MnO_4: A New p-Type Transparent Conducting Oxide by Computational Materials Design. *Adv. Funct. Mater.* 2013, 23, 5267–5276.

111. Hautier, G.; Miglio, A.; Waroquiers, D.; Rignanese, G.-M.; Gonze, X., How Does Chemistry Influence Electron Effective Mass in Oxides? A High-Throughput Computational Analysis. *Chem. Mater.* 2014, 26, 5447–5458.

112. Ueda, K.; Inoue, S.; Hirose, S.; Kawazoe, H.; Hosono, H., Transparent p-type semiconductor: LaCuOS layered oxysulfide. *Appl. Phys. Lett.* 2000, 77, 2701–2703.

113. Scanlon, D. O.; Buckeridge, J.; Catlow, C. R. A.; Watson, G. W., Understanding doping anomalies in degenerate p-type semiconductor LaCuOSe. *J. Mater. Chem. C* 2014, 2, 3429–3438.

114. Cava, R. J.; Zandbergen, H. W.; Ramirez, A. P.; Takagi, H.; Chen, C. T.; Krajewski, J. J.; Peck, W. F.; Waszczak, J. V.; Meigs, G.; Roth, R. S.; Schneemeyer, L. F. $LaCuO_{25+x}$ and $YCuO_{2.5+x}$ Delafossites — Materials with Triangular $Cu^{2+\delta}$ Planes. *J. Solid State Chem.* 1993, 104, 437–452.

115. Katayama-Yoshida, H.; Koyanagi, T.; Funashima, H.; Harima, H.; Yanase, A., Engineering of nested Fermi surface and transparent conducting p-type Delafossite $CuAlO_2$: possible lattice instability or transparent superconductivity? *Solid State Commun.* 2003, 126, 135–139.

116. Ingram, B. J.; Gonzalez, G. B.; Mason, T. O.; Shahriari, D. Y.; Barnabe, A.; Ko, D. G.; Poeppelmeier, K. R., Transport and defect mechanisms in cuprous delafossites. 1. Comparison of hydrothermal and standard solid-state synthesis in $CuAlO_2$. *Chem. Mater.* 2004, 16, 5616–5622.

117. Nagarajan, R.; Draeseke, A. D.; Sleight, A. W.; Tate, J., p-type conductivity in $CuCr_{1-x}Mg_xO_2$ films and powders. *J. Appl. Phys.* 2001, 89, 8022–8025.

118. Kandpal, H. C.; Seshadri, R., First-principles electronic structure of the delafossites ABO_2 (A=Cu, Ag, Au; B=Al, Ga, Sc, In, Y): evolution of $d^{10}–d^{10}$ interactions. *Solid State Sci.* 2002, 4, 1045–1052.

119. Tate, J.; Jayaraj, M. K.; Draeseke, A. D.; Ulbrich, T.; Sleight, A. W.; Vanaja, K. A.; Nagarajan, R.; Wager, J. F.; Hoffman, R. L., p-Type oxides for use in transparent diodes. *Thin Solid Films* 2002, 411, 119–124.

120. Snure, M.; Tiwari, A., $CuBO_2$: A p-type transparent oxide. *Appl. Phys. Lett.* 2007, 91, 092123.

121. Yanagi, H.; Hase, T.; Ibuki, S.; Ueda, K.; Hosono, H., Bipolarity in electrical conduction of transparent oxide semiconductor $CuInO_2$ with delafossite structure. *Appl. Phys. Lett.* 2001, 78, 1583–1585.

122. Liu, M.-L.; Huang, F.-Q.; Chen, L.-D.; Wang, Y.-M.; Wang, Y.-H.; Li, G.-F.; Zhang, Q., p-type transparent conductor: Zn-doped $CuAlS_2$. *Appl. Phys. Lett.* 2007, 90, 072109.

123. Chan, G. H.; Liu, M. L.; Chen, L. D.; Huang, F. Q.; Bugaris, D. E.; Wells, D. M.; Ireland, J. R.; Hersam, M. C.; Van Duyne, R. P.; Ibers, J. A., Syntheses, crystal structures, and physical properties of $La_5Cu_6O_4S_7$ and $La_5Cu_{6.33}O_4S_7$. *Inorg. Chem.* 2008, 47, 4368–4374.

124. Paudel, T. R.; Zakutayev, A.; Lany, S.; d'Avezac, M.; Zunger, A., Doping Rules and Doping Prototypes in A_2BO_4 Spinel Oxides. *Adv. Funct. Mater.* 2011, 21, 4493–4501.

125. Paudel, T. R.; Lany, S.; d'Avezac, M.; Zunger, A.; Perry, N. H.; Nagaraja, A. R.; Mason, T. O.; Bettinger, J. S.; Shi, Y.; Toney, M. F., Asymmetric cation nonstoichiometry in spinels: Site occupancy in Co_2ZnO_4 and Rh_2ZnO_4. *Phys. Rev. B* 2011, 84, 064109.

126. Perkins, J. D.; Paudel, T. R.; Zakutayev, A.; Ndione, P. F.; Parilla, P. A.; Young, D. L.; Lany, S.; Ginley, D. S.; Zunger, A.; Perry, N. H.; Tang, Y.; Grayson, M.; Mason, T. O.; Bettinger, J. S.; Shi, Y.; Toney, M. F., Inverse design approach to hole doping in ternary oxides: Enhancing p-type conductivity in cobalt oxide spinels. *Phys. Rev. B* 2011, 84, 205207.

127. Shi, Y.; Ndione, P. F.; Lim, L. Y.; Sokaras, D.; Weng, T.-C.; Nagaraja, A. R.; Karydas, A. G.; Perkins, J. D.; Mason, T. O.; Ginley, D. S.; Zunger, A.; Toney, M. F., Self-Doping and Electrical Conductivity in Spinel Oxides: Experimental Validation of Doping Rules. *Chem. Mater.* 2014, 26, 1867–1873.

128. Windisch, C. F.; Ferris, K. F.; Exarhos, G. J., Synthesis and characterization of transparent conducting oxide cobalt–nickel spinel films. *J. Vac. Sci. Technol. A* 2001, 19, 1647–1651.

129. Zakutayev, A.; Paudel, T. R.; Ndione, P. F.; Perkins, J. D.; Lany, S.; Zunger, A.; Ginley, D. S., Cation off-stoichiometry leads to high p-type conductivity and enhanced transparency in Co_2ZnO_4 and Co_2NiO_4 thin films. *Phys. Rev. B* 2012, 85, 085204.

130. Nagaraja, A. R.; Perry, N. H.; Mason, T. O.; Tang, Y.; Grayson, M.; Paudel, T. R.; Lany, S.; Zunger, A., Band or Polaron: The Hole Conduction Mechanism in the p-Type Spinel Rh_2ZnO_4. *J. Am. Ceram. Soc.* 2012, 95, 269–274.

131. Kaur, J.; Bethge, O.; Wibowo, R. A.; Bansal, N.; Bauch, M.; Hamid, R.; Bertagnolli, E.; Dimopoulos, T., All-oxide solar cells based on electrodeposited Cu_2O absorber and atomic layer deposited ZnMgO on precious-metal-free electrode. *Sol. Energy Mater. Sol. Cells* 2017, 161, 449–459.

6

Complex Molybdenum–Vanadium Oxide Bronzes and Suboxides as Catalysts for Selective Oxidation and Ammoxidation of Light Hydrocarbons

Douglas J. Buttrey*, Douglas A. Blom[†] and Thomas Vogt[‡]

*Center for Catalytic Science and Technology, Department of Chemical and Biomolecular Engineering, University of Delaware, Newark, DE 19716, USA
[†]NanoCenter, Department of Chemical Engineering, University of South Carolina, Columbia, SC 29208, USA
[‡]NanoCenter, Department of Chemistry and Biochemistry, University of South Carolina, Columbia, SC 29208, USA

Introduction

Heterogeneous catalysis is of central importance to the chemical industry, particularly as we seek efficient and sustainable new processes for the transformation of naturally abundant hydrocarbons to valuable intermediates that can be used as platform feedstocks for the synthesis of monomer precursors for synthetic fibers, plastics, coatings, and other consumer products.[1] Among the desired chemical reactions, the selective oxidation and ammoxidation reactions are central as they account for roughly 25% of critical industrial organic products and intermediates needed for the manufacturing of many industrial products and consumer goods. Of particular importance are ethylene, propylene oxide, acrolein, acrylic acid, acrylonitrile, methacrylic acid, maleic anhydride, phthalic anhydride, and methyl tert-butyl ether.[2]

The major challenge in selective oxidation reactions is that they are in competition with the more thermodynamically favored complete oxidation of hydrocarbons to CO_2 and water. The objective in selective oxidation chemistry is to identify catalysts, almost always metal oxides, that produce the desired product with high yield at temperatures sufficiently low to avoid free radical formation. For most of the industrially important oxidation products, complex mixed-metal oxides, such as multicomponent bismuth molybdates or uranium antimonates for producing acrolein and acrylonitrile from propylene, or vanadyl pyrophosphate (VPO) for efficient conversion of *n*-butane to maleic anhydride have been used. Multistep reactions are involved, for which multifunctional active site configurations are wanted.

Multifunctional selective heterogeneous metal oxide oxidation catalysts have some general features in common. In a seminal paper by R. K. Grasselli,[3] these common traits were enumerated as seven principles ('pillars') underpinning the properties of selective heterogeneous oxidation catalysis: (i) available lattice oxygen (Mars–van Krevelen mechanism[4]), (ii) intermediate metal–oxygen bond strengths, (iii) stable and suitable host structure, (iv) multiple redox-active species, (v) multifunctional active sites, (vi) isolated active site configurations, and (vii) when required, cooperation with secondary promoter phases. These catalysts are redox active structurally and compositionally complex materials in which the active sites are spatially isolated.[1-3,5,6] The materials must be sufficiently stable against reduction so that the catalyst structure does not degrade significantly during catalysis. Furthermore, these catalysts must at least temporarily accommodate either oxygen vacancy formation, or the formation of more complex defects associated with oxygen deficiency.[7,8] The structural and compositional complexity required to accommodate these 'seven design pillars' presented researchers with 'six head winds' when structurally characterizing the complex catalytic system such as the M1/M2 catalyst: (1) in the case of M1 over 200 independent crystallographic parameters to describe the unit cell in *Pba2*, (2) varying mixed Mo/V site occupancies, (3) partial Te occupancies in the heptagonal channels, the presence of (4) multiple phases and (5) complex grain boundaries, and finally (6) the unavailability of single crystals. In this chapter, we examine complex molybdenum–vanadium suboxides and bronzes engineered to accommodate the 'seven pillars' and describe in more detail the challenges met during the structural characterization of these compositionally and structurally complex oxides.

Introduction to transition metal oxide bronzes and related suboxides

Oxide bronzes are a classification of non-stoichiometric materials, known for almost two centuries, which contain transition metal elements with different oxidation states. The name 'bronze' was given to these oxides due to their lustrous metallic-like appearance. The most common materials are those containing transition metal sites with some displaying the maximum valence (the Group number) and others reduced by one electron, leading to mixed d^0 and d^1 configurations. The original examples, reported by Wöhler in 1824,[8] were tungsten bronzes derived from the parent host structure WO_3, but with the tungsten partially reduced from W(VI) to a mixture of W(V) and W(VI) by incorporating an additional cation. Typically, the additional cation may be an alkali, alkaline-earth, or lanthanide ion; examples of tungsten bronzes are Na_xWO_3 ($x \approx 0.3$), K_xWO_3 ($x \approx 0.33$), Ba_xWO_3 ($x \approx 0.12$), and Tm_xWO_3 ($x \approx 0.10$). Other ionic species such as H^+, In^+, Cu^+, Sn^{2+}, Pb^{2+} (see Ref. 9), Sb^{3+} (see Ref. 10), and Bi^{3+} may be incorporated into the framework.[11] Similarly, bronzes may be derived from other fully oxidized transition metal oxides such as MoO_3, Nb_2O_5, and Ta_2O_5 which can form partially reduced compounds such as $Na_{0.9}Mo_6O_{17}$[12], $Sr_{0.7}NbO_3$[13], and $Ba_3Ta_5O_{15}$ (see Ref. 14) that contain d^0 and d^1 transition metal states.

Some of these structures consist of interlinked networks of corner-sharing MO_6 octahedra that can be thought of as simple intercalated variants of the cubic ReO_3 structure, or when the additional intercalated ion approaches full occupancy, as a cubic perovskite. Other bronzes belong to a tetragonal family made up of 3-, 4-, and 5-member rings of connected MO_6 octahedra where the 5-membered rings serve as sites for the intercalation of the additional ions and are known as *Tetragonal I* phases (see Figure 1).

Orthorhombic and *Tetragonal II* bronzes are distorted variants of the cubic forms. Finally, some of these bronzes form phases with the *Hexagonal Tungsten Bronze* (HTB) structure type, in which hexagonal rings

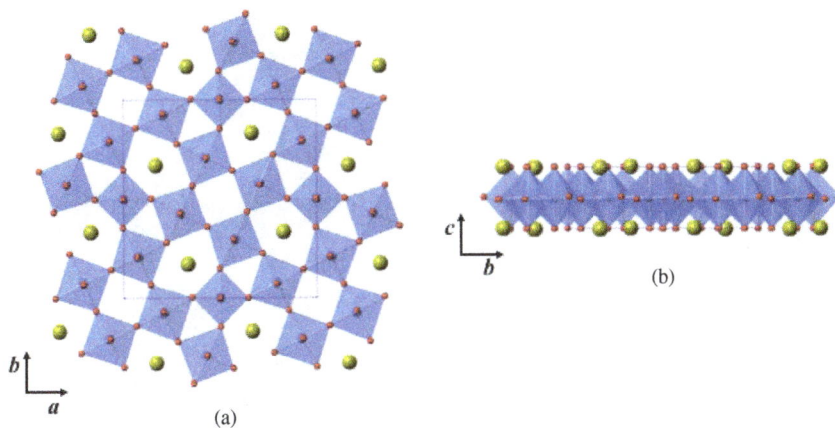

Figure 1. Ideal tetragonal structure NaW_3O_9. (a) [001] projection and (b) [100] projection.

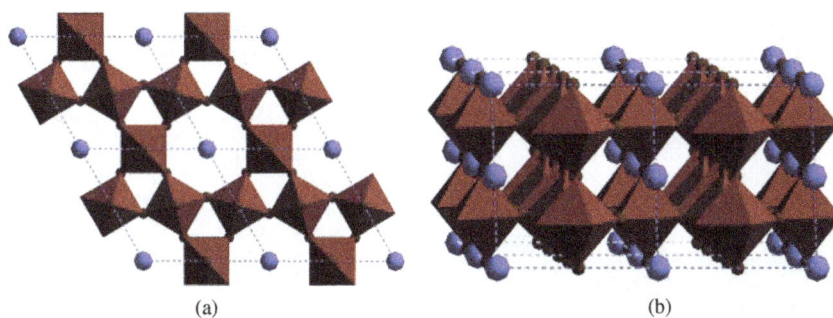

Figure 2. Ideal HTB structure KW_3O_9. (a) [001] projection and (b) clinographic lateral view.

of corner-shared MO_6 octahedra enclose channels that intercalate the additional cations. An illustration of an HTB-type bronze structure, exemplified by KW_3O_9, is shown in Figure 2.

In addition to these bronzes, intermediate suboxides with mixed valence states can be formed by mild reduction without inserting additional cations. In the tungsten family, these are exemplified by $W_{17}O_{47}$ (see Ref. 15) $W_{18}O_{49}$[16–17], and so-called crystallographic shear structures in the Magneli series W_nO_{3n-2} (*n*: integer), such as $W_{20}O_{58}$, $W_{24}O_{70}$[18], and $W_{40}O_{118}$.[19] Similarly, suboxides of molybdenum with mixed d^0 and d^1 Mo states such as Mo_4O_{11}, Mo_5O_{14}, and $Mo_{17}O_{47}$ can be formed by mild reduction reactions. Niobium and tantalum seem not to form analogous forms. In some cases, these suboxide structures are similar to the bronzes discussed above, but lacking the second metal ion species, leaving open channels. Examples with a pentagonal ring unit forming networks with open hexagonal channels are shown in Figure 3.

Among the bronzes and suboxides are a class of materials that contain pentagonal subunits with a pentagonal biprismatic center made of M_6O_{21} units. Suboxides in this family include $W_{18}O_{49}$, $Mo_{17}O_{47}$, $Nb_8W_9O_{47}$, and Mo_5O_{14}. Examples of niobium-based structures containing Nb_6O_{21} units with intercalated alkali metal ions are presented in Figure 4. These materials are derived from heteropoly oxometalate Keggin-type compounds.[20] While any of these bronzes or suboxides may provide redox activity needed for selective oxidation reactions, the most promising examples offering substantial site isolation are the ones with pentagonal subunits, as they tend to form larger unit cell structures in which certain sites are spatially better separated than others. Our focus from this point onward will be primarily on these types of materials.

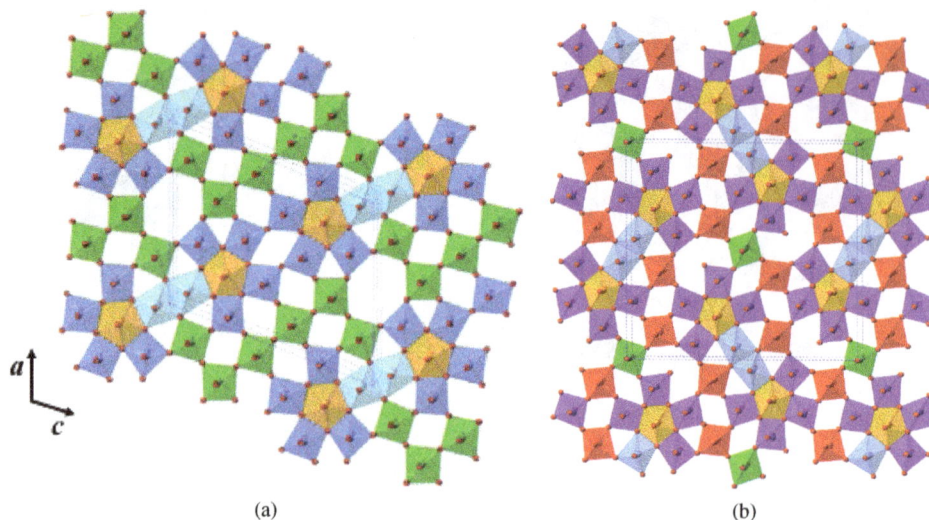

Figure 3.   The structures of (a) $W_{18}O_{49}$, and (b) $Mo_{17}O_{47}$, both featuring M_6O_{21} pentagonal ring units interconnected by corner-shared octahedra. The colors highlight the different sites which can be substituted by W^{5+}, W^{6+}, Mo^{5+}, Mo^{6+}, V^{4+}, and V^{5+} as highlighted in the later part of the chapter.

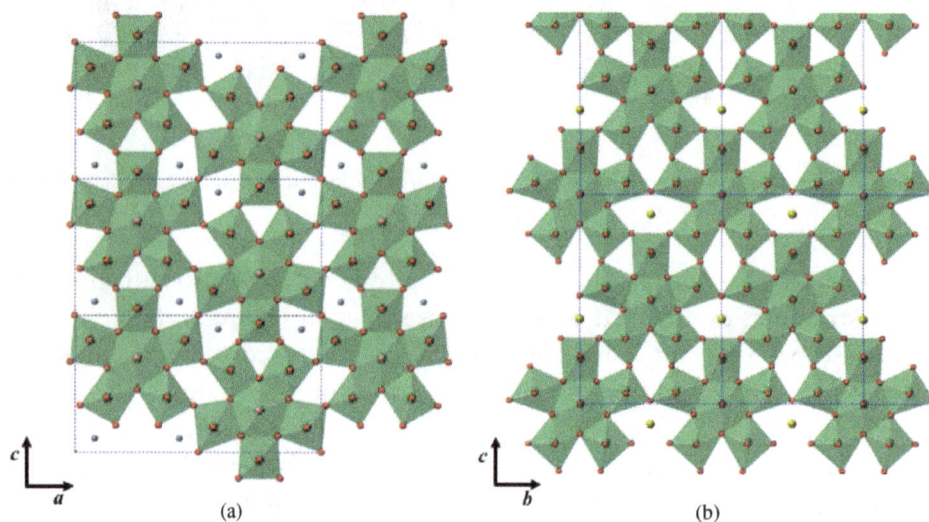

Figure 4.   The structures of (a) $LiNb_6O_{15}$, and (b) $NaNb_5O_{15}$, both featuring corner-shared pentagonal ring units with somewhat similar framework patterns, but subtle differences result in different nanopores and intercalation patterns for the alkali species.

Vanadium-containing oxides for paraffin activation in selective oxidation catalysis

Several vanadium (5+)-containing materials have shown promise as light paraffin (amm)oxidation catalysts due to the active nature of the vanadium moieties for alkane activation. To date, the most successful heterogeneously catalyzed light alkane oxidation process is the conversion of *n*-butane to maleic anhydride using vanadyl pyrophosphate (VPO) — its structure is depicted in Figure 5. The vanadium sites are paired by edge-sharing on the active surface and site-isolated by phosphate tetrahedra that surround these vanadium pairs. The V^{5+} sites in VPO are produced in an otherwise V^{4+} bulk phase by oxygen non-stoichiometry near the surface of the catalyst. Unfortunately, VPO suffers from poor mobility of oxygen in the lattice.

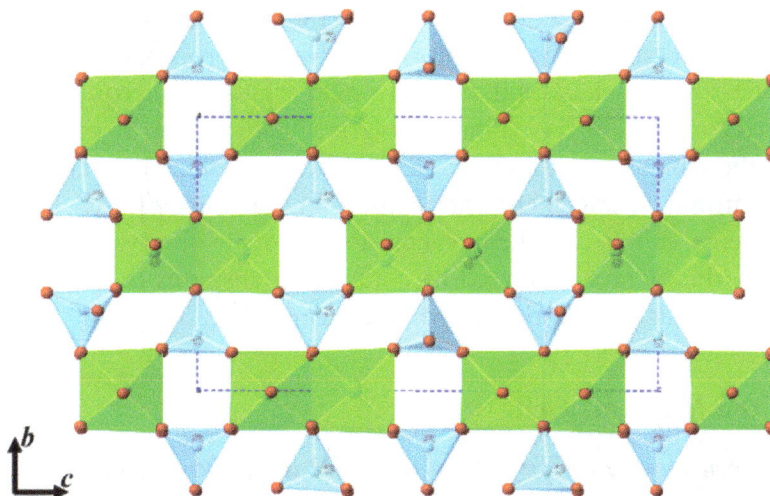

Figure 5. [100] projection of the vandyl pyrophosphate structure with composition $(VO)_2P_2O_7$. Vanadium octahedral pairs (green) are proposed to have mixed 4+ and 5+ valences at the catalyst surface, and tetrahedral phosphate linking groups (blue) provide site isolation.

Other systems with mixed-metal oxides such as Mo–V–O, Mg–V–O[21], Sb–V–O[22] and Sb–V–P–O (see Ref. 23) materials have also shown potential as alkane-selective oxidation or ammoxidation catalysts for the production of acrylic acid or acrylonitrile. Recent attention has turned toward the incorporation of one or more additional metals into these systems in efforts to enhance the catalytic conversion and selectivity. Although evidence for catalytic activity in these systems has been well established, detailed structural characterization of the active phases and studies on the catalytic mechanisms have been forthcoming only since the early 2000s. A variety of multicomponent oxide bronze and suboxide systems are being actively investigated in efforts to improve catalytic performance and to better understand the underlying fundamental steps during the catalysis. Apart from the case of the VPO catalyst, product yields are still not sufficient for widespread capital investment in order to modify existing commercial processes or to construct new plants; however, a small-scale plant for the production of acrylonitrile from propane has been successfully operated by Ashahi Chemical in Thailand in recent years using a Mo–V–Nb–Te–O catalyst.[24] The detailed characterization of this Mo–V–Nb–Te–O catalyst system and related forms will be the subject of the remainder of this chapter. We emphasize points that all too often are nowadays consigned to 'supplementary materials' despite their paramount importance.

Catalytically active molybdenum–vanadium suboxides and bronzes

The Mo–V–Nb–Te–O catalyst formulations for propane (amm)oxidation typically contain one primary phase (M1), but may also contain a second phase (M2) as a promoter. The focus on these phases is motivated by the presence of vanadium and the expectation that the structures will be sufficiently complex to provide the required multifunctionality and isolated active sites. Identification of these phases was initially accomplished by combining some basic selected area electron diffraction and phase-contrast imaging studies combined with a simultaneous refinement of high-resolution synchrotron X-ray and neutron powder diffraction data. Advances in aberration-corrected scanning transmission electron microscope (STEM) imaging in the past decade or so has added a new level of detail to the complexity of the structural models for these materials. Based on the elucidation of these structural *and* compositional details, it has been possible to propose mechanistic models for the active surface or surfaces.

Transmission electron microscopy (TEM) was used as the initial tool to identify the compositions and unit cell constants of individual catalyst phases. The extremely small probe sizes used in TEM allow the direct observation of nanometer-sized particles as individual single crystals. Electron diffraction is routinely used to determine the unit cell constants of crystalline or polycrystalline phases. Selected area electron diffraction pattern is used to assess the symmetry within the unit cell. Well-crystallized materials produce phase-contrast images from which the atomic arrangement within the unit cell can be often at least partially deduced, although the low-Z elements, in this case the oxygens, are not easily observable. Chemical analysis in TEM can be performed using either X-rays generated by core state transitions or the associated electron energy loss (EELS) produced as a result of inelastic electron scattering.

Characterization of the M1 Unit Cell

Preliminary selected area electron diffraction (SAED) and phase-contrast (HREM) imaging

The [001] zone selected area electron diffraction (SAED) for the M1 phase is shown in Figure 6. The two unit cell directions in reciprocal space within this zone, a^* and b^*, are immediately evident in this pattern, as well as the associated systematic absence conditions. These absences reveal the presence of either two principle glide planes or two screw axes. The spacing between the spots can be used to quantify the a and b

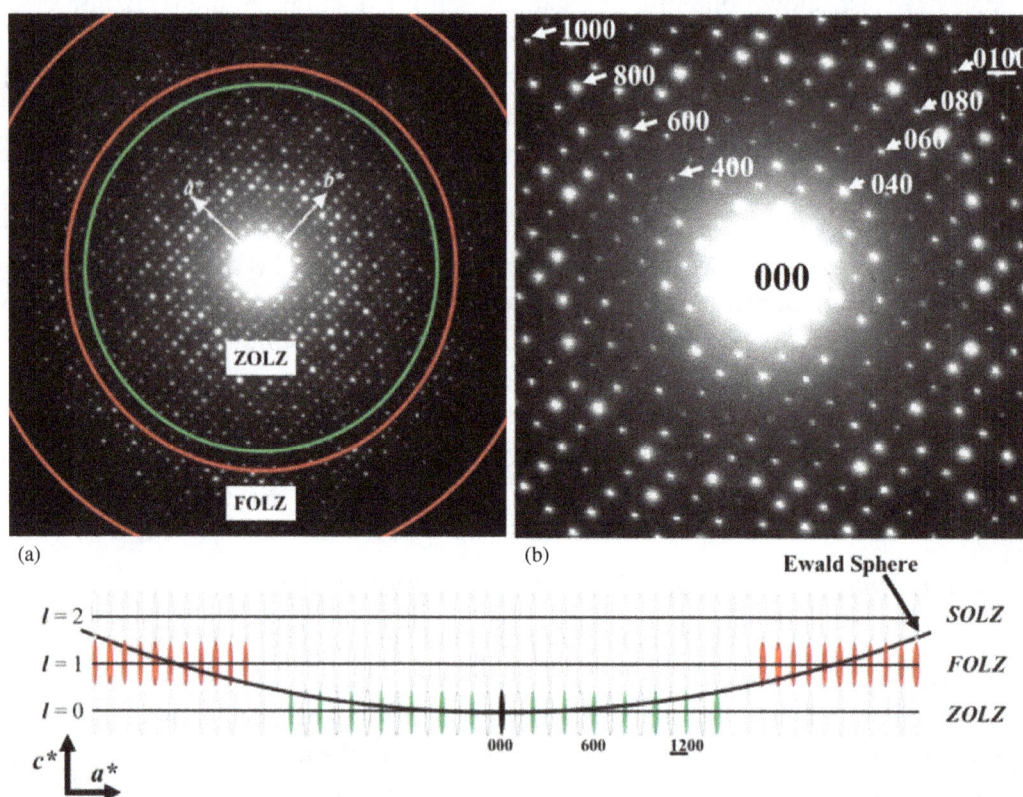

Figure 6.   Interpretation of the [001] SAED pattern for the M1 phase. (a) Full [001] pattern showing the ZOLZ band from the caustic spot (direct beam) out to the green ring. The red rings represent the inner and outer limits of the FOLZ band. (b) Enlargement of the ZOLZ with indexing along the principal directions showing systematic absences. Lower view schematic illustrates the intersection of the Ewald sphere and the reciprocal lattice rods. Those shown in green are those observed in the ZOLZ, in red are those observe in the FOLZ, in gray are ones that do not make contact with the Ewald sphere, and dotted ones are absent by symmetry.

lattice constants to within the resolution of these patterns; the approximate values obtained are $a = 2.11$ nm, $b = 2.67$ nm. The ring of diminished intensity just beyond the green ring in Figure 6(a) indicates the outer boundary of the zero-order Laue zone (ZOLZ). Reflections within the ZOLZ have the general indices $h k 0$. This boundary represents the limit of contact of the Ewald sphere (or sphere of diffraction) with the reciprocal lattice rods associated with the $h k 0$ reflections. The elongation of the reciprocal lattice rods is inversely proportional to the thickness of the crystallite along the electron beam direction. The high-order diffraction spots visible in the ZOLZ indicate that this particular crystal is just a few nanometers thick.

The higher intensity ring of diffraction spots just beyond the ZOLZ and diminished intensity band is called the first-order Laue zone (FOLZ). Reflections in the FOLZ band have the general indices $h k 1$. The diameter of the FOLZ ring depends on the wavelength of the electrons and the modulus of the lattice vector parallel to the electron beam, in this case c. The FOLZ ring diameter in Figure 5 is consistent with a unit cell with a c-parameter that is only about 0.4 nm, a typical MO_6 octahedral spacing.

In order to determine the c-axis dimension directly and learn more about the unit cell symmetry, it is necessary to search for another M1 crystallite that can be oriented to a different principle zone axis, either [100] or [010]. In Figure 7, the [010] zone pattern is shown, along with enlargements of the ZOLZ and FOLZ regions of the pattern as panels on the right and left of the figure. The vertical spot pattern matches the spacings along a^* from Figure 6, and the perpendicular c^* spacing can be used to determine the c parameter as close to 0.40 nm. Comparison of the spacings in the ZOLZ and FOLZ patterns reveal that a systematic absence related to symmetry is present along a^*, but not along c^*. In this zone, the ZOLZ reflections are of the form $h0l$ and those in the FOLZ are indexed as $h1l$. These results, together with those from Figure 6, are consistent with a pair of glide planes, but inconsistent with a pair of screw axes parallel to a^* and b^*. This combination is only found in two of the 230 space groups, determining the orthorhombic M1 symmetry as being either *Pbam* or *Pba2*. This means that there is either a binary axis parallel to c or a mirror plane normal to it. Given the very large lattice constants in the ab-plane, it is highly unlikely that all atoms lie within the mirror planes. It is therefore reasonable to conclude that the correct space group is *Pba2*.

An HREM image was obtained from the [001] zone alignment shown in the diffraction pattern in Figure 5. In this image, phase-contrast transfer associated with the metal sites is observed, but the oxygens

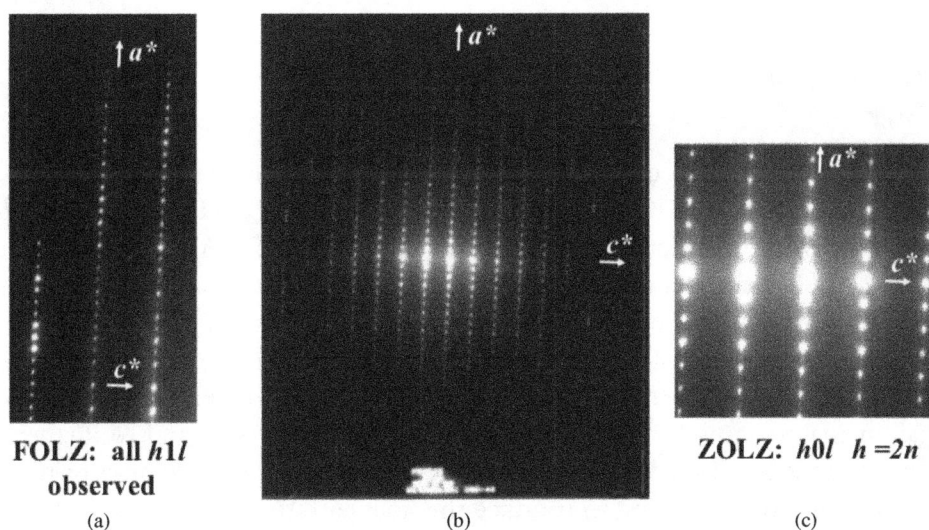

FOLZ: all $h1l$ observed

(a)

(b)

ZOLZ: $h0l$ $h = 2n$

(c)

Figure 7.   The [010] zone SAED pattern from a very thin crystallite of the M1 phase in which the principle unit cell directions in reciprocal space are labeled as a^* and c^*. The enlarged view of the ZOLZ (right panel) shows a special condition $h0l$: $h = 2n$ which is not observed in the FOLZ (left panel).

Simulated HREM Image   **Experimental HREM Image**

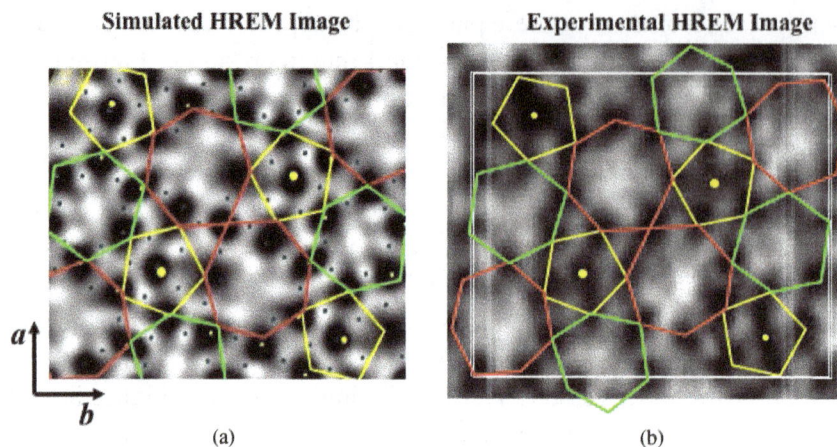

(a)   (b)

Figure 8.   (a) Simulation and (b) experimental [00-1] zone HREM images from the same very thin crystallite of the M1 phase from which Figure 6 was obtained. The color overlays indicate the presence of centered pentagonal ring units (yellow), and linking sites that form a network of hexagonal (green) and heptagonal (red) nanochannels.

contribute too little to be observed. Also, we found that during electron beam irradiation most of the tellurium sublimed from the structure. Tellurium is expected to enter the lattice as an intercalated species — an additional metal ion species that makes the material a bronze. The [00-1] HREM image is shown in Figure 8 along with a computer simulated image based on our preliminary model for the metal framework of the M1 structure with the Te omitted. We have annotated both the simulated and experimental images, indicating the presence of pentagonal rings with a metal atom in the center in pentagonal biprismatic coordination (yellow), hexagonal rings (green), and heptagonal rings (red), each enclosing empty nanochannels. The unit cell boundaries are outlined in white on the experimental image. Based on the HREM image, we can obtain estimates for the x, y coordinates of the metal sites, apart from those for Te. Comparing with the pentagonal ring bronzes considered earlier in Figures 3 and 4, we can guess likely z coordinates for the metals and can also suggest likely x, y, z coordinates for the oxygens; however, the distribution of Mo, V, and Nb among the framework metal sites and the location of Te remain to be determined.

Using synchrotron X-ray diffraction and neutron diffraction for refinement of the M1 structure

Synchrotron X-ray powder diffraction data (S-XPD) were collected using the beamline X7A at the National Synchrotron Light Source (NSLS) at Brookhaven National Laboratory (BNL) in Long Island, NY. Due to the presence of Mo, the X-ray wavelength was selected to be ~0.70577 Å and then carefully set using a temperature-controlled (111) channel-cut single crystal of silicon as a double monochromator on the incident side of the sample and calibrated using a CeO_2 standard. The specimen was ground and sieved to <400 mesh and then loaded into a glass capillary with a diameter of 0.30 mm that was mounted horizontally and rotated rapidly about its axis. Slits were set to discriminate against divergent X-rays. A germanium (220) crystal was used as an analyzer to select against inelastically scattered X-rays and to further monochromate the scattered beam. Data were collected by frame-stepping a 4096-channel position-sensitive detector (PSD) with a 4° window in 2θ and then merging to produce the final pattern.

Neutron powder diffraction (NPD) data were collected using the 32-detector BT-1 diffractometer at the National Institute of Standards and Technology's NIST Center for Neutron Research, with the sample mounted in a vanadium can. A Cu (311) monochromator with a 90° takeoff angle, $\lambda = 1.5398$ Å, and in-pile collimation of 15 min of arc were used. Data were collected over the range of $3 \leq 2\theta \leq 168°$ with a step size of 0.05°.

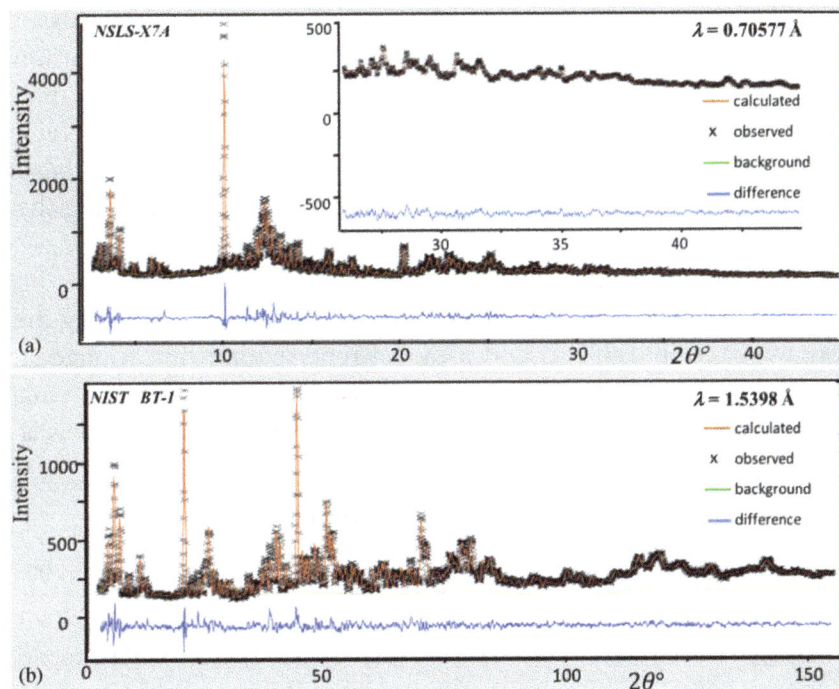

Figure 9. Refined model fit [27] to (a) synchrotron X-ray powder diffraction pattern data from NSLS beamline X7A powder diffractometer at $\lambda = 0.70577$ Å collected with a 4° PSD with inset showing the high-angle data and (b) neutron powder diffraction data from NIST beamline BT-1 using Cu (311) K_α at $\lambda = 1.5398$ Å.

The S-XPD and NPD data (see Figure 9) were simultaneously refined using GSAS[25] with the graphical user interface program EXPGUI.[26] The refinement of each model was started by first considering all framework octahedral sites and the pentagonal bipyramidal sites fully occupied by Mo. The background and histogram scale factors were refined, followed by unit cell parameters, and then the instrumental zero-point corrections and wavelength of the S-XPD data. The synchrotron wavelength was refined to account for variations in wavelength calibration and sample environment, such as temperature variations between the S-XPD and NPD instruments. At this point, peak shapes were refined for both data sets. The peak shape parameters were then fixed and the atomic coordinates of all metal and oxygen sites were refined. The complexity of the model did not allow refinement of individual atomic thermal displacement parameters.

Instead, to limit the number of parameters involved, just two isotropic thermal parameters were used: one for the metal atoms and the second for the oxygen atoms. The site occupancies for molybdenum vs. species were then refined to allow for the presence of vanadium, niobium, and tellurium. The relative weightings of the neutron and synchrotron data were adjusted so that the weighted sum of squared differences of the two data sets were the same. Difference Fourier maps were generated from both the S-XPD and NPD fits to check for missing or incorrectly assigned atoms. The combined use of S-XPD and NPD data sets in generating these maps provides a check on possible discrepancies in metal and/or oxygen coordinates. Small features in the difference Fourier maps may be attributed to several possible sources; in particular, they may originate from instrumental noise, particularly in the NPD data where the intensity-to-background ratio and q-range is not as high as in the S-XPD data. One small and poorly defined feature observed in the Fourier maps indicated that there is a small unaccounted random distribution of electronic and nuclear density within the heptagonal nanopores. This was subsequently assigned as intercalated tellurium present at very low occupancies.[27] Based on the occupancies initially determined, the average composition of M1 was determined to be $Mo_{7.5}V_{1.5}NbTeO_{29}$. We later reported on an improvement to this initial refinement,[28] after accounting for trace contaminant phases (Mo_5O_{14} and Te metal) in the analysis and attributing the scattering from the heptagonal pores to the low-level, disordered occupancy of tellurium; with this improvement, the M1 composition was revised to $Mo_{7.8}V_{1.2}NbTe_{0.937}O_{28.9}$. This formula unit can be rewritten to more explicitly indicate the tellurium intercalation sites by using the equivalent form $(TeO)_{0.937}(Mo_{7.8}V_{1.2}Nb)$ O_{28}. The Te occupancy in the heptagonal channel was not accounted for in the initial analysis[27] because the impurity phases had been excluded from the refinement, making attempts at refining the small Te occupancy in the heptagonal channels statistically insignificant; only with the trace impurities accounted for was the additional Te significant in the refinement.

The resulting atomic coordinates, thermal parameters, and occupancies (omitting the disorder site within the heptagonal pores) are listed in Table 1, and a model representing this average structure is shown in Figure 10. Note that the tellurium intercalated as Te–O chains in the hexagonal channels is represented in this figure, but the disordered, low-level occupancy of Te in the heptagonal channels is not shown.

Table 1. Atomic coordinates for the Ml phase (from Ref. 28).

Site	Wyckoff no.	Site sym.	x	y	z	U_{iso} 100	Occ
Mo1	2a	2(001)	0	0	0.5	1.453(30)	0.8
V1	2a	2(001)	0	0	0.5	1.453(30)	0.2
Mo2	2b	2(001)	0	0.5	0.641(11)	1.453(30)	0.2
V2	2b	2(001)	0	0.5	0.641(11)	1.453(30)	0.8
Mo3	4c	1	0.1174(8)	0.2273(6)	0.480(5)	1.453(30)	0.5
V3	4c	1	0.1174(8)	0.2273(6)	0.480(5)	1.453(30)	0.5
Mo4	4c	1	0.1762(5)	0.4824(4)	0.533(4)	1453(30)	1
Mo5	4c	1	0.2149(6)	0.3405(5)	0.671(4)	1.453(30)	1
Mo6	4c	1	0.2842(5)	0.2084(4)	0.638(5)	1.453(30)	1
Mo7	4c	1	0.3801(7)	0.0993(6)	0.473(5)	1.453(30)	0.5
V7	4c	1	0.3801(7)	0.0993(6)	0.473(5)	1.453(30)	0.5
Mo8	4c	1	0 4584(5)	0.2250(4)	0.651(5)	1453(30)	1
Nb9	4c	1	0.3597(6)	0.3185(4)	0.518(4)	1.453(30)	1
Mo10	4c	1	0.0005(6)	0.1354(4)	0.667(5)	1.453(30)	1
Mo11	4c	1	0.3442(5)	0.4377(4)	0.655(5)	1.453(30)	1
Te12	4c	1	0.5438(4)	0.1018(3)	0.554(4)	1.453(30)	1
O1	2a	2(001)	0	0	0.026(13)	0.911(31)	1
O2	2b	2(001)	0	0.5	0.055(13)	0.911(31)	1
O3	4c	1	0 1167(14)	0.2283(11)	0.043(9)	0.911(31)	1
O4	4c	1	0.1827(12)	0.4786(9)	0.110(11)	0.911(31)	1
O5	4c	1	0.2171(12)	0.3412(10)	0.098(9)	0.911(31)	1
O6	4c	1	0.2852(14)	0.2132(8)	0.088(13)	0.911(31)	1
O7	4c	1	0.3831(14)	0.1052(10)	0.040(10)	0.911(31)	1
O8	4c	1	0.4513(12)	0.2261(11)	0.072(10)	0911(31)	1
O9	4c	1	0.3515(13)	0.3150(10)	0.082(10)	0.911(31)	1
O10	4c	1	0.0117(13)	0.1384(9)	0.074(12)	0.911(31)	1
O11	4c	1	0.3439(13)	0.4385(11)	0.101(12)	0.911(31)	1

(*Continued*)

Table 1. (*Continued*)

Site	Wyckoff no.	Site sym.	x	y	z	U_{iso} 100	Occ
O12	4c	1	0.5337(12)	0.1156(10)	0.012(9)	0.911(31)	1
O13	4c	1	0.5184(13)	0.4279(11)	0.542(10)	0.911(31)	1
O14	4c	1	0.5764(12)	0.3354(11)	0.576(13)	0.911(31)	1
O15	4c	1	0.0452(14)	0.2658(10)	0.625(10)	0.911(31)	1
O16	4c	1	0.5843(13)	0.0302(10)	0.657(9)	0.911(31)	1
O17	4c	1	0.6944(14)	0.2978(11)	0.571(14)	0.911(31)	1
O18	4c	1	0.7755(14)	0.2167(11)	0.592(12)	0.911(31)	1
O19	4c	1	0.6709(13)	0.0937(9)	0.573(12)	0.911(31)	1
O20	4c	1	0.9595(14)	0.4341(11)	0.607(10)	0.911(31)	1
O21	4c	1	0.8041(14)	0.3527(11)	0.556(11)	0.911(31)	1
O22	4c	1	0.7940(12)	0.1251(10)	0.582(12)	0.911(31)	1
O23	4c	1	0.7648(12)	0.0254(10)	0.585(11)	0.911(31)	1
O24	4c	1	0.8661(13)	0.2585(10)	0.587(12)	0.911(31)	1
025	4c	1	0.9026(13)	0.1149(11)	0.586(13)	0.911(31)	1
O26	4c	1	0.9067(13)	0.0112(10)	0.572(11)	0.911(31)	1
O27	4c	1	0.8204(12)	0.4484(10)	0.603(10)	0 911(31)	1
O28	4c	1	0.9410(12)	0.3407(11)	0.582(12)	0.911(31)	1
O29	4c	1	0.9529(13)	0.1998(11)	0.595(13)	0.911(31)	1
O30	4c	1	0.1531(14)	0.2982(11)	0.578(12)	0.911(31)	1

Notes. Space Group Pba2 (#32); $a = 21.134(2)$ Å, $b = 26.658(2)$ Å, $c = 4.0146(3)$ Å with $z = 4$.

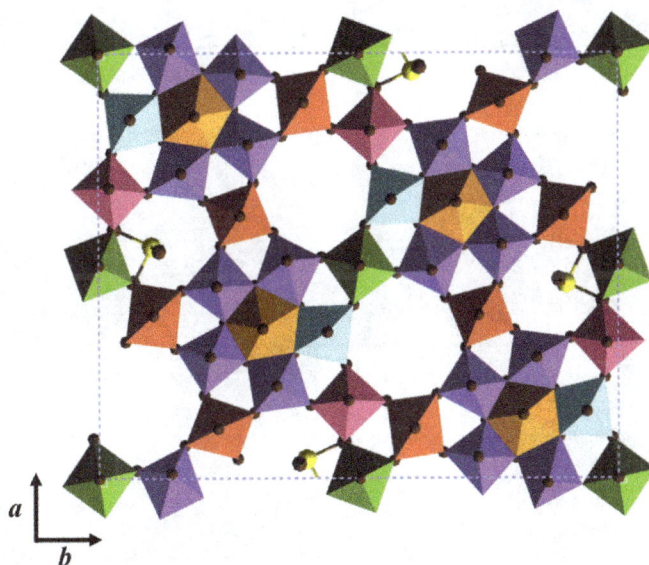

Figure 10. Model for the average structure of M1 with composition $Mo_{7.8}V_{1.2}NbTe_{0.937}O_{28.9}$. Note that the low-level disordered occupancy of Te in heptagonal channels is omitted here. The color coding for the different sites is as follows: Mo^{6+}/Mo^{5+} (red), Mo^{5+}/V^{4+} (green), Mo^{5+} (purple), Nb^{5+} (orange), Mo^{5+} (teal), Mo^{6+}/Mo^{5+} (pink), and Te^{4+} (yellow) based on preliminary assessment of likely valences.

Characterization of the M2 structure and composition using SAED, HREM imaging, and combined S-XPD and NPD

We established that the M2 structure is a derivative of the HTB family, idealized earlier in Figure 2, but it exhibits a slight distortion that is accompanied by crystallographic twinning along the equivalent a_n directions of the ideal hexagonal setting. As a result, superstructure reflections appear in SAED patterns from the principal zone, [001]; these additional reflections are evidence of an orthorhombic distortion of the HTB unit cell. The observations of superstructure in this case can initially be confusing since twinning make the additional reflections appear along all combinations of a_n^* if the selected area chosen is large enough or the twins are small enough so that the SAED encompasses all three equivalent domains. Evidence from the SAED patterns in which the superstructure appears only along one or two of the a_n^* directions from the parent hexagonal setting reveals orthorhombic symmetry as the true symmetry of the unit cell.

Figures 10(b) and 10(c) appear to suggest that there are systematic absences along a^* for $h00$; $h = 2n$; however, structure factors for these reflections are not zero, but very small (1000–$5000 \times$ smaller than Fmax) and may not be observable at the exposure times used to obtain conventional SAED patterns. Diffuse

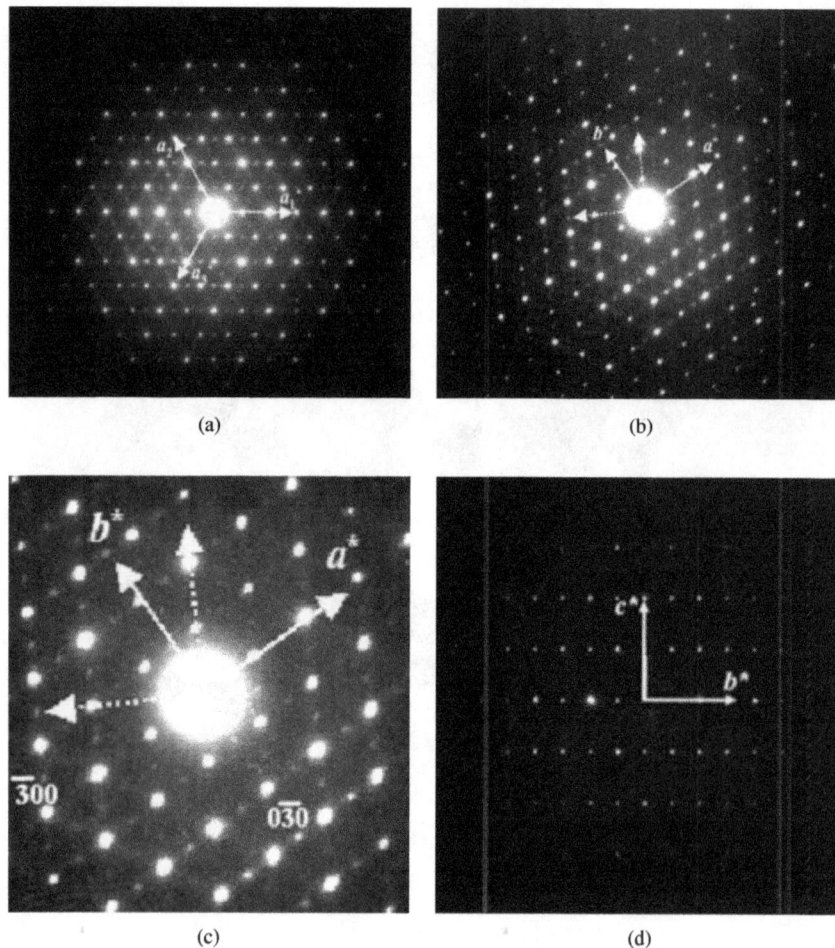

Figure 11.   SAED patterns from the M2 phase. (a) [001] zone showing a twinned superstructure along a_1^*, a_2^*, and a_3^* (hexagonal) and (b) along a^* and b^* of two orthorhombic domains rotated 60° to one other. (c) Enlarged schematic of reflections around the caustic (000). (d) SAED pattern of M2 in the [100] zone.

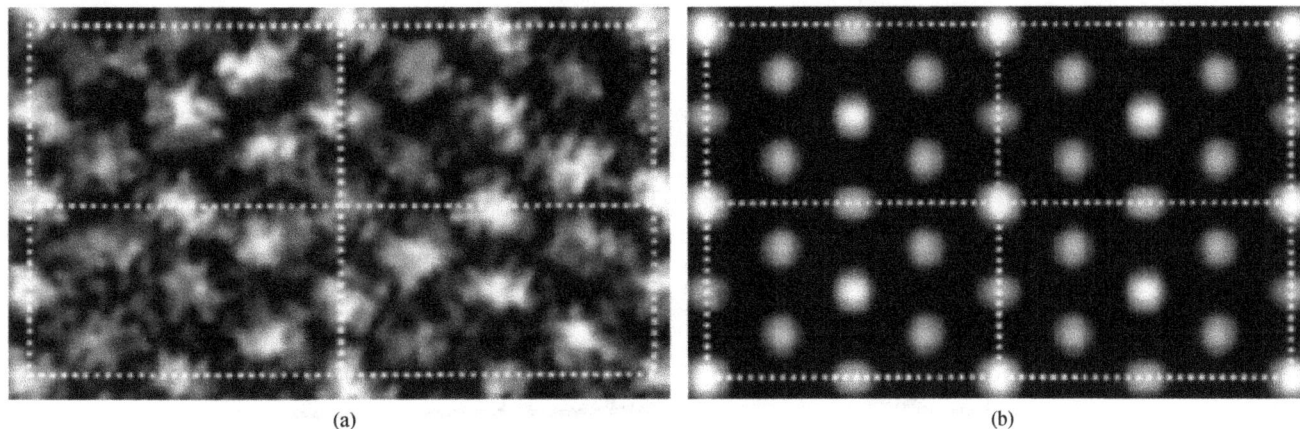

(a)   (b)

Figure 12.   M2 phase (a) 2 × 2 unit cell HREM image in the [001] zone and (b) the corresponding [001] HREM image simulation with $\Delta f = -43$ nm and thickness $t = 9.2$ nm.

scattering in the [001] zone that appears to be uniformly distributed suggests that the twinning is weakly short-range correlated and close to random. Since no systematic absences are observed in the [100] zone (Figure 10(d)), only two space groups are possible for this unit cell: *P222* (#16) or *Pmm2* (#25). Using symmetry considerations similar to those exercised for the M1 analysis lead to the conclusion that the unit cell of M2 is most probably *Pmm2* with unit cell parameters $a \cong 12.6$ Å, $b \cong 7.29$ Å, $c \cong 4.01$ Å. Phase-contrast HREM imaging confirms the distorted HTB-type structure (Figure 11).

S-XPD and NPD data (see Figure 13) were collected in a similar manner to the M1 phase and were simultaneously refined, again using GSAS[25] with the graphical user interface-program EXPGUI;[26] results are shown in Table 2. The initial model was based on the HTB-type parent structure and the evidence from SAED and HREM imaging for the orthorhombic distortion. It was found that the intercalated tellurium sites were statistically distributed and that they tend to orient toward the more vanadium-rich octahedral sites in the hexagonal rings as illustrated in Figure 14. The displacement of tellurium toward vanadium is associated with the stronger repulsion between Mo^{6+} and Te^{4+} centers in combination with the relatively short V^{4+}–O bond distances. Due to the inequivalence of the framework sites surrounding each hexagonal channel, two crystallographically distinguishable types of Te^{4+} sites exist. Vanadium preferentially occupies two of the three framework sites, resulting in Te being displaced further on average from the channel center Te(2) surrounded by these two sites. This is in contrast with the Te(1) site that is adjacent to a single vanadium site, with the other site containing no vanadium. The Te(1) channel contains a distorted TeO_4E trigonal bipyramid (E = lone pair) with extended TeO chains along *c*, whereas the Te(2) channel contains a distorted TeO_3E tetrahedron with only a weak interaction between TeO units along *c*. The arrangement of Mo, Nb, Te, and V in M2 would seem to provide potential olefin oxidation sites, but the absence of pentavalent vanadium in this structure would seem to render it inactive for paraffin activation.

Initial STEM HAADF investigations of the MoVNbTeO M1 phase

A well-characterized M1 powder sample previously used for the combined X-ray and neutron structural refinement described in DeSanto *et al.*[28] was investigated using the aberration-corrected JEOL-2100F STEM at the University of South Carolina set up in the high-angle annular dark-field (HAADF) mode. The high-resolution HAADF image provided in Figure 15 shows an M1 crystallite oriented along the [00$\bar{1}$]

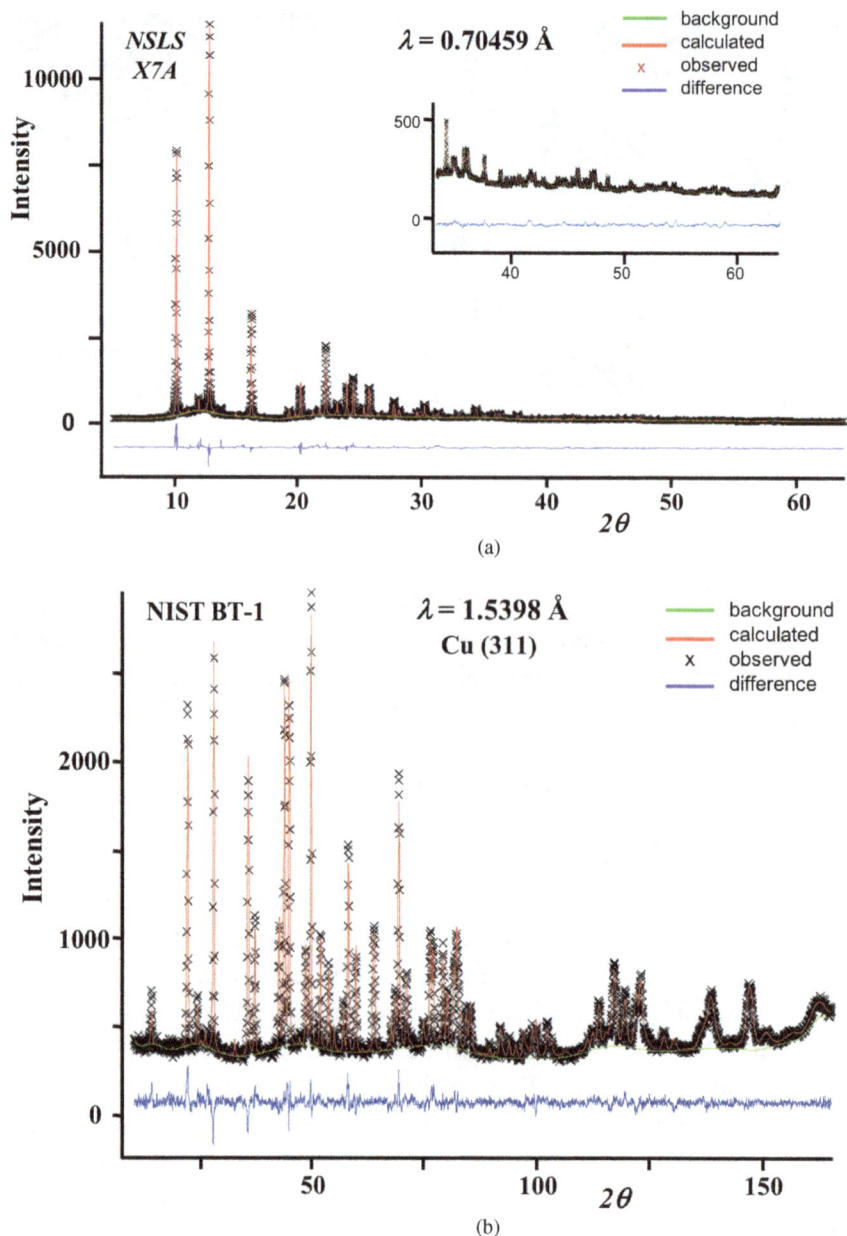

Figure 13. Refined model fit[28] to (a) synchrotron X-ray powder diffraction pattern data from NSLS beamline X7A powder diffractometer at $\lambda = 0.70459$ Å collected with a 4° PSD with inset showing the high-angle data and (b) neutron powder diffraction data from NIST beamline BT-1 using Cu-(311) K_α at $\lambda = 1.5398$ Å.

direction. The improved resolution using an instrument with aberration–correction is immediately apparent as all of the atomic columns are now well resolved. It is also clear from the image that adjacent atomic columns have different contrast. This initial observation was encouraging since the occupancies of the crystallographically distinct atomic sites, when refined against neutron and X-ray powder diffraction data, revealed different mixed Mo and V amounts. The image in Figure 15 shows both a polyhedral and wire-frame rendering of the unit cell based on the Rietveld model scaled with a constant aspect ratio and superimposed on the HAADF image. A good qualitative agreement between the superimposed unit cells and the HAADF image is observed.

Table 2. Atomic coordinates for the M2 phase (from Ref. 28).

Site	Wyckoff	x	y	z	Occ.
Te1	4i	0.0237(26)	0.0621(18)	0.609(5)	0.237(5)
Te2	4i	0.5070(39)	0.5754(19)	0.624(5)	0.218(4)
Mo3	1c	0.5	0	0.577(14)	0.54(4)
V3	1c	0.5	0	0.577(14)	0.46(4)
Mo4	4i	0.2505(8)	0.2463(11)	0.583(6)	0.78(1)
V4	4i	0.2505(8)	0.2463(11)	0.583(6)	0.22(1)
Mo5^c	1b	0	0.5	0.594(9)	0.5
Nb5^c	1b	0	0.5	0.594(9)	0.5
O1	4i	0.1047(10)	0.3098(12)	0.589(4)	1
O2	2f	0.2935(13)	0.5	0.583(5)	1
O3	4i	0.3965(10)	0.1913(12)	0.596(4)	1
O4	2e	0.7933(13)	0	0.565(5)	1
O5	1c	0.5	0	0.086(6)	1
O6	1b	0	0.5	0.065(6)	1
O7	4i	0.0387(18)	0.0425(36)	0.103(7)	0.237(5)
O8	4i	0.2514(13)	0.2445(18)	0.076(4)	1
O9	4i	0.4778(19)	0.4855(54)	−0.004(10)	0.218(4)

Notes. Space Group *Pmm2*: $a = 12.6294(6)$ Å, $b = 7.29156(30)$ Å, $c = 4.012010(7)$ Å with $Z = 4$.

^{a}NPD: $R_{wp} = 4.05\%$, $R_p = 3.31\%$.

^{b}SPD: $R_{wp} = 5.54\%$, $R_p = 4.20\%$.

^{c}Fractional occupancy not refined.

Analysis of the HAADF images explored the presence and occupancy of Te sites in the center of the hexagonal (site S12) and heptagonal (site S13) channels. In the DeSanto *et al.* model, the refined S12 site within the hexagonal channel was shifted away from the center toward the S2 site.[28] The S12 site is indicated by arrows in the enlarged portion of Figure 16. The intensity line scan shown at the top of Figure 16 was taken from the atomic columns outlined with a white rectangle. The peaks in this line scan confirm the DeSanto model,[28] which showed that the S12 site is shifted away from the center of the channel toward the S2 site.

In addition to confirming the Te position in the hexagonal channels, we were furthermore able to detect an appreciable contrast in two of the four heptagonal channels enclosed in the white ovals drawn in the bright-field and HAADF STEM images as shown in Figure 17. The inverted contrast between the bright- and dark-field images provides a strong corroboration of the location of Te. The occupancy of Te in the heptagonal channels was estimated by comparing the number of occupied channels to the number of unoccupied channels. Occupation was determined through visual inspection of the contrast in the center of the channel. The results from this counting study are shown in Figure 18 and indicate that ≈ 26% of the heptagonal channels exhibited measurable filling in the HAADF image. This is remarkably consistent with the Rietveld model predicting approximately 20%.[28]

Due to the good contrast variability between the individual atomic columns observed in the dark-field aberration-corrected images, we extracted quantitative information in order to compare these results with the model derived from X-ray and neutron structural refinements. The model of the DeSanto refinement of the M1 phase uses ~ 200 adjustable parameters.[28] High-quality diffraction data and a suitable starting model

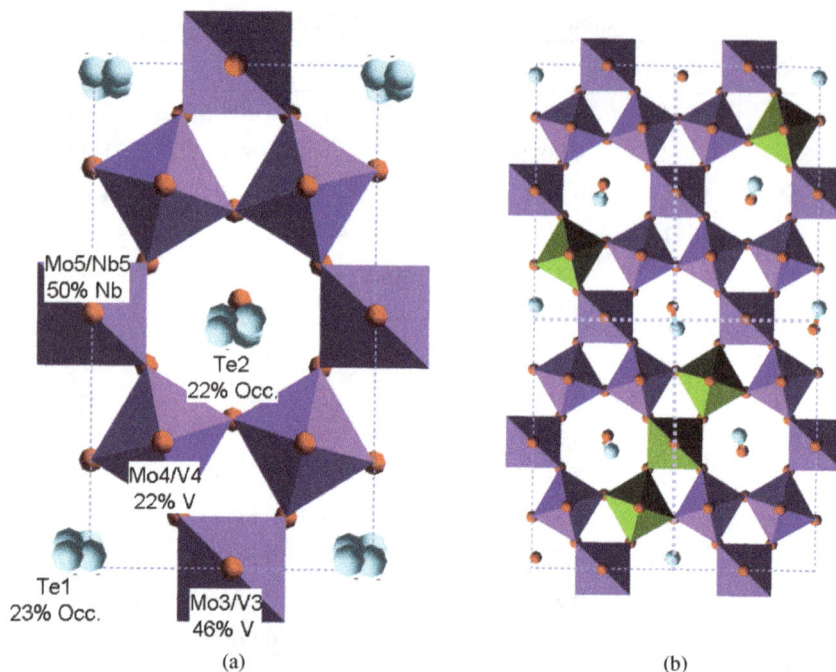

Figure 14. Rendering of the M2 structure based on the Rietveld refinement from Ref. 28. (a) Single unit cell with the statistically distributed Te distortions exaggerated and the oxygen atoms associated with the disordered Te sites at the corners of the unit cell omitted for clarity. No distinction is made here between Mo, V, and Nb in the octahedral sites. (b) 2 × 2 unit cell rendering showing one possible representation of statistically distributed vanadium octahedra (green) and molybdenum or niobium octahedra (purple). Note that the Te displacements are shifted toward vanadium sites in the hexagonal channels. Niobium and molybdenum cannot be distinguished from these refinements, so are assumed to be randomly mixed.

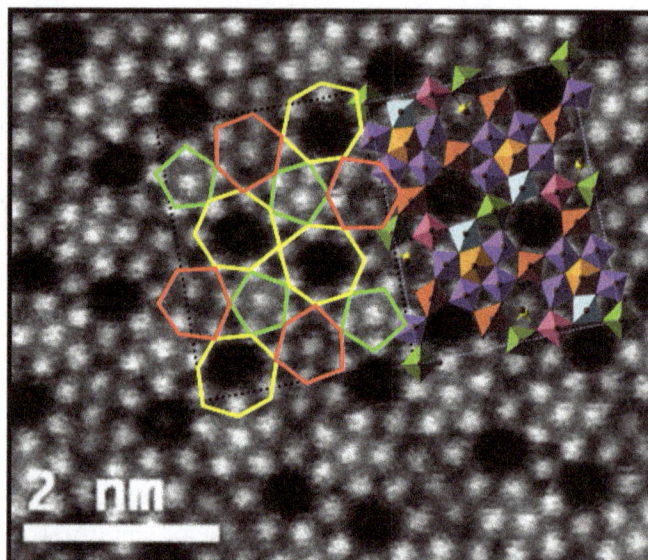

Figure 15. Magnified portion of the [00−] projection showing a scaled model unit cell superimposed on the high-resolution HAADF STEM image. Note that the images have not been processed or altered. Color coding is given in Figure 9.

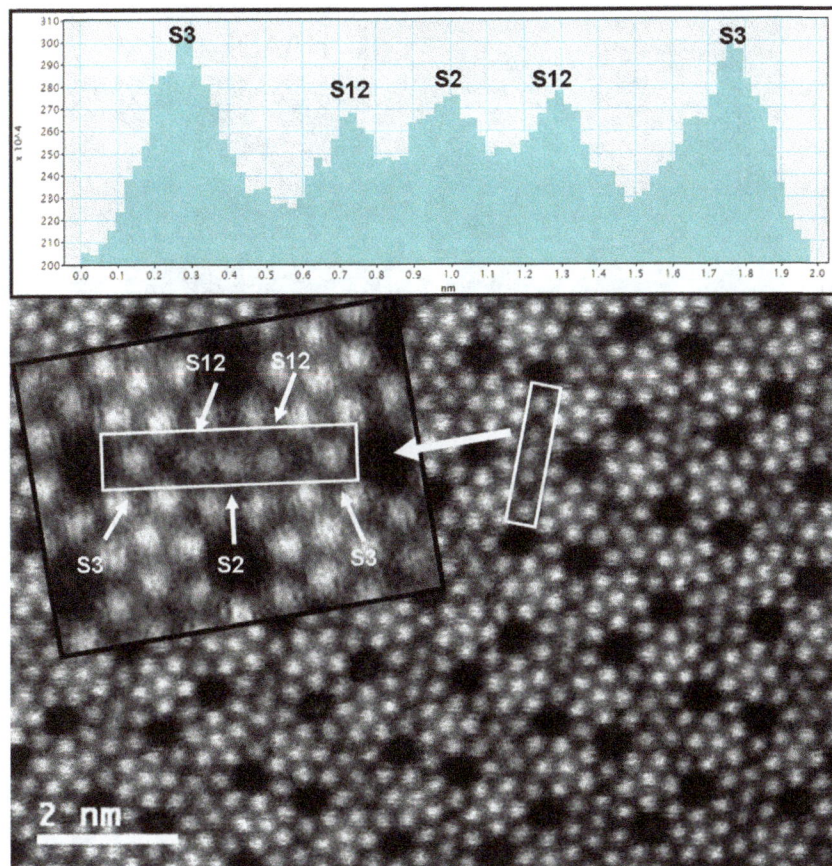

Figure 16.   A HAADF image of the S–M1–MoVNbTe sample showing the S3–S12–S2–S12–S3 sites in the enlarged inset. The line scan at the top of the image is from the region enclosed in the white rectangle.

with imposed constraints are needed to ensure a stable and converging refinement using least-squares methods. Aberration-corrected HAADF imaging can become a critical first step to derive such initial model parameters. The sub-Ångstrom resolution allows the precise measurements of atomic site positions in projections and the approximate Z^2 relationship between the atomic number and the image contrast permits an initial estimation of the atomic site compositions. However, the downside to this straightforward analysis is an inability to resolve low Z atoms such as oxygen which require bright-field imaging where the simple Z^2 relationship between the atomic number and the image no longer holds. Extensive frozen-phonon multislice simulations are needed to then test structural models. See Ref. 29 for a recent example involving a new apatite-type oxide ion conductor where both dark- and bright-field imaging were used in conjunction with synchrotron and neutron powder diffraction.

A two-pronged approach depicted in Figure 19 was used for the initial interpretation of HAADF images and the extraction of image-based structural models. The top part outlines the determination of the atomic coordinates and the bottom part the estimation of the atomic site composition based on the image contrast. To verify the validity of the method, the image-based MoVNbTeO model was compared to the initial structural model of the same catalyst developed by DeSanto *et al.* based on the refinement of high-resolution X-ray and neutron powder diffraction data.[28]

In the <001> projections, the M1 structure has no overlapping metal sites, which allows for the direct extraction of their fractional *x* and *y* atomic coordinates. Atomic coordinates were determined by superimposing

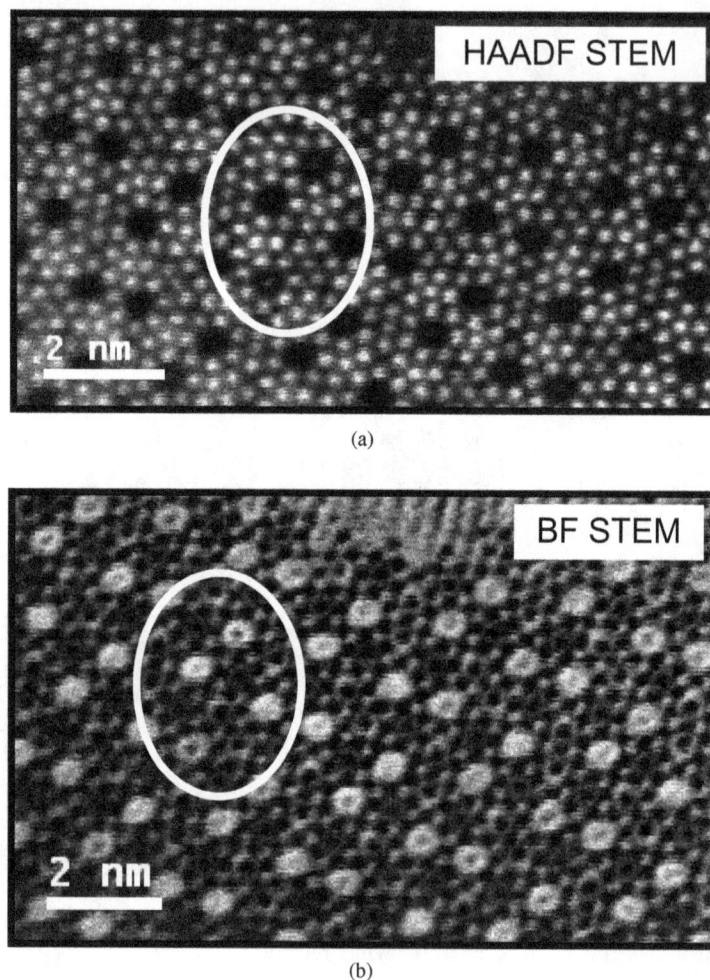

Figure 17. (a) High-resolution HAADF STEM image showing areas with occupied heptagonal channels within the white oval, (b) high-resolution BF STEM image of the same area showing the same filled heptagonal channels. Note that the images have not been processed or altered.

an adjustable grid on top of the unit cell. The adjustable grid was needed to compensate for rastering and sample drift distortions. To minimize the influence of these distortions, each metal coordinate site within the first quadrant of a chosen unit cell was measured and averaged for four different reference points [i.e. (0,0), (0,0.5), (0.5,0), (0.5,0.5)], corresponding to projections of the binary axes in the *Pba2* (Space Group No. 32) symmetry.[31] The reference points are labeled in Figure 20 (red circles), and an example of the mesh applied to each of the reference points for the measurement of a quadrant within a single unit cell is shown in Figure 21. By using four reference points fixed by symmetry and only a single quadrant for the measurement of atomic coordinates, we significantly reduce the maximum distance that any given atomic column lies from the nearest point of reference. By sampling with respect to each of the four nearest reference points, the systematic errors due to rastering distortions and other environmental disturbances (e.g. stray electrical fields, temperature drift, vibrations) cancel approximately and the average of the measured fractional x, y coordinate best reflects the atomic position within the unit cell.

To quantify the atomic composition of each column seen as a projection, we made the following assumptions: (i) the observed scattering follows the Z^2 Rutherford relationship[32–34]; (ii) within a single unit cell,

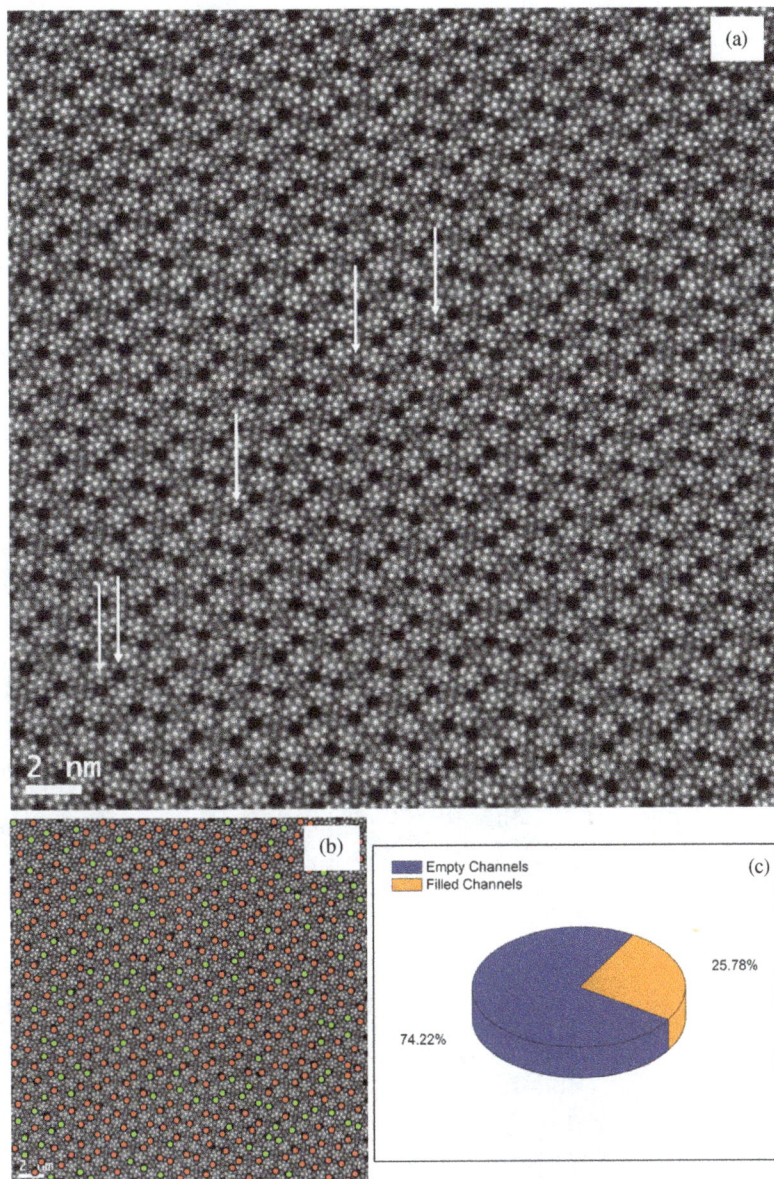

Figure 18. (a) HAADF STEM image showing a large area of the M1-MoVNbTe sample. The arrows point toward some of the occupied heptagonal channels. (b) A map of occupied (green) versus unoccupied (red) heptagonal channels (c) A pie chart showing the percentage of occupied vs unoccupied heptagonal channels.

the thickness of the crystal was constant; (iii) scattering contributions from oxygen were only considered at positions in the <001> projections which are superimposed with metal site columns; (iv) a constant integration area was appropriate for all metal framework sites; (v) the background was constant throughout the portion of the image being analyzed; and (vi) the intensity of each atomic column is independent and does not contain contributions from neighboring atomic sites.

Using these assumptions, the total intensity for each atomic column was calculated by integrating the individual pixel intensity for each framework site. The background was calculated by averaging the integrated intensity of several empty heptagonal channels near the unit cell of interest. The raw intensity for each site was determined by subtracting the background from the total intensity of each atomic column. Finally, the

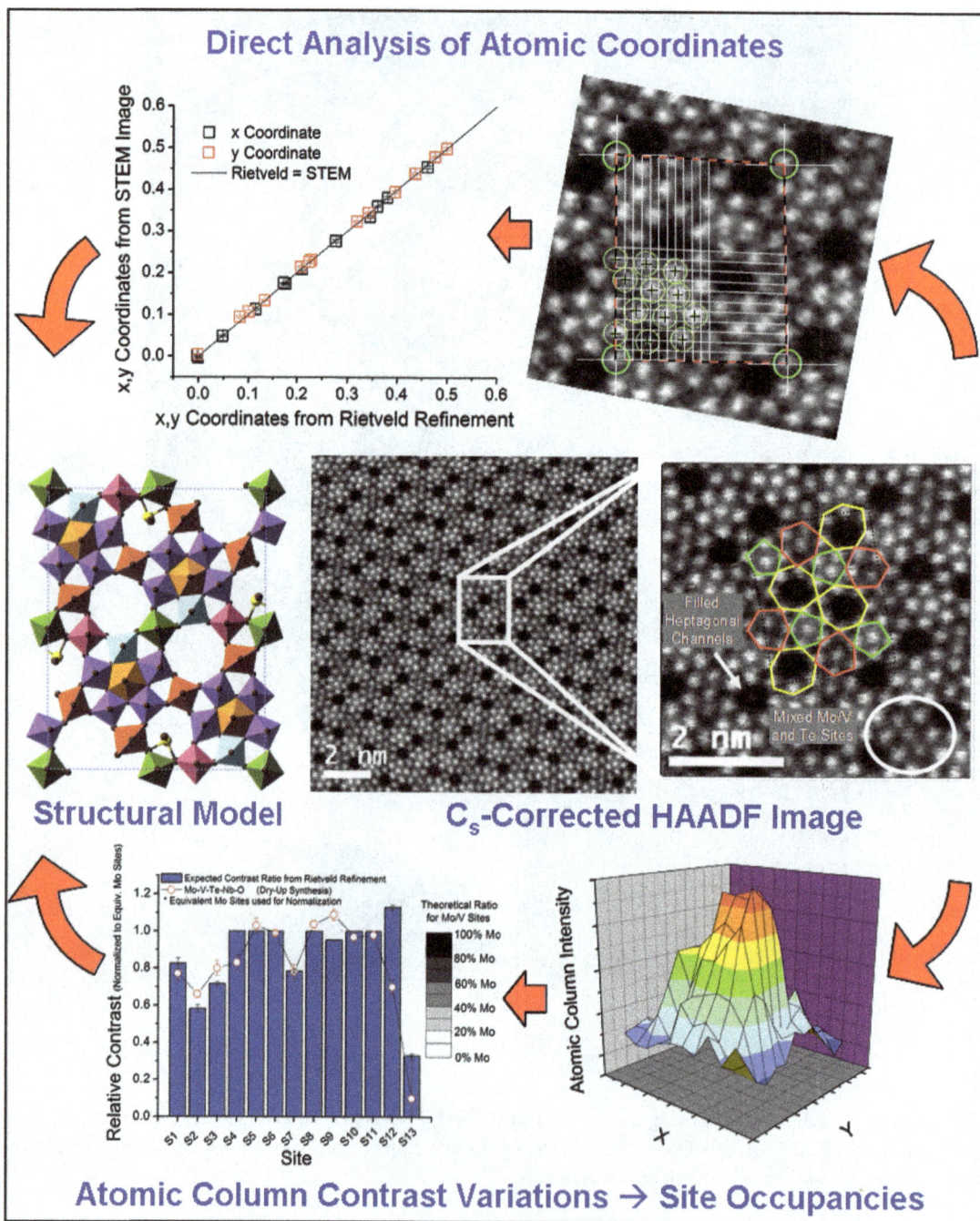

Figure 19. Schematic of the HAADF STEM image analysis algorithm.[30]

raw intensities of each atomic site were converted to a contrast ratio through normalization of the site intensity by the average intensity of the five octahedral members of the pentagonal ring. The normalization process removes the majority of the thickness contributions. Any remaining variability in the contrast ratio of each atomic site within the unit cell should be the result of composition differences. The approach of using the average of five pentagonal ring sites for normalization was chosen because we expect the occupancy of these octahedral sites to be at or very close to 100% Mo based on the prior refinements of the MoVNbTeO and MoVTaTeO M1 phases.[28,35] If this assumption proves valid, then it should be possible to quantify the

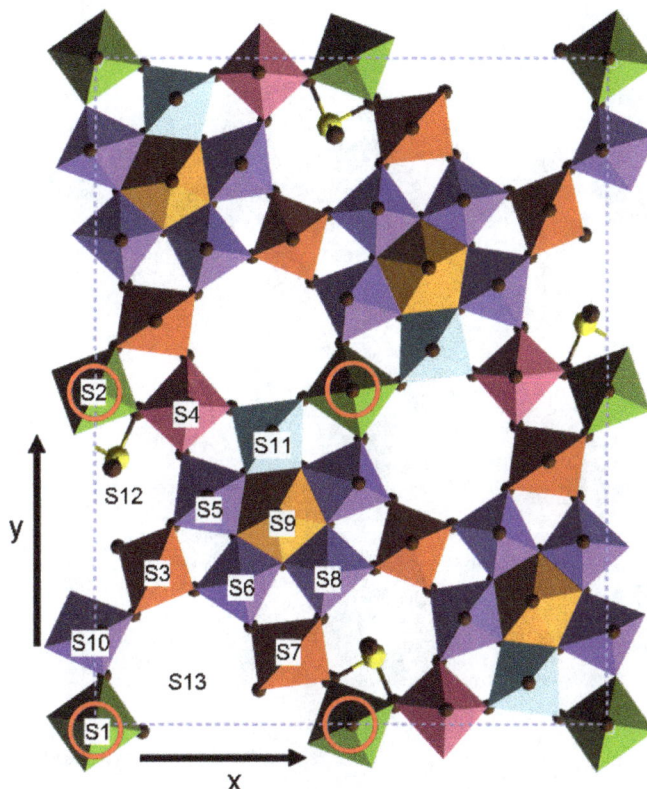

Figure 20.   Idealized M1 model showing the reference points used for the measurement of the atomic coordinates (red circles) within a single quadrant of the unit cell. Color coding is same as in Figure 10.

composition of each site using the contrast ratio. To test this assumption for each unit cell, the contrast ratio for each pentagonal site can be compared with the expected value of unity. The variability in the contrast ratio of these positions should provide an estimate for the expected uncertainty of the remaining atomic columns in the framework.

Using the HAADF image interpretation algorithm described in the previous section, the MoVNbTe sample was quantitatively analyzed. The plot in Figure 22 provides a comparison between the x, y coordinates from the Rietveld model[28] and HAADF image along with a 45° line as a reference to indicate a perfect match. The comparison of the HAADF-based x, y coordinates with those from the Rietveld-refined model reveals good agreement. Quantification of the differences between the two models is shown in Figure 22. Most deviations are less than 0.15 Å which is well below the 1 Å point-to-point resolution of the microscope. The high level of precision afforded by HAADF-derived coordinates should enable their direct insertion into a starting model for a complete structural refinement. In this case, the agreement between the two models indicates that no revisions to the atomic coordinates are necessary for the existing Rietveld model.

Comparison of the expected contrast ratios from the refined M1 model and HAADF images are presented in Figure 23. Overall, the majority of measured intensities follow the site occupancy trends predicted by the Rietveld model; however, the contrast for site S4 was anomalously low, as will be discussed shortly. The measured contrast ratio for each of the five sites that form the pentagonal ring (S5, S6, S8, S10, S11) is very close to the expected value of unity and was bound in a range between 0.97 and 1.03. This range represents a precision of ±4% for the Mo content and provides a benchmark for the expected level of uncertainty for the measurement of atomic column compositions using HAADF images. This initial result is important since it validates the normalization assumption corresponding to a pure Mo site occupancy.

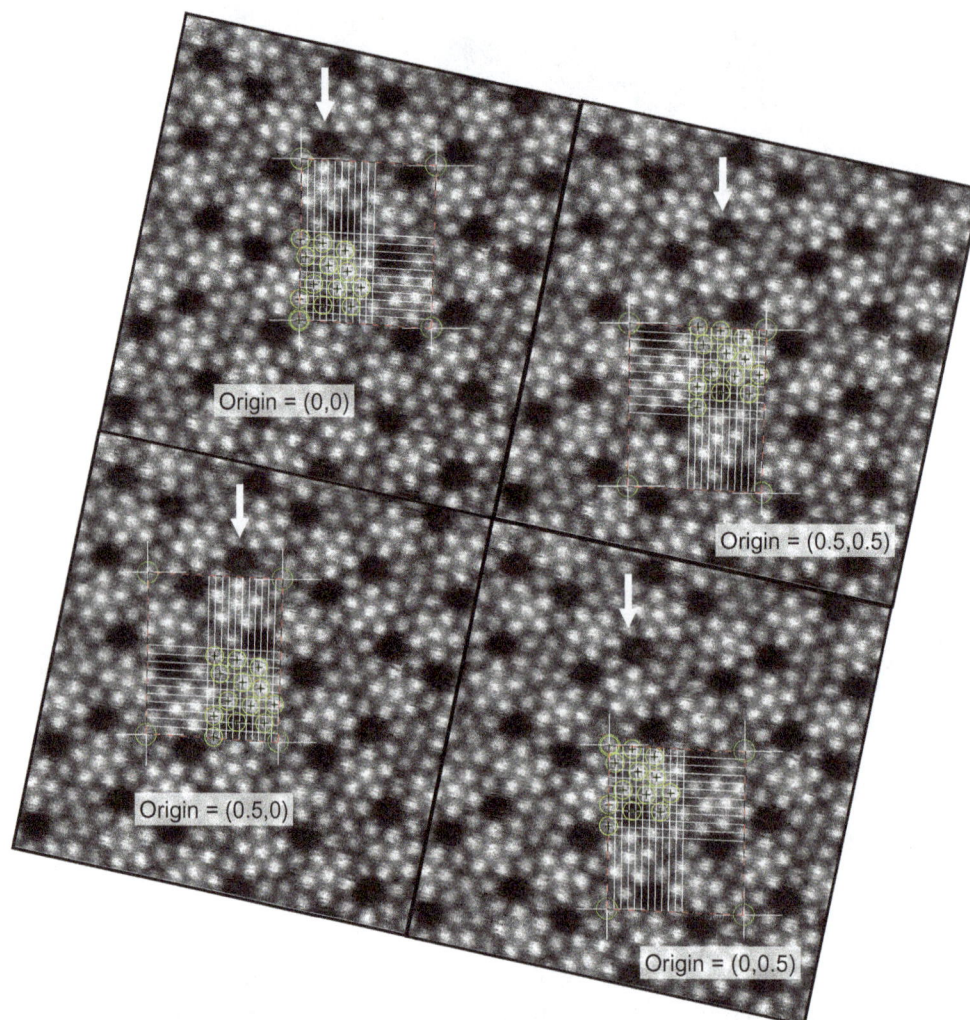

Figure 21.   An example of the adjustable mesh applied to each of the four reference points for a single unit cell. The white arrow points toward the same feature in each case.

A second important observation is the measured contrast ratios of the mixed Mo/V occupancy sites (S1–S3, S7). The depressed values for the measured contrast ratios for these sites were consistent with V substitution and were in reasonable agreement with the expected contrast ratios derived from the refined model based on powder diffraction. The agreement between HAADF model and the refined model was ~10% for these sites. However, some larger deviations between the refined model and the HAADF model were observed for sites S4, S9, S12, and S13.

For site S4, the measured contrast ratio was significantly lower than expected and indicated some vanadium occupancy. This is in contrast to the refined model that predicted full Mo occupancy. It is believed that the refined site occupancy is incorrect for two reasons. First, a subsequent model developed by Murayama et al.[36] suggested an alteration to the DeSanto model[28] in which site S4 should contain partial V occupancy. Second, looking back at the structural model in Figure 10, it is reasonable to expect that site S4 would have a similar occupancy as S7, due to their geometric similarity — site S7 shows an ~30% V occupancy. The measured contrast ratios for sites S4 with S7 support this contention since these sites have similar values.

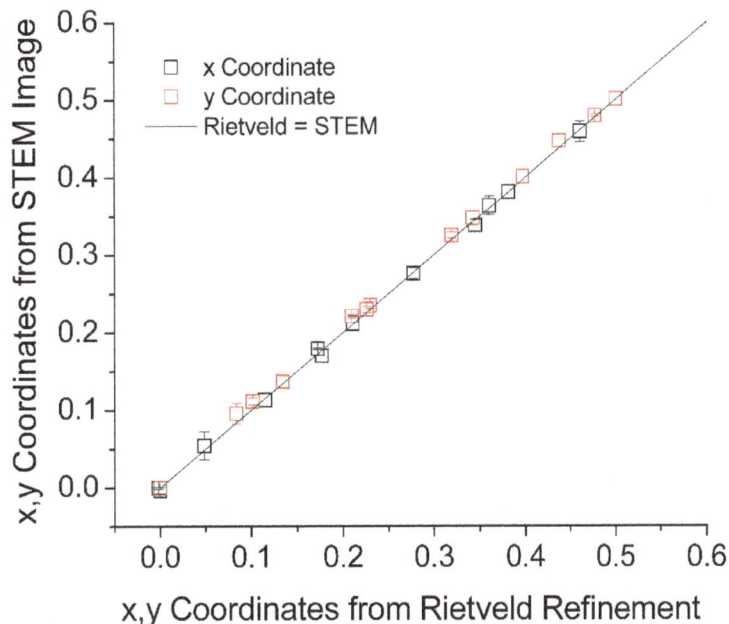

Figure 22. Plot showing the fractional coordinates of the refined model[28] plotted against the experimentally determined coordinates measured directly from raw high-resolution HAADF STEM images for M1-MoVNbTe. Note that the standard errors fall within the size of the plot symbols.

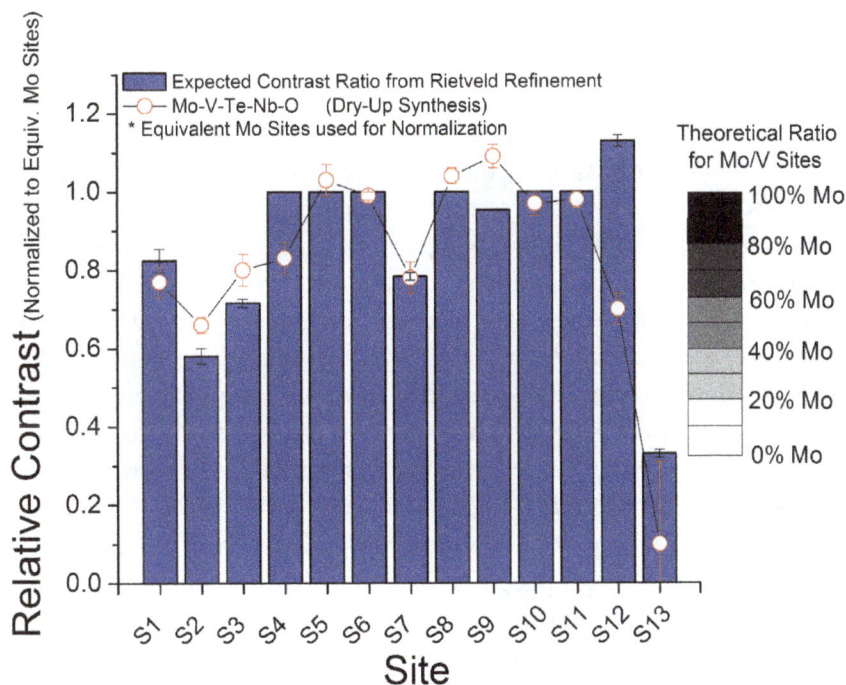

Figure 23. Comparison of occupancies derived from HAADF-STEM with those obtained by Rietveld refinement for M1-MoVNbTe. All sites were normalized by assuming full Mo occupancy of the five pentagonal ring sites. Discrepancies for the Te occupancies are attributed to electron beam induced sublimation.

For site S9, the measured contrast ratio was significantly higher than the expected contrast ratio for a site fully occupied by Nb. The contrast ratio for a site fully occupied by Nb is expected to be just below unity (~ 0.95). As can be seen in Table 1, the isotropic thermal displacement parameter, U_{iso}, for Nb was refined together with the other metals assuming a common value so as to limit the number of adjustable parameters. Based on the unusually high contrast for Nb in the STEM HAADF imaging, it seems that the thermal parameter for Nb in the pentagonal biprismatic center may actually be significantly smaller than for the edge-shared Mo octahedra forming the pentagonal rings, and this is likely to be the cause for the elevated contrast observed for this S9 site. This suggests that the thermal parameter for the S9 site should be refined separately from those for the other sites.

Finally, for the Te sites S12 and S13, the HAADF image contrast is significantly diminished compared to what would be expected for the Te occupancy based on the refined model based on powder diffraction data. The low Te occupancy in the HAADF model is consistent with partial sublimation during exposure to the electron beam.[27] An additional factor that could complicate the quantification of atomic species in sites S12 and S13 is the cation alignment along the direction of the beam. It is possible that the Te atoms in the Te–O chains (in the case of MoVNbTeO) exhibit a degree of statistical variability of the exact position of the Te atom in the ab-plane. If this is the case, the atomic contrast could be depressed due to a disruption of the electron channeling down the atomic columns due to the x and y displacements of the Te atoms in the Te–O chains running in the z-direction. An alternative factor that could contribute to a diminished contrast in this position is the incomplete occupancy of the Te–O chains. If a Te–O chain begins a few nanometers below the surface of the framework, then this position would appear to be out of focus with respect to the surface of the crystal and would lead to a somewhat blurred feature with lower contrast in the HAADF image.

The Nb site location

In the analysis of the HAADF images of the MoVNbTeO M1 phase, the location and quantification of Nb was neglected. In the DeSanto *et al.* model, Nb is *assumed* to occupy the position in the center of the pentagonal ring.[28] The reason for this omission is that the isoelectronic Nb^{5+} is indistinguishable from Mo^{6+} using standard X-ray, neutron, or electron diffraction methods. Usually, such problems with X-ray and electron contrast from similar or isoelectronic elements are overcome by using neutron diffraction, for which the nuclear scattering mechanism is isotope dependent. In this instance, unfortunately, the natural mix of Mo isotopes (^{92}Mo–^{100}Mo) leaves us with insufficient contrast differences against ^{93}Nb; the coherent neutron scattering cross-section for Nb is 6.23 barns and the weighted average for natural Mo is 5.80 barns.[37] To approach this problem, we hypothesized that substitution of a Group V transition element into the M1 framework structure may provide some insights. The two options for this substitution are V and Ta. Naturally, Ta would be the better option since V is already included in the M1 structure and additional V would be unlikely to solve the problem. Furthermore, Ta is chemically very similar to Nb, being isoelectronic, of similar radius, and forming similar compounds. Using the high Z of Ta as a site-specific substitution for Nb should provide an indirect probe to test the proposed location of Nb in the M1 framework since substitution of Nb by Ta will provide significant contrast for electron, X-ray, and neutron diffraction techniques. In an earlier study, the structural changes of Ta substitution for Nb were explored by simultaneous Rietveld refinement of X-ray and neutron data.[38] The result of that study indicated a successful site-specific substitution of Ta in the structure, and suggested that the Ta occupied the expected position in the center of the pentagonal ring, though with less than full occupancy and leaving residual Ta_2O_5 as an exsolved phase.[35]

A representative [001] high-resolution HAADF image for a Nb- and Ta-containing M1 phase is shown in Figure 24. In both images, the atomic columns are well resolved and exhibit the orthorhombic framework. At first glance, it is clear that the atomic columns in the pentagonal centers of the MoVTaTe specimen

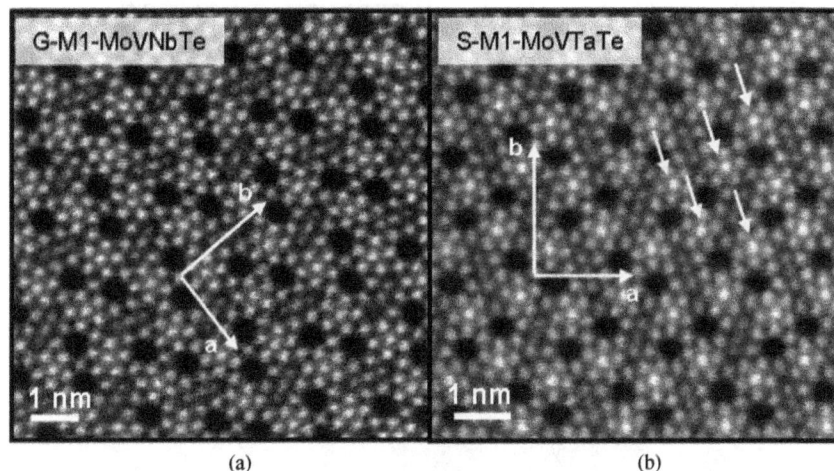

Figure 24.   HAADF images of (a) M1-MoVNbTe, and (b) M1-MoVTaTe catalysts viewed in [001] projection. The a- and b-axis vectors represent the edges of single unit cells in each case. Note that the images shown have not been processed or altered in any way.

in Figure 24(b) (white arrows) display very high contrast compared to those in the MoVNbTe catalyst. This high contrast indicates higher average Z, consistent with this site preferentially containing Ta. Finding Ta in the pentagonal center is consistent with the model developed from diffraction data by DeSanto *et al.* for the same Ta-containing catalyst.[35] This also reinforces the proposed position of Ta by Yamazoe and Kihlborg in an earlier study of Ta-doped Mo_5O_{14}; the structure of Mo_5O_{14} is related to M1 in that it is built upon a network of similar pentagonal units,[39] but with different connectivity. Using this indirect argumentation, we conjecture that the location of Nb in the M1-MoVNbTe structure is likely in the pentagonal center, as has been previously assumed, provided that an isomorphic substitution of Ta for Nb did occur. The match of valence, crystal radius, and typical crystal chemistry suggests that Nb and Ta would exhibit similar site preferences. Contrast within the octahedral sites surrounding the pentagonal center is nearly constant for each sample. This is a second indication of an isomorphic substitution into the pentagonal center since substitution of Ta into a Mo site would be easily detected in the HAADF images.

Characterization of Rotational Twinning in the MoVTeNbO M2 Phase Using HAADF

The M2 phase has been found to serve as a promoter in the best M1-containing catalysts for the selective (amm)oxidation of propane to either acrylic acid or acrylonitrile. As shown earlier, SAED studies on M2 revealed superstructure reflections that in some cases reflections were consistent with hexagonal symmetry[28,40] and in other cases, electron diffraction patterns displayed superlattice reflections along only one of the apparent hexagonal directions, indicating the presence of a subtle distortion.[40] The proposed explanation for the distortion was that the distorted M2 unit cell has orthorhombic symmetry and that the crystallites are comprised of multiple twins displaced by 60° rotation from one another. We will now revisit this using aberration-corrected HAADF imaging.

A typical image of M2 in the [001] orientation is shown in Figure 25. A unit cell is outlined (black rectangle) for identification purposes. The image in Figure 26(a) shows the diffractogram of the image in Figure 25, and the schematic in Figure 26(c) provides the expected diffractogram for the orthorhombic M2 phase. An overlay of these figures is provided in Figure 26(b). From the superimposed images, it is clear that the image in Figure 25 is comprised of three different 60° twins since the expected diffraction pattern only accounts for a third of the superlattice reflections. If the orthorhombic schematic is replicated twice and

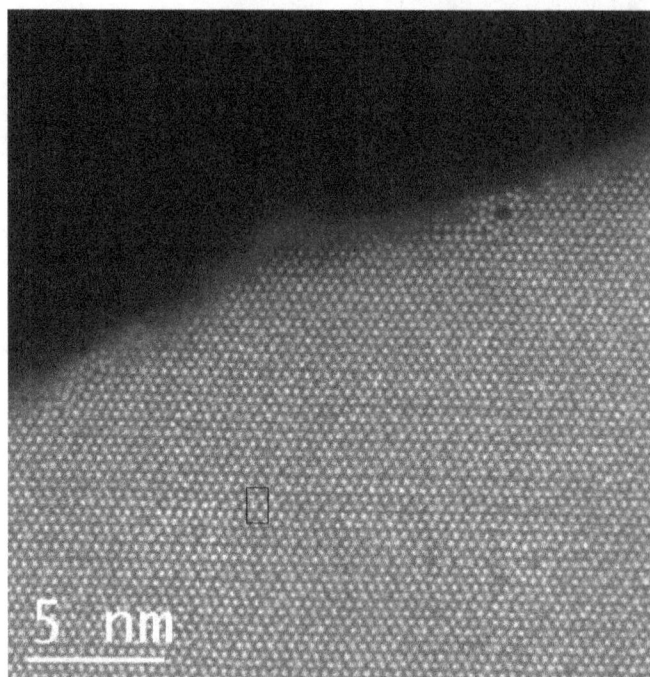

Figure 25. HAADF STEM image of M2 phase in the [001] zone axis orientation. The outline shows the orthorhombic unit cell of the material.

rotated by 60° and 120°, respectively, as shown in sequential images in Figures 26(d) and 26(e), then super-lattice reflections observed in the diffractogram are well described as demonstrated in Figure 26(f) by super-imposing the three orthorhombic diffraction patterns on top of the diffractogram.

To illustrate and isolate the individual twin domains further, Fourier image processing techniques were applied to the data. The {120} reflections in the diffractogram (shown circled in Figure 26(a)) for one of the possible twin orientations were isolated using masking. The inverse FFT of the masked image highlights the regions in the original image responsible for those particular reflections. These reflections were selected for the other two domain orientations, followed by subsequently combining the three IFFT images to create the RGB composite shown in Figure 27. This image shows how the three different twin orientations exist in very small domains that are typically only a few unit cells in size. This twin analysis validates the previous findings of DeSanto et al.[28] and Garćía-Gońzalez et al.[40]

Note the defect near the crystallite boundary in the M2 image in Figure 25 and the more extensive defect region near the edge of the M2 crystallite in Figure 28. The defective edge seen in Figure 28 displays multiple pentagonal rings and heptagonal channels that are common to the M1 phase, though there is no evidence of longer range ordering or regular registry with the M2 lattice. These are M1-like features present only as near-surface defects, suggesting that M1-like arrangements are energetically unfavorable in M2. This, in turn, suggests that any synergy between M1 and M2 in catalysis is not associated with the well-ordered intergrowth of M2 as a promoter, but perhaps shared disordered or defect-rich interfaces. Woo et al.[38] report that the proposed promotor effect of M2 is not evident in their studies and that M2 is much less active than M1 for propylene conversion, challenging the hypothesis of M2 as a good promotor.

Revisiting the rietveld refinement to improve the M1 model

The first STEM HAADF experiments suggested the need to improve the structural model of the M1 phase initially obtained from the combined refinement of X-ray and neutron diffraction data. In addition, new

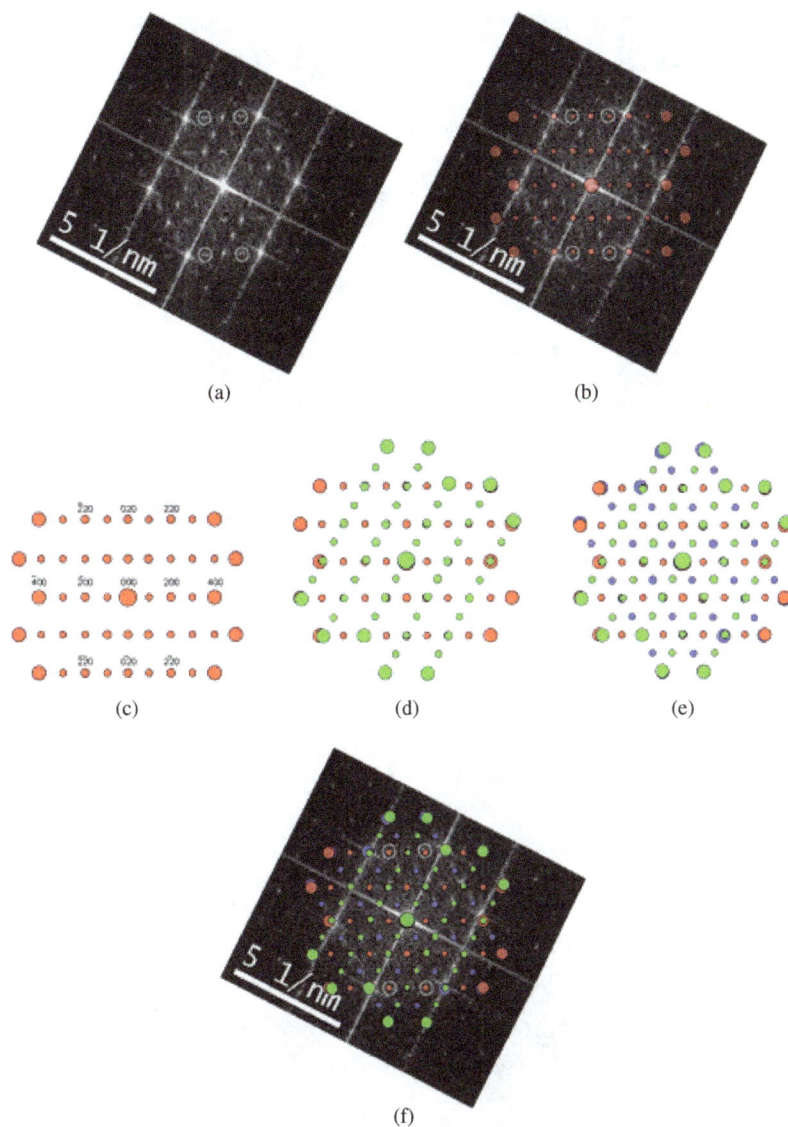

Figure 26. (a) Fast Fourier transform (FFT) of the HAADF image shown in Figure 24. (b) The FFT shown with an overlay of an ideal diffraction pattern for the M2 phase. Note that some of the reflections are unaccounted for. (c) An idealized [001] diffraction pattern for the M2 phase. Additional diffraction patterns are superimposed and rotated by (d) 60° and (e) 120°. (f) All three patterns are shown superimposed on the FFT. The FFT pattern and analysis were adapted from Ref. 41.

evidence for the presence of other low-level impurities needed to be incorporated to give us higher confidence in bond distances and angles, as well as occupancies and valences on which mechanistic models are based on.

As outlined above, the most notable difference between the Rietveld and STEM models for M1 were associated with the S4 site. The STEM HAADF analysis appears to indicate mixed Mo/V occupancy similar to that of site S7, whereas the Rietveld model showed full Mo occupancy at S4, but mixed occupancy at S7. In the Rietveld refinement of complex structures, it is necessary to use chemical constraints to limit the number of adjustable parameters. One of the constraints applied in the DeSanto *et al.* models was to lock the S4 molybdenum occupancy at 100% early in the refinement process when this site appeared to contain only molybdenum. The characteristics of the S4 and S7 sites are of special interest, since these have been associated with the complex active center of M1 in proposed mechanistic models.[2,5,42–44] The HAADF

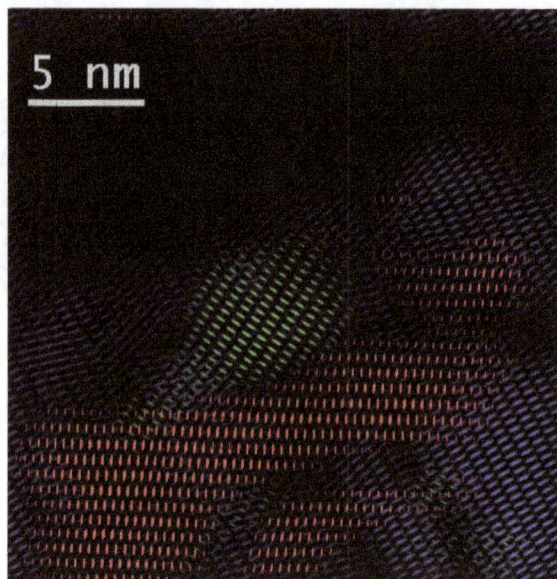

Figure 27. RGB composite image reconstructed from Figure 25 using the inverse FFTs of masked diffractograms. Each of the three twin orientations corresponds to a different color.

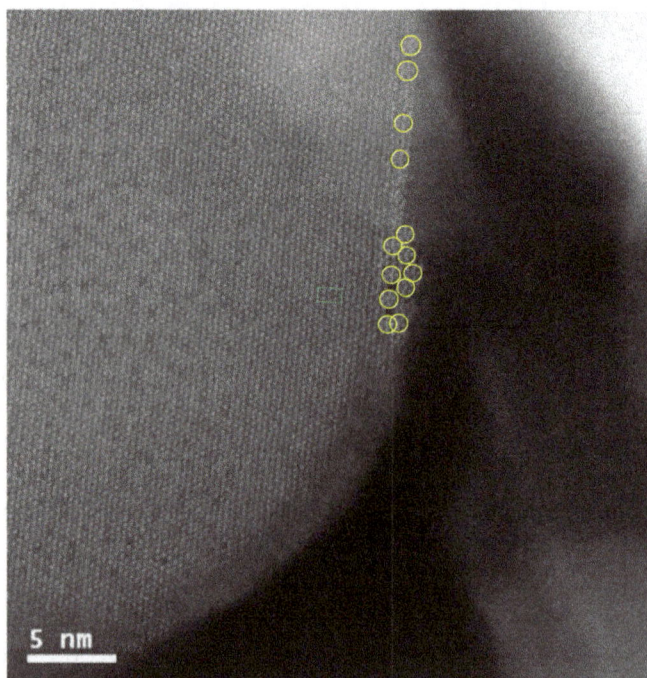

Figure 28. HAADF STEM image of M2 phase in the [001] zone axis orientation. The outline in green shows the orthorhombic M2 unit cell and the yellow circles enclose pentagonal rings indicating the presence of nascent M1-like features as defects near the crystallite edge. These pentagonal rings are not well ordered, but do show beginnings of formation of two heptagonal channels reminiscent of the M1 structure.

STEM contrast results for site S4 suggest mixed Mo,V occupancy, so performing the Rietveld refinement with variable S4 occupancy is expected to further improve the goodness of fit.

A second opportunity to improve the M1 model is to account for a V-substituted Mo_5O_{14}-type impurity phase.[45] Furthermore, another previously unidentified impurity, a Mo-substituted V_2O_5-type structure, was

Figure 29. HAADF STEM image of a crystallite of Mo–V_2O_5 impurity within the M1 specimen used for refinement. The insert shows a simulation of [010] V_2O_5. Note that the distortion in the image is an extrinsic artefact from the STEM rastering.

subsequently also included in an updated Rietveld refinement. Figure 29 shows a STEM HAADF image of this Mo-substituted V_2O_5-type; this was found within the M1 preparation used for the refinement data collection, yet was not readily apparent from the diffraction data as a result of the diffraction peaks of its small unit cell being masked by the much more complex M1 pattern. Additionally, in this new refinement, the sensitivity of the refinement to the groupings of isotropic thermal parameters was explored.

The same synchrotron X-ray and neutron diffraction data sets as used previously were simultaneously refined and provided improved metal coordinates and occupancies, as well as oxygen coordinates.[46–48] The neutron data were weighted by a factor of 10 with respect to the X-ray data due to the substantially weaker intensity of the neutron source and the need to accurately determine oxygen coordinates. Distinguishing vanadium and molybdenum mixed occupancies can also be more confidently done using neutron data due to the larger difference of their coherent cross-sections ($b_V = 7.6(6)$ fm vs $b_{Mo} = 6.715(20)$ fm).

The refinement procedure was as follows: histogram scale factors were refined initially, along with background parameters and the phase fractions. Shifted Chebyshev polynomials with 24 and 16 terms were used for the background fitting in the S-XPD and NPD data sets, respectively. The unit cell parameters for M1 and the impurity phases were refined. While continuing to refine all of the above parameters, the angular zero points for each instrument and the synchrotron X-ray wavelength were refined.[28] Next, the peak shape parameters for the S-XPD and NPD data sets were refined. Peak shape profiles of the S-XPD data were fitted using pseudo-Voigt functions; whereas, simple Gaussian functions were used to fit the NPD peak shapes. The associated peak shape parameters were then locked, and the atomic coordinates of all metal and oxygen sites in M1 refined. The initial model included fractional occupancies of V for sites S1–S11, except for the pentagonal bipyramidal S9 site which is assumed to be fully occupied by Nb.[35] Tellurium was initially located in both hexagonal and heptagonal channel sites and refined with separate fractional occupancies. Thermal parameters were refined as isotropic and in common groupings in order to minimize the number of adjustable parameters in the model. A common isotropic thermal parameter was refined for the Mo and V octahedral sites. A separate isotropic thermal parameter was used for the pentagonal bipyramidal Nb site (S9),

Table 3. Phase fractions from simultaneous S-XPD and NPD refinement.[48]

Phase	Earlier refinement[28]	New refinement
M1	89 wt.%	82 wt.%
Te^0	4 wt.%	5 wt.%
Mo_5O_{14}-type	7 wt.%	7 wt.%
$VMoO_5$-type	na	6 wt.%

another for each of the Te sites (S12 and S13), and three more groupings for the oxygen sites, depending on their polyhedral location. The motivation for selecting three isotropic thermal parameter groupings for oxygen will be addressed later. Fractional occupancies were subsequently refined for each distinguishable metal site, along with the refinement of other instrumental and structural parameters. Lastly, the peak profile parameters were refined once again to further improve the fit to the two data sets.

The phase fractions for M1 and the impurity phases are presented in Table 3; these differ little from the earlier refinement apart from the inclusion of the $VMoO_5$ phase that had not been observed previously.

The results differ somewhat from the earlier refinement of Ref. 28. Now the three impurity phases are present in similar weight percent amounts, and the dominant M1 phase represents about 82 wt.% of the specimen.[48] The presence of four coexisting phases satisfies the Gibbs phase rule. It is also useful to note that the impurity phases have small unit cells and have therefore not been observed in powder diffraction data as their diffraction peaks are masked by the peaks of the large cell M1 phase. Including the additional $VMoO_5$-type impurity significantly improves the refinement statistics as shown in Table 4. The new model is similar to the initial DeSanto one in terms of the general features of the coordination environments of each metal site; however, there is a significant improvement in all metrics of the goodness-of-fit to both the synchrotron and neutron data (see Table 5) which gives us a much higher confidence in the coordinates and occupancies. This will allow a better assessment of the elemental distribution and the associated metal valence states. Much of the improvement is due to the inclusion of the $VMoO_5$-type impurity phase. This phase has also been reported by others in synthetic studies and is particularly prevalent in slurry syntheses involving excess amounts of vanadium.[49–51] This phase has been reported as having a cation-deficient formula $V_{0.95}Mo_{0.97}O_5$ (JCPDS: 77-0649) by Plyasova *et al.*[52,53] However, there appears to be no clear evidence to support cation deficiency in this structure, so we have assumed this material to be stoichiometric $VMoO_5$, while using the coordinates reported by Plyasova *et al.*[52,53] As shown in Table 3, this phase accounts for about 6 wt.% of the sample and was not recognized in any previous refinements. Another change observed in the new model is that the Nb site, S9, shows a thermal parameter that is lower by a factor of 3 compared to the original refinement; this significant drop was anticipated based on the unusually high contrast for this site observed in STEM HAADF images.

Determination of valence states using bond valence analysis

The bond valence sum method was used to obtain oxidation state estimates for each site. This method involved first determining the individual bond valence contributions, v_{ij}, using

$$v_{ij} = \exp\left\{\frac{R_0 - d_{ij}}{B}\right\}$$

Table 4.   Refinement results.[48]

Site	Wyckoff	x	y	z	$U_{iso} \cdot 100$	Occupancy
Mo1	2a	0	0	0.5	1.98(9)	0.699(31)
V1	2a	0	0	0.5	1.98(9)	0.301(31)
Mo2	2b	0	0.5	0.620(9)	1.98(9)	0.421(27)
V2	2b	0	0.5	0.620(9)	1.98(9)	0.579(27)
Mo3	4c	0.1175(7)	0.2284(6)	0.466(7)	1.98(9)	0.573(20)
V3	4c	0.1175(7)	0.2284(6)	0.466(7)	1.98(9)	0.427(20)
Mo4	4c	0.1767(6)	0.4774(5)	0.513(7)	1.98(9)	0.804(22)
V4	4c	0.1767(6)	0.4774(5)	0.513(7)	1.98(9)	0.196(22)
Mo5	4c	0.2119(6)	0.3427(4)	0.627(7)	1.98(9)	0.954(20)
V5	4c	0.2119(6)	0.3427(4)	0.627(7)	1.98(9)	0.046(20)
Mo6	4c	0.2800(5)	0.2118(4)	0.646(7)	1.98(9)	0.883(21)
V6	4c	0.2800(5)	0.2118(4)	0.646(7)	1.98(9)	0.117(21)
Mo7	4c	0.3841(6)	0.1018(5)	0.491(7)	1.98(9)	0.760(20)
V7	4c	0.3841(6)	0.1018(5)	0.491(7)	1.98(9)	0.240(20)
Mo8	4c	0.4594(5)	0.2278(4)	0.635(7)	1.98(9)	1
Nb9	4c	0.3591(5)	0.3177(4)	0.493(6)	0.56(21)	1
Mo10	4c	0.0014(6)	0.1329(4)	0.660(6)	1.98(9)	1
Mo11	4c	0.3427(5)	0.4409(4)	0.642(6)	1.98(9)	0.948(25)
V11	4c	0.3427(5)	0.4409(4)	0.642(6)	1.98(9)	0.052(25)
Te12	4c	0.5498(7)	0.1042(4)	0.544(8)	3.34(51)	0.711(10)
Te13	4c	0.6852(32)	0.4118(28)	0.317(17)	3.34(51)	0.146(7)
O1	2a	0	0	0.018(9)	0.111	1
O2	2b	0	0.5	0.061(10)	0.111	1
O3	4c	0.1192(8)	0.2297(6)	0.032(7)	0.111	1
O4	4c	0.1767(7)	0.4748(5)	0.063(8)	0.111	1
O5	4c	0.2123(7)	0.3385(5)	0.026(8)	0.111	1
O6	4c	0.2830(7)	0.2146(5)	0.085(8)	0.111	1
O7	4c	0.3866(7)	0.1062(5)	0.061(8)	0.111	1
O8	4c	0.4520(7)	0.2279(7)	0.021(7)	0.111	1
O9	4c	0.3561(7)	0.3199(6)	0.058(8)	0.111	1
O10	4c	0.9944(7)	0.1355(5)	0.084(7)	0.111	1
O11	4c	0.3439(7)	0.4398(6)	0.052(9)	0.111	1
O12	4c	0.5506(11)	0.1122(7)	0.022(8)	0.183	0.711(10)
O13	4c	0.5224(7)	0.4300(5)	0.550(9)	0.111	1
O14	4c	0.5743(7)	0.3340(5)	0.553(9)	0.111	1
O15	4c	0.0425(7)	0.2692(5)	0.581(8)	0.111	1
O16	4c	0.5801(8)	0.0354(7)	0.564(9)	1.45(20)	1
O17	4c	0.7015(9)	0.2958(7)	0.630(7)	0.111	1
O18	4c	0.7760(7)	0.2149(5)	0.562(8)	0.111	1
O19	4c	0.6677(7)	0.0956(5)	0.568(9)	0.111	1

(Continued)

Table 4. (*Continued*)

Site	Wyckoff	x	y	z	$U_{iso} \cdot 100$	Occupancy
O20	4c	0.9631(8)	0.4335(7)	0.574(10)	1.45(20)	1
O21	4c	0.8147(6)	0.3514(6)	0.582(7)	0.111	1
O22	4c	0.7984(6)	0.1243(5)	0.542(8)	0.111	1
O23	4c	0.7700(6)	0.0302(6)	0.576(8)	0.111	1
O24	4c	0.8687(7)	0.2559(5)	0.545(9)	0.111	1
O25	4c	0.9065(6)	0.1150(6)	0.565(8)	0.111	1
O26	4c	0.9087(7)	0.0164(6)	0.531(7)	0.111	1
O27	4c	0.8334(8)	0.4552(6)	0.603(7)	1.45(20)	1
O28	4c	0.9457(7)	0.3411(6)	0.565(8)	0.111	1
O29	4c	0.9511(7)	0.1986(6)	0.568(9)	0.111	1
O30	4c	0.1508(7)	0.2995(6)	0.543(8)	0.111	1
O31	4c	0.682(5)	0.433(4)	0.975(27)	0.183	0.146(7)

Notes. Space group: *Pba2* (No. 32); $a = 21.134(1)$ Å, $b = 26.647(1)$ Å, $c = 4.0140(2)$ Å with Z = 4. Formula unit: $\{TeO\}_{0.86(1)} \cdot Mo_{7.48(6)} V_{1.52(6)} NbO_{28}$.

Table 5. Comparison of previous and new refinement statistics.[48]

Parameters	New refinement	Previous refinement[27]
R_{wp} (S-XPD)	5.95%	7.20%
R_p (S-XPD)	4.76%	5.80%
R_{wp} (NPD)	3.92%	5.19%
R_p (NPD)	3.36%	4.46%
χ^2	6.54	10.42

where d_{ij} is the experimentally refined distance between the two atoms, i and j, R_0 is the 'ideal' distance expected between two bonded atoms based on their formal oxidation states, and B is an empirical constant for which the value is most commonly taken to be 0.37.[54–56] For all but one case, a value of $B = 0.37$ was used since it is considered as typical; however, for the case of Mo^{5+}, Brown *et al.* suggested that a value of $B = 0.29$ is more appropriate. This change for Mo^{5+} has little influence on valence results, as compared with the usual choice of $B = 0.37$.

Bond valence sums were obtained by summing over all coordinating neighbors, j, for the ith site as:

$$V_i = \sum_{j=1}^{N} v_{ij}$$

Semiempirical relationships between bond strengths and bond lengths have been shown to provide robust and reliable assessments of the contribution from each neighbor in the coordination environment, although a bond valence sum model does not take into account any covalency in metal–oxygen bonds.[56] For each framework site, we show results for anticipated valence states corresponding to both d^0 and d^1 configurations in Table 6.

The bond valence sum results indicate that all sites in the pentagonal ring perimeter sites (Mo5/V5, Mo6/V6, Mo8, Mo10, and Mo11), as well as the pentagonal biprismatic ring center site, Nb9, have d^0 configurations, i.e. these metals are in the highest oxidation state (+6 for Mo and +5 for V and Nb), independent

Table 6. Bond valence sum and electron configuration results for framework metal sites.

Site type	Site ID	BVS using $R_0(Mo^{6+})/R_0(Mo^{5+})$	BVS using $R_0(V^{5+})/R_0(V^{4+})$	Mo and V e configuration
B	Mo1/V1	5.324/4.578	4.019/3.818	Both d^1
B	Mo2/V2	5.291/4.616	4.104/3.898	Both d^1
L	Mo3/V3	5.495/4.898	4.149/3.941	Mixed d^0/d^1
L	Mo4/V4	5.520/4.887	4.168/3.959	Mixed d^0/d^1
PR	Mo5/V5	6.117/5.809	5.042/4.789	Both d^0
PR	Mo6/V6	6.286/5.882	5.056/4.803	Both d^0
L	Mo7/V7	5.798/5.245	4.111/3.905	Mixed d^0/d^1
PR	Mo8	6.517/6.318	—	d^0
PR	Mo10	6.222/5.826	—	d^0
PR	Mo11/V11	6.142/5.755	5.532/5.504	Both d^0

Site type	Site ID	BVS using $R_0(Nb^{5+})$		e configuration
PB	Nb9	5.432		d^0

Notes: B = Binary axis octahedra; L = Linking Octahedra; PR = Pentagonal Ring Octahedra; PB = Pentagonal Biprism.

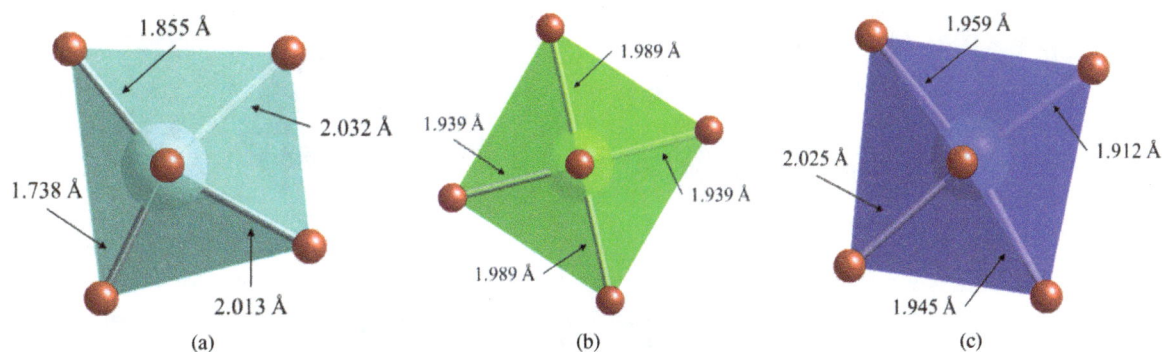

Figure 30. Representative octahedra in [001] projection showing different levels of asymmetry. (a) Mo11, (b) Mo1/V1, and (c) Mo7/V7.

of whether occupancies are mixed or not, although it should be noted that the V substitution is very low in the ring perimeter sites. As corroborating evidence, we observe strong out-of-center distortions, typical of a second-order Jahn–Teller effect expected for d^0 octahedra[57,58] in each case. To illustrate this, Figure 30(a) shows an example of the pronounced out-of-center distortion for site Mo11; this is typical of all the pentagonal ring sites. Similar pentagonal ring units in Mo_5O_{14} and $Mo_{17}O_{47}$ are also found to contain Mo^{6+}, as evidenced by bond valence sums and octahedral distortions.

In contrast, the octahedra residing on the binary axes, sites Mo1/V1 and Mo2/V2, clearly show evidence of d^1 configuration, i.e. these mixed-occupancy sites are comprised of Mo^{5+} and V^{4+} based on both the bond valence sums and the nearly symmetric coordination environments. The Mo1/V1 site geometry is represented in Figure 30(b). Guliants *et al.* used DFT methods to show that the Bader charges for vanadium at the Mo2/V2 site should be +4.[59]

The remaining octahedral sites are those that link together the pentagonal rings, mixed occupancy sites Mo3/V3, Mo4/V4, and Mo7/V7. These site valences are more difficult to interpret with respect to their valence sums and octahedral distortions, suggesting that they represent a mix of d^0 and d^1 configurations.

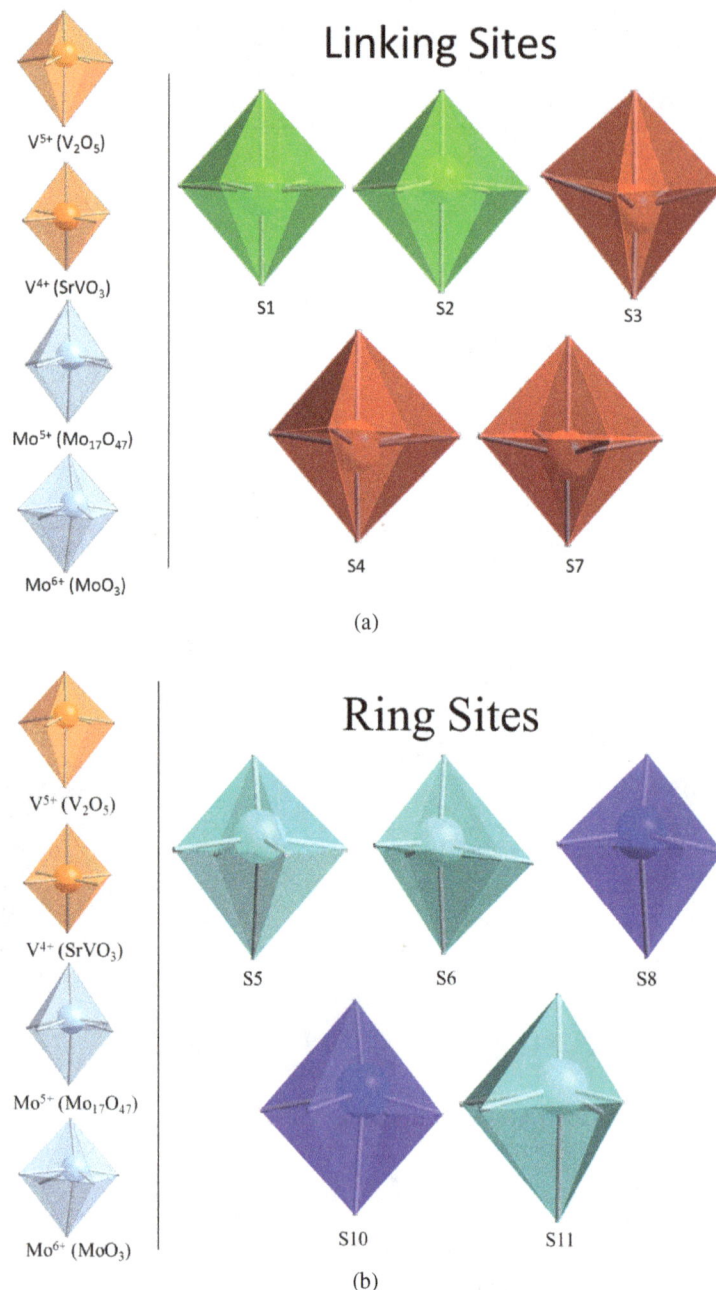

Figure 31.   3D rendering of the average metal octahedral geometries showing different levels of asymmetry compared with model octahedra: (a) the linking sites; (b) the pentagonal ring sites. Model octahedra are adapted from: V_2O_5: Enjalbert *et al.*,[26] SrVO₃: Rey *et al.*,[36] $Mo_{17}O_{47}$: Kihlborg,[37] MoO_3: Kihlborg.[38] Note that the site labelings here use S for site identification, rather than explicitly indicating Mo and/or V.

The geometry of site Mo/V7 is illustrated in Figure 30(c); this site is noticeably less distorted than the Mo11 site shown in Figure 30(a), yet is not as symmetric as in case of then Mo1/V1 site represented in Figure 30(b). The coordination environments for these sites seem to be intermediate between the two limiting configurations. We can also see this by comparing 3D illustrations for the linking sites in Figure 31(a) with those of the pentagonal ring octahedra shown in Figure 31(b). We propose that these linking octahedra, i.e. those not on the binary axes, are of mixed valence. The bond length chart provided in

Figure 32 provides further evidence, where the variation in the bond distances can be compared among the various octahedral sites and can be related to representative model octahedra for molybdenum and vanadium compounds.

Identification of the valence states of each site is important for mechanistic modeling, although it is important to recognize that they represent the bulk valence states rather than those on the reactive surfaces. Overall, the M1 crystallite must satisfy the electroneutrality requirement. If we accept the evidence for valence states for sites 1, 2, 5, 6, and 8–11, and if we take the Te sites to be 4+ with capping oxygens, then we can make inferences concerning the remaining valences for sites 3, 4, and 7. To review what we have: (i) The two binary axis sites, Mo1/V1 and Mo2/V2, have d^1 configurations with occupancies as determined from the Rietveld refinement, (ii) all pentagonal ring perimeter sites and the pentagonal biprismatic center have d^0 configurations with the occupancies as indicated (including Nb9), and (iii) the tellurium is present as Te^{4+} with capping oxygens. We can now determine the total cationic charge associated with the remaining sites to within our experimental uncertainties. Based on the refined occupancies, we have the formula unit $\{TeO\}_{0.86(1)} \cdot Mo_{7.48(6)}V_{1.52(6)}NbO_{28}$. The mixed valence cation sites (sites Mo3/V3, Mo4/V4, and Mo7/V7) must account for an overall charge contribution of +14.95(10) to maintain electroneutrality. This means that the average valence on these sites is +5 to within the random variability of our results. If these three octahedral sites were exclusively d^1, we would find occupancies where the net cationic valence contribution would be +14.14(6); whereas, if these sites were exclusively d^0, the net cationic valence contribution would be +17.14(6). To achieve the electroneutrality, the balance must involve a linear combination where about 73%

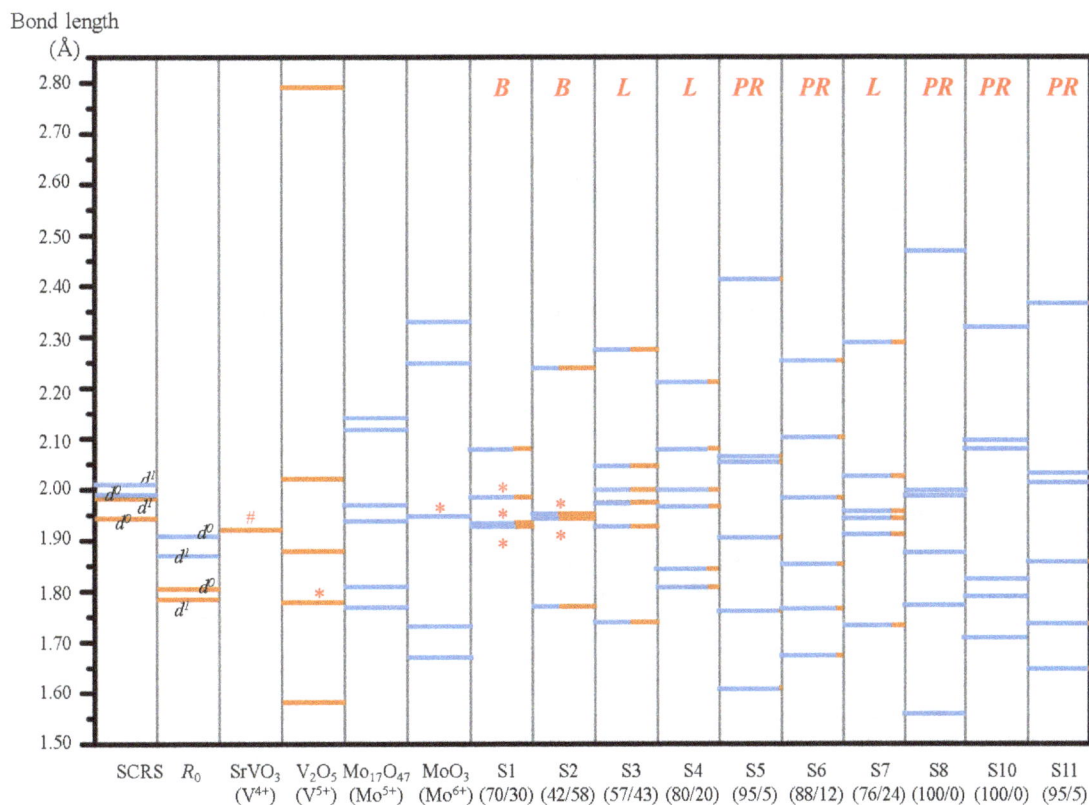

Figure 32. Comparison of bond lengths among the M1 octahedral sites and with representative examples for d^0 and d^1 octahedral molybdenum and vanadium sites in model compounds. Blue bars represent Mo–O bonds while orange bars are V–O bonds. Bars mixed with blue and orange are scaled to the specific site occupancies. The hash notation (#) represents the six identical bond lengths. The asterix notation (*) represents two matching bond lengths. SCRS: Shannon Crystal Radii Sum,[60] R_0: ideal bond length from Brown *et al.*,[54] SrVO3: Rey *et al.*,[61] V2O5: Enjalbert *et al.*,[62] Mo17O47: Kihlborg,[63] MoO3: Kihlborg.[64]

of these sites are d^1 and the rest are d^0 with a random variability of about ±3%. We have yet to determine how the individual valence states are distributed among these three sites to give this net result.

Having this linear combination of cationic valences distributed among the Mo3/V3, Mo4/V4, and/or Mo7/V7 sites would suggest that the geometric distortions on some or all of these sites should be, at least on average, intermediate in magnitude. The extent of octahedral distortion should be consistent with the extent of valence mixing. Since each of these three sites exhibits distortions that are intermediate between the d^0 and d^1 extremes, it would seem that all three are of mixed valence. It is interesting to observe that if we propose that all d^0 species are vanadium (therefore V^{5+}) and all d^1 species are molybdenum (Mo^{5+}), then the refined occupancies indicate that we would have 71.2% d^1 and 28.8% d^0, a combination very closely matching the charge balance (i.e. 73% and 27%). Of course, this is not the only permutation that would achieve electroneutrality, but it does suggest that significant d^0 vanadium may exist in three linking sites. An alternative scenario satisfying electroneutrality would be to propose 27% Mo^{6+} (d^0), 46% Mo^{5+} (d^1), and 27% V^{4+} (d^1); in this case, we have a mix of Mo and V reduced sites with no d^0 vanadium present. It should be noted that the extent of tellurium intercalation can vary such that if a specimen is rendered deficient in tellurium, then the balance of charge must redistribute, and the cationic valences for the three octahedral sites in question would have to exhibit higher d^0 content, if the Mo/V occupancies remain the same.

We can examine how the observed geometries for the three mixed occupancy/mixed valence sites can be explained by considering how these arise from a simple linear combination of typical Mo^{5+} and V^{5+} coordination environments. In other words, by assuming the scenario in which all Mo sites are d^1 and all V sites are d^0, we can see if the observed average geometry is reasonable. In Figure 33, we show a typical d^1 coordination

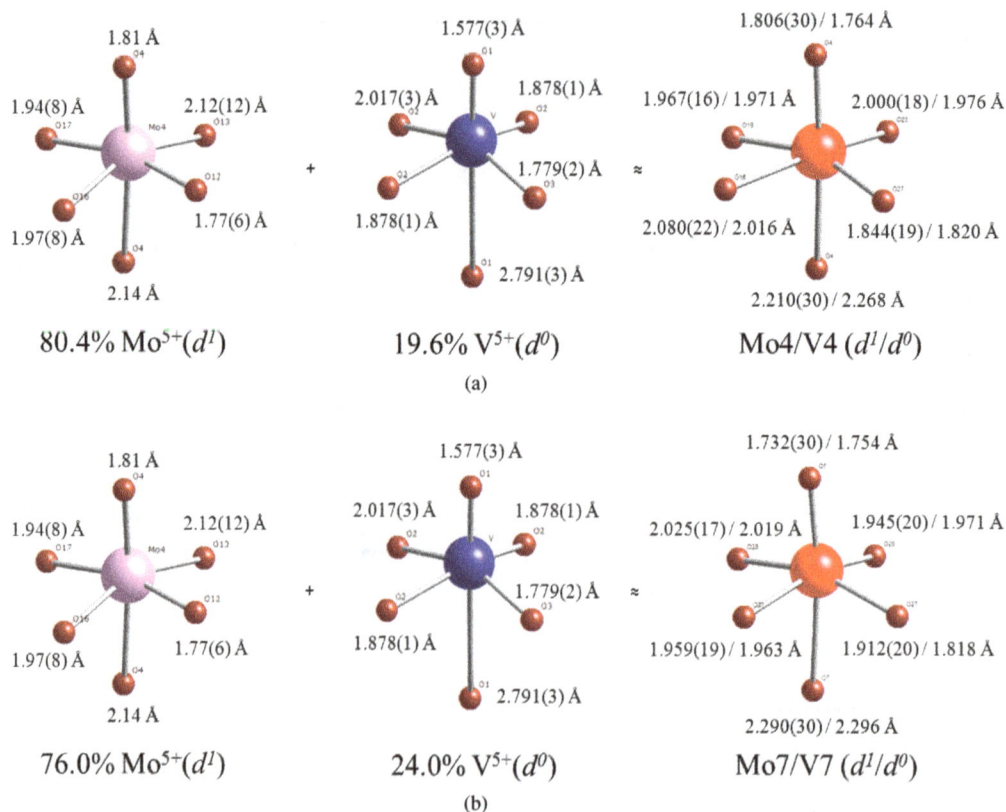

Figure 33. Linear combinations of representative $Mo^{5+}(d^1)$ and $V^{5+}(d^0)$ octahedra are similar to refined coordination environments of (a) Mo4/V4 and (b) Mo7/V7. Average bond distance values for the mixed combinations (right side) in front of the slash are the refined bond lengths, and after the slash are those calculated by linear combination.

environment for molybdenum (showing only minor distortions) and a typical d^0 coordination environment for vanadium (showing significant out-of-center distortion) for the cases of sites Mo4/V4 and Mo7/V7. The representative case for V^{5+} geometry has been selected from the structure of V_2O_5 as reported by Enjalbert et al.[62] For the typical Mo^{5+} case, we have chosen the $Mo_{17}O_{47}$ structure refined by Kihlborg,[63,65] selecting an octahedral linking site that clearly has a bond valence sum very close to +5.0. Linear combinations based on these model cases yield mixed occupancy averages that are quite similar to the refined results. The deviations between refined and calculated bond lengths are generally less than 0.05 Å.

Looking again at the bond valence sums for vanadium in Table 6, we see that the values for vanadium suggest 4+ rather than 5+ states that we proposed for the three mixed configuration sites. It is important to recognize, however, that vanadium is present in mixed occupancy as a minor element in these sites. As a consequence, the average bond distance in these sites is longer than expected for V–O due to the blending with Mo–O, so the BVS suggests a lower valence. If we combine the BVS values for Mo^{5+} and V^{+5}, and weight these linearly by the occupancy, we obtain weighted bond valence sum values of +4.58, +4.75, and +4.97 for sites Mo3/V3, Mo4/V4, and Mo7/V7, respectively. Since bond valence sums are generally considered to be good to within about ±10%, the results are consistent with pentavalent vanadium, and most convincing for Mo4/V4 and Mo7/V7 cases. This argument is somewhat less convincing for the Mo3/V3 site, which may indicate that this site has mixed valences for both Mo and V, in other words, having a more complex mix of Mo^{5+}, Mo^{6+}, V^{4+}, and V^{5+}.

Recently, Lunkenbein et al.[66] reported on using annular bright-field (ABF) STEM imaging to directly measure the octahedral distortions in the related MoVOx catalyst. ABF STEM allowed for the probing of localized metal oxygen distortion, and therefore localized valence states.

High-Temperature and *In Situ* HAADF Measurements

Upon heating to 780 K in vacuum, we observed via HAADF STEM that the pentagonal ring units are stable, but the linking metal octrahedra significantly disordered.[67] Additionally, loss of Te after prolonged exposure to high vacuum at elevated temperature was observed. Aouine et al.[68] performed *in situ* STEM on the MoVTeO phase. They report that at 350°C, under 1 mbar of 30/15/55 $O_2/C_2H_6/N_2$ no significant disordering of the linking octahedral occurs. The Te in the hexagonal channels is reduced under these conditions and displaced closer to the Mo2/V2 site. Additionally, for the specimen they measured, two different populations of Te-containing hexagonal channels were evident in the STEM micrographs. More recently, *in situ* measurements of MoVNbTeO M1 were reported at oxidative dehydrogenation (ODH) conditions.[69] Changes in the HAADF contrast for the Te sites over time led the authors to propose a mechanism where a portion of the intercalated Te–O chains are reduced to Te metal creating oxygen radicals in the process. Other structural changes were observed which depended on the composition of gas in the *in situ* experiment.

Multislice Frozen-Phonon Image Simulation

We performed multislice frozen-phonon image simulations to quantitatively compare the experimental HAADF image contrast with the computed contrast for the improved model structure.[70] The Nb site contrast was larger in the image simulations than experimentally observed, suggesting that the atomic displacement parameter from the new Rietveld refinement is smaller than it should be. The image simulation for the Mo2/V2 site was significantly dimmer than observed experimentally. Similar comparisons between image simulations and experimental HAADF images to deduce quantitative compositional information have been published for related oxide bronzes MoVTeO, MoVTeNbO, and MoVTeTaO.[71–73]

The proposed catalytic mechanism

We will now consider the possible relevance of the valence distribution findings above to the models for the selective (amm)oxidation mechanisms proposed by Grasselli *et al.*[2,5,6,42,44] and Gryzbowska *et al.*[21] The Grasselli mechanism invokes a synergistic interplay of sites Mo2/V2, Mo4/V4, and Mo7/V7 with neighboring Te–O in the hexagonal channel. The exchange between two resonance structures (V^{5+}=O $\leftrightarrow$ $^{4+}V^{\cdot}$–O$^{\cdot}$) through thermal excitation gives the surface terminal oxygen a partial radical character and therefore allows it to easily abstract a methylene–H from an approaching propane molecule to activate it. In the Gryzbowska mechanism, the bridging oxygen in the V^{5+}–O–V^{4+} linkage was proposed to abstract the first hydrogen from paraffin, due to high nucleophilicity.[21] Obviously, for either of these proposed mechanisms, the presence of V^{5+} is critical. Our structural model with the proposed valence distribution provides plausible evidence for having vanadium with +5 valence in the required Mo4/V4 and Mo7/V7 linking sites, and this is consistent with either of the proposed mechanisms. A rendering of the likely active site arrangement within the M1 structure is highlighted in Figure 34. It should again be emphasized that powder diffraction methods provide average bulk structural information, so the refinement results do not necessarily indicate the surface occupancies or valences; however, V^{5+} has been independently reported in surface studies, such as an investigation by Wachs *et al.* using methanol adsorption in temperature-programmed surface reaction (TPSR).[74]

One of the characteristics specified in the 'seven pillars' of heterogeneous selective oxidation catalysis that we have not yet addressed is that of site isolation. The proposed active site is bounded by a set of six M_6O_{21} pentagonal rings that join to form an elongated hexagon pattern, and includes one of the ring perimeter sites, Mo5, in one of these six rings. The cluster of linking sites enclosed within the set six pentagonal rings, and the intercalated Te–O chains are unique among all of the known pentagonal ring frameworks, i.e. this motif is not found in any of the other frameworks in the class that were presented at the beginning of this

Figure 34.   Highlighted rendering of the local environment with the Mo4/V4 and Mo7/V7 sites adjacent to the Te12 site, which is (001) feature believed to be the active site. Color coding is given in Figure 10.

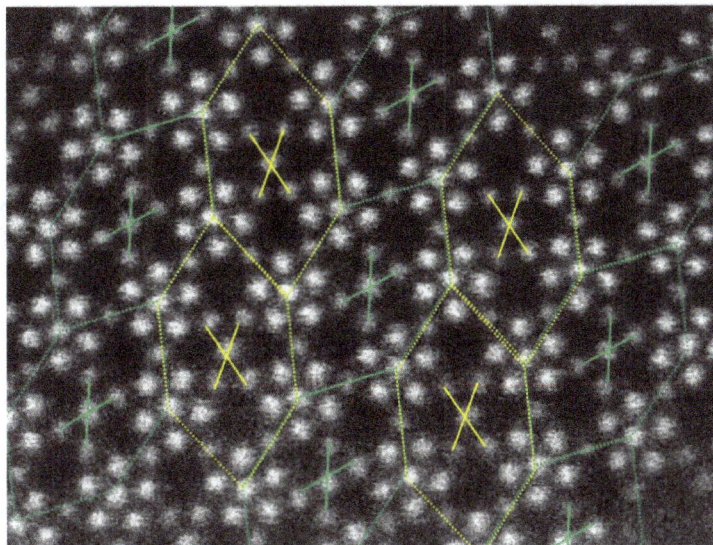

Figure 35. [001] HAADF-STEM image showing site isolation created by the close packing of symmetry-related sets of pentagonal rings enclosing the unique cluster of linking sites (2 × S4, 2 × S7, and S2) as marked with an X. Note that this specimen was prepared without tellurium, so the intercalation sites are vacant. The green and yellow annotated pattern illustrates the glide symmetry in this structure.

chapter. These clusters are separated from one another by the M_6O_{21} pentagonal units and are present in the closest packed arrangement possible. In Figure 35, a HAADF-STEM image of an M1 specimen prepared without the intercalated Te–O chains is annotated to show the periodically repeating sets of six M_6O_{21} pentagonal rings (with symmetry-related repeats marked in green and yellow), enclosing the unique linking site motif filling space in the (001) plane.

Conclusions

Rietveld refinement of both synchrotron X-ray and neutron powder data in combination with HAADF-STEM imaging provides a powerful combination of methods from which a significant structural insight can be obtained. The final Rietveld refinement clearly indicates that vanadium preferentially occupies the linking sites in M1, while the pentagonal ring perimeter sites are mostly molybdenum and the pentagonal biprismatic centers favor niobium. The refined coordinates, bond valence sums evaluated under the constraint of overall electroneutrality provide compelling evidence that the mixed occupancy sites, Mo3/V3, Mo4/V4, and Mo7/V7 have mixed d^0/d^1 configurations and the Mo4/V4 and Mo7/V7 pair seems to be comprised of mixed $Mo^{5+}(d^1)$ and $V^{5+}(d^0)$. We find that these sites show out-of-center moderately weak Jahn–Teller distortions, as we would expect with partial d^0 occupancies. These findings are consistent with suggested propane activation mechanisms.

In the beginning, we recognized the set of 'seven pillars' proposed by Robert K. Grasselli[3] as important attributes characterizing effective multifunctional selective heterogeneous metal oxide oxidation catalysts. These were (i) available lattice oxygen (Mars–van Krevelen mechanism[4]), (ii) intermediate metal–oxygen bond strengths, (iii) stable and suitable host structure, (iv) multiple redox-active species, (v) multifunctional active sites, (vi) isolated active site configurations, and (vii) when required, cooperation with secondary promoter phases. The M1 catalyst does seem to satisfy all of these specifications, except perhaps the last one, which was classified by Grasselli as not necessarily essential.

References

1. Thomas, J. M.; Thomas, W. J., *Principles and Practice of Heterogeneous Catalysis*, VCH, Weinheim, 1997, ISBN 3-527-29239-X.
2. Grasselli, R. K., *Catal. Today* 1999, 49, 141.
3. Grasselli, R. K., *Top. Catal.* 2001, 15, 93.
4. Mars P. van Krevelen, D. W., *Chem. Eng. Sci. Suppl.* 1954, 3, 41.
5. Grasselli, R. K., Burrington, J. D., Buttrey, D. J., DeSanto Jr., P., Lugmair, C. G., Volpe Jr., A. F., Weingand, T., Multifunctionality of Active Centers in (Amm)oxidation Catalysts: from Bi-Mo-O_x to Mo-V-Nb-(Te,Sb)-O_x. *Top. Catal.* 2003, 23, 5.
6. Grasselli, R. K., Buttrey, D. J., DeSanto Jr., P., Burrington, J. D., Lugmair, C. G., Volpe Jr., A. F., Weingand, T., Active Centers in Mo-V-Nb-(Te,Sb)-O_x (Amm)oxidation Catalysts. *Catal. Today* 2004, 91–92, 251.
7. Grasselli, R. K., Buttrey, D. J., Burrington, J. D., Andersson, A., Holmberg, J., Ueda, W., Kubo, J., Lugmair C. G., Volpe, Jr., A. F., Active Centers, Symbiosis and Redox Properties of MoV(Nb,Ta)TeO Ammoxidation Catalysts. *Top. Catal.* 2006, 38, 7.
8. Wöhler, F., *Ann. Physik*, 1824, 2, 350.
9. Haydon, S. K., Jefferson, D. A., *J. Solid State Chem.* 2002, 168, 306.
10. Triantafyllou, S. T., *et al.*, *J. Solid State Chem.* 1997, 134, 344.
11. Jefferson, D. A., Uppal, M. K., Smith, D. J., *J. Solid State Chem.* 1984, 53, 101.
12. Greenblatt, M., Molybdenum Oxide Bronzes with Quasi-Low-Dimensional Properties. *Chem. Rev.* 1988, 88 (1), 31–53; *Low-Dimensional Electronic Properties of Molybdenum Bronzes and Oxides*, Ed. C. Schlenker, Kluwer Academic Publishers, 1989.
13. Ridgley, D., Ward, R., *J. Am. Cheme. Soc.* 1955, 77, 6132.
14. Feger, C. R., Ziebarth, R. P., *Chem. Mater.* 1995, 7, 373.
15. Dobson, M. M., Tilley, R. J., *Acta Crystallogr. B* 1988, 44, 474.
16. Magneli, A., *Arkiv Kemi* 1950, 1, 223.
17. Viswanathan, K., Brandt, K., Salie, E., *J. Solid State Chem.* 1981, 36, 45.
18. Sundberg, M., *J. Solid State Chem.* 1980, 35, 120.
19. Mohammad, A. A., Gillet, M., *Thin Solid Films* 2002, 408, 302.
20. *Polyoxometalate Chemistry — Some Recent Trends*, Ed. F. Sécheresse, World Scientific Series in Nanoscience and Nanotechnology, Vol. 8, World Scientific. 2013. https://www.worldscientific.com/worldscibooks/10.1142/8749
21. Grzybowska-Swierkosz, B., Active centres on vanadia-based catalysts for selective oxidation of hydrocarbons. *Appl. Catal. A* 1997, 157, 409.
22. Nilsson, J., Landa-Canovas, A., Hansen, S., Andersson, A., *Catal. Today* 1997, 33, 97; Olga Guerrero-Perez, M., Pena, M. A., Fierro, J. L. G., Banares, M., A Study about the Propane Ammoxidation to Acrylonitrile with an Alumina-Supported Sb-V-O Catalyst. *Ind. Eng. Chem. Res.* 2006, 45, 4537–4543.
23. Grasselli, R. K., Agaskar, P., Buttrey, D. J., White, B. D., Some Structural and Catalytic Aspects of NASICON-Structured Mixed Metal Phosphates. In *3rd World Congress of Oxidation Catalysis: Studies in Surface Science and Catalysis* 1997, 110, 219–226.
24. https://www.chemicals-technology.com/projects/ptt-asahi-chemical-acrylonitrile-methyl-methacrylate/ (accessed June 18, 2018).
25. Larson, A. C., von Dreele, R. B., LANSCE MS-H805, LANL 2000.
26. Toby, B. H., *J. Appl. Cryst.* 1981, 14, 149.
27. DeSanto, P., Buttrey, D. J., Grasselli, R. K., Lugmair, C. G., Volpe, A. F., Toby, B. H., Vogt, T., Structural Characterization of the Orthorhombic Phase M1 in MoVNbTeO Ammoxidation Catalyst. *Top. Catal.* 2003, 23, 23–38.
28. DeSanto, P., Buttrey, D. J., Grasselli, R. K., Lugmair, C. G., Volpe Jr., A. F., Toby, B. H., Vogt, T., Structural Aspects of the M1 and M2 Phases in MoVNbTeO Propane Ammoxidation Catalysts. *Z. Krist.* 2004, 219, 152.
29. Tate, M. L., Blom, D. A., Avdeev, M., Brand, H. E. A., McIntyre, G. J., Vogt, T., Evans, I. R., New apatite-type oxide ion conductor, $Bi_2La_8[(GeO_4)_6]O_3$: Structure, properties and direct imaging of low-level interstitial oxygen atoms using aberration-corrected scanning transmission electron microscopy. *Adv. Funct. Mater.* 2017, 27, 1605625.
30. Pyrz, W. D., Blom, D. A., Vogt, T., Buttrey, D. J., *Angew. Chem. Intl. Edn.* 2008, 47, 2788.
31. *International Tables for Crystallography*, Vol. A, Ed. T. Hahn, D. Reidel Publishing Company: 1983.
32. Williams, D. B., Carter, C. B., *Transmission electron microscopy: a textbook for materials science*, Plenum Press: New York, 1996.
33. Abe, E., Pennycook, S. J., Tsai, A. P., *Nature* 2003, 421, 347.
34. Varela, M., Lupini, A. R., van Benthem, K., Borisevich, A. Y., Chisholm, M. F., Shibata, N., Abe, E., Pennycook, S. J., *Annu. Rev. Mater. Res.* 2005, 35, 539.

35. DeSanto, P., Buttrey, D. J., Grasselli, R. K., Pyrz, W. D., Lugmair, C. G., Volpe, A. F., Vogt, T., Toby, B. H., *Top. Catal.* 2006, 38, 31.

36. Murayama, H., Vitry, D., Ueda, W., Fuchs, G., Anne, M., Dubois, J. L., *Appl. Catal. A-Gen.* 2007, 318, 137.

37. NIST Center for Neutron Research www.nist.gov/ncnr

38. Woo, J., Sanghavi, U., Vonderheide, A., Guliants, V. V., A study of M1/M2 phase synergy in the MoVTe(Nb,Ta)O catalysts for propane ammoxidation to acrylonitrile. *Appl. Catal. A-Gen.* 2016, 515, 179.

39. Yamazoe, N., Kihlborg, L., *Acta Cryst. B – Str. Sci.* 1975, 31, 1666.

40. Garćia-Gonzalez, E., Nieto, J. M. L., Botella, P., Gonzalez-Calbett, J. M., *Chem. Mater.* 2002, 14, 4416.

41. Blom, D. A., Pyrz, W. D., Vogt, T., Buttrey, D. J., Aberration-corrected STEM investigation of the M2 phase of MoVNbTeO selective oxidation catalyst. *J. Electron Micr.* 2009, 58, 193. (Special Issue on Advanced Electron Microscopy in Materials Physics).

42. Grasselli, R. K., Burrington, J. D., *Handbook of Heterogeneous Catalysis*, 2nd ed. Vol. 7, Eds. Ertl, G., Knözinger, H., Schüth, F., Weitkamp, J., Wiley-vch Verlag GmbH & Co. KgaA: 2008, 3479–3489. Grasselli, R. K., Tenhover, M. A., *Handbook of Heterogeneous Catalysis*, 2nd ed., Vol. 7, Eds. Ertl, G., Knözinger, H., Schüth, F., Weitkamp, J., Wiley-vch Verlag GmbH & Co. KgaA. 2008, 3489–3503.

43. Grasselli, R. K., Buttrey, D. J., DeSanto Jr., P., Burrington, J. D., Lugmaird, C. G., Volpe Jr., A. F., Weingand, T., *Catal. Today* 2004, 91–92, 25.

44. Grasselli, R. K., *Top. Catal.* 2002, 21, 79.

45. Pyrz, W. D., Blom, D. A., Sadakane, M., Kodato, K., Ueda, W., Vogt, T., Buttrey, D. J., *Chem. Mater.* 2010, 22, 2033.

46. McCusker, L. B., Von Dreele, R. B., Cox, D. E., Louër, D., Scardi, P., *J. Appl. Cryst.* 1999, 32, 36.

47. Von Dreele, R. B., *Combined X-ray and Neutron Rietveld Refinement*, The Rietveld Method, Eds. Young, R.A., Oxford University Press: New York, 1993, 227–235.

48. Li, X., Buttrey, D. J., Blom, D. A., Vogt, T., Improvement of the structural model for the M1 phase Mo-V-Nb-Te-O propane (amm)oxidation catalyst, *Top. Catal.* 2011, 54, 614.

49. Baca, M., Pigamo, A., Dubois, J. L., Millet, J-M. M., *Top. Catal.* 23 (2003) 39.

50. Solsona, B., López-Nieto, J. M., Oliver, J. M., Gumbau, J. P., *Catal. Today* 2004, 91–92, 247.

51. Sanfiz, A. C., Hansen, T. W., Teschner, D., Schnörch, P., Girgsdies, F., Trunschke, A., Schlögl, R., Looi, M. H., Abd Hamid, S. B., *J. Phys. Chem. C* 2010, 114, 1912.

52. Plyasova, L. M., Solov'eva, L. P., Tsybulya, S. V., Kryukova, G. N., Zabolotnyi, V. A., Olen'kova, I. P., *J. Struc. Chem.* 1991, 32, 110.

53. Molchanov, V. V., Plyasova, L. M., Goidin, V. V., Lapina, O. B., Zaikovskii, V. I., *Inorg. Mater.* 1995, 31, 1121.

54. Brown, I. D., Altermatt, D., *Acta Cryst B.* 1985, 41, 244.

55. Brese, N. E., O'Keeffe, M., *Acta Cryst B.* 1991, 47, 192.

56. Brown, I. D., *The Bond-Valence Method: An Empirical Approach to Chemical Structure and Bonding, Structure and Bonding in Crystals II.* Eds. O'Keeffe, M., Navrotsky, A., Academic Press: 1981, 1–30.

57. Martin, K., Brown, I. D., *J. Solid State Chem.* 1995, 115, 395.

58. Pearson, R. G., *Proc. Nat. Acad. Sci.* 1975, 72, 2104.

59. Govindasamy, A., Muthukumar, K., Yu, J., Xu, Y., Guliants, V. V., *J. Phys. Chem. C* 2010, 114, 4544.

60. Shannon, R. D., *Acta Cryst A* 1976, 32, 751.

61. Rey, M. J., Dehaudt, P. H., Joubert, J. C., Lambert-Andron, B., Cyrot, M., Cyrot-Lackmann, F., *J. Solid State Chem.* 1990, 86, 101.

62. Enjalbert, R., Galy, J., *Acta Cryst C* 1986, 42, 1467.

63. Kihlborg, L., *Acta Chem. Scand.* 1960, 14, 1612.

64. Kihlborg, L., *Ark. Kemi.* 1963, 21, 257.

65. Kihlborg, L., *Acta Chem. Scand.* 1963, 17, 1485.

66. Lunkenbein, T., Girgsdies, F., Wernbacher, A., Noack, J., Auffermann, G., Yasuhara, A., Klein-Hoffmann, A., Ueda, W., Eichelbaum, M., Schlogl, Trunsehke, A. Schlogl, R., Willinger, M. G., *Angew. Chem. Int. Ed.* 2015, 54(23), 6838–6831.

67. Blom, D. A,. Vogt, T., Allard, L. F., Vogt, T., Buttrey, D. J., *Top. Catal.* 2014, 57, 1138–1144.

68. Aouine, M., Epicier, T., Millet, J-M. M., *ACS Catal.* 2016, 6, 4775.

69. Zhu, Y., Sushko, P. V., Melzer, D., Jensen, E., Kovarik, L., Ophus, C., Sanchez-Sanchez, M., Lercher, J. A., Browning, N. D., *J. Am. Chem. Soc.* 2017, 139, 12,342.

70. Blom, D. A., Li, X., Mitra, S., Vogt, T., Buttrey, D. J., STEM HAADF Image Simulation of the Orthorhombic M1 Phase in the Mo-V-Te-Nb-O Propane Oxidation Catalyst. *ChemCatChem* 2011, 3, 1028.

71. Epicier, T., Aouine, M., Nguyen, T. T., Millet, J-M M., *ChemCatChem* 2017, 9, 3526.
72. Woo, J., Borisevich, A., Kock, C., Guliants, V. V., *ChemCatChem* 2015, 7, 3731.
73. Woo, J., Guliants, V. V., *Appl. Catal. A* 2016, 512, 27.
74. Wachs, I. E., Jehng, J., Ueda, W., *J. Phys. Chem. B* 2009, 109, 2275.

7

New Crystalline Complex Metal Oxide Catalysts with Porous, Acidic, and Redox Properties

Satoshi Ishikawa*, Zhenxin Zhang*, Toru Murayama[†], Masahiro Sadakane[‡] and Wataru Ueda*,[§]

*Department of Material and Life Chemistry, Faculty of Engineering, Kanagawa University,
Rokkakubashi, Kanagawa-ku, Yokohama-shi, Kanagawa 221-8686, Japan
[†]Department of Applied Chemistry, Graduate School of Urban Environmental Sciences,
Tokyo Metropolitan University, 1-1 Minami-osawa, Hachioji, Tokyo 192-0397, Japan
[‡]Department of Applied Chemistry, Graduate School of Engineering, Hiroshima University,
1-4-1 Kagamiyama, Higashi Hiroshima 739-8527, Japan
[§]uedaw@kanagawa-u.ac.jp

Introduction

Solid-state catalysts have been utilized not only in industrial chemical processes,[1] but also in various new areas like environmental processes, new energy generation, chemical production based on biomass and natural gases, and so forth. In addition, the solid catalysis methodology is expected to expand to other fields, for example medicine synthesis, nuclear power generation, cosmo system, and marine area. This trend is clearly based on the fact that solid catalysis is principally a phenomenon with extremely low or practically no impact on nature. In spite of that, catalysis is one of the key technologies for developing a sustainable society. There are still a lot of problems in developing new solid catalysts and their application technology. One of the main issues in this respect is the lack of widely applicable synthetic methodology of solid catalysts in rational ways. Although there are many examples of catalytic materials, catalytic reactions, and catalytic systems that have been developed from tremendous elaborative researches, the technological knowledge accumulated during the developments are not applicable in a straightforward to developing new other catalysts. This is simply due to the fact that a new catalytic reaction needs new catalytic functions based on new catalytic materials. One could imagine a new catalytic system, where the system would usually consist of variety of constitutional elements in a complex arrangement, so that synthetic realization in a desired direction is extremely difficult.

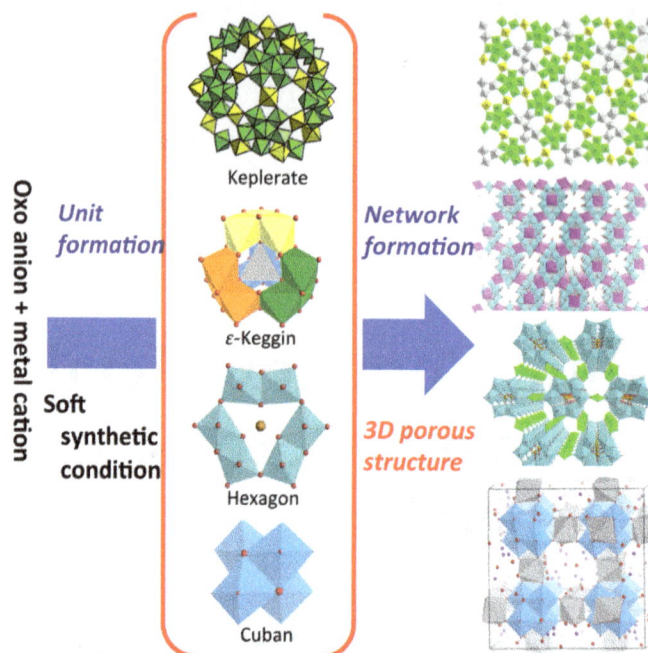

Figure 1.   Four new types of crystalline complex metal oxide catalysts with 3D structure.

One often suggests, therefore, that analytical understanding of the solid catalyst complexity will help to create a new rational synthetic protocol of solid catalysts. Although this approach is absolutely important, we are still on the way and far from realization because there are always many unpredictable and uncontrollable factors in catalytic material synthesis, even in analyses-driven directions. Instead, pure inorganic synthetic approach that can control elemental positions and introduce 3D structure in solid crystals seems more realistic if we take into account that solid catalysts work by collaborative action of constituent elements in atomic scale. It will be more desirable if elemental complexity and structural high dimensionality are introduced in solid catalyst materials with inorganic synthetic methods. In this context, the most important criterion is that the synthesized material has to be catalytic. There are many examples of inorganic solid materials in 3D structure and with tuned elemental positions, but most of them cannot be catalytic materials due to the lack of design methodology for attaining a highly efficient interaction field for the reactant molecule and solid catalysts. The introduction of 3D pore structure in solid catalysts is most desirable for this purpose. This approach is not totally new, but many of these achieved materials are thermally unstable and so not well applicable as catalysts.[2]

Here in this chapter, four new members of solid complex metal oxide catalysts based on group 5 and 6 elements are introduced, all of which were created using the unit synthesis protocol and have extremely high catalytic performance based on the 3D porous structure formed by the unit-network arrangement of metal octahedra as summarized in Figure 1.

Crystalline Microporous $Mo_3VO_{11.2}$ Oxides

Structure and unit assembly

Crystalline orthorhombic and trigonal $Mo_3VO_{11.2}$ (Orth-MoVO and Tri-MoVO, respectively) are comprised of network arrangement based on pentagonal $\{Mo_6O_{21}\}^{6-}$ unit and $\{MO_6\}$ octahedral, and forms a highly organized crystalline structure containing a hexagonal channel and a heptagonal channel in the crystal structure.[3-5] Among the channels, the heptagonal channel works as a micropore capable of adsorbing small molecules including N_2, CO_2, CH_4, C_2H_6, etc.[6-8] Since these catalysts show outstanding catalytic activity for selective oxidations for ethane and acrolein, many efforts have been devoted for clarifying the catalytically active site over these catalysts.[9-13]

Crystalline microporous $Mo_3VO_{11.2}$ is formed by hydrothermal synthesis using a precursor solution containing polyoxomolybdate, $[Mo_{72}V_{30}O_{282}(H_2O)_{56}(SO_4)_{12}]^{36-}$ ($\{Mo_{72}V_{30}\}$), which is comprised of 12 pentagonal $\{Mo_6O_{21}\}^{6-}$ units and 30 $[V=O]^{2+}$ units.[5,14] Crystal formation of Mo_3VO_x materials is considered to proceed via the structural unit assembly based on the pentagonal $\{Mo_6O_{21}\}^{6-}$ unit supplied from $\{Mo_{72}V_{30}\}$ and the single octahedral unit with occluding counter cation inside the heptagonal channel under hydrothermal condition to form a highly organized crystal structure. The structural unit assembly process is susceptible to the pH of the precursor solution, and in fact, Orth-MoVO and Tri-MoVO can be separately obtained by changing the pH of the precursor solution.

Besides Orth-MoVO and Tri-MoVO, two other crystalline Mo_3VO_x materials, tetragonal (Tet-MoVO) and amorphous (Amor-MoVO),[15] have been reported to date. Tet-MoVO is formed by an appropriate heat treatment of Orth-MoVO. Amor-MoVO is obtained by hydrothermal synthesis using the highly concentrated precursor solution of $\{Mo_{72}V_{30}\}$. The structural models and the HAADF-STEM images of these four materials are shown in Figure 2.[16,17]

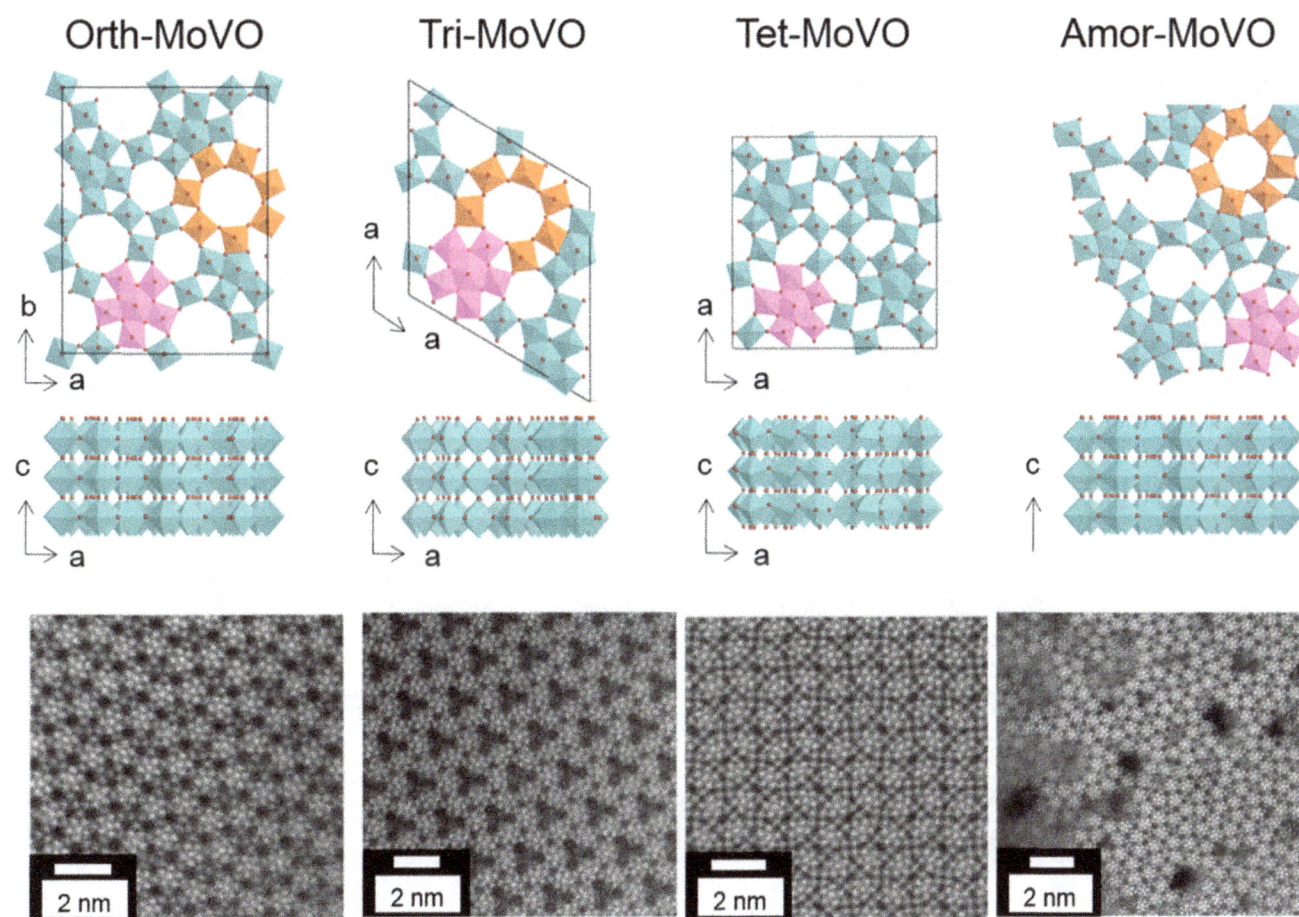

Figure 2. Structure images and HAADF-STEM images of orthorhombic, trigonal, tetragonal, and amorphous $Mo_3VO_{11.2}$. Pink, pentagonal $\{Mo_6O_{21}\}^{6-}$ unit; orange, heptagonal channel. White spots in HAADF-STEM images represent elements and black spots represent void spaces. These images were taken on a JEOL-2100F at the University of South Carolina by Douglas A. Blom.

All these materials are rod-shaped crystals derived from the stacking of the octahedral units and have the same elemental composition ($Mo_3VO_{11.2}$). The cross-section of the rods is constituted by the pentagonal $\{Mo_6O_{21}\}^{6-}$ unit and the $\{MO_6\}$ (M = Mo, V) octahedral, while the structural arrangement was different depending on the crystal phases. The arrangement of the pentagonal $\{Mo_6O_{21}\}^{6-}$ unit and the $\{MoO_6\}$ octahedral forms a hexagonal channel and a heptagonal channel in the crystal structure of Orth-MoVO, Tri-MoVO, and Amor-MoVO, while Tet-MoVO has no heptagonal channel in the crystal structure.

Catalysis field for selective oxidation

The selective oxidations of ethane and acrolein over these four catalysts were carried out, and the obtained results are listed in Table 1.[9–13,18–21] Orth-MoVO, Tri-MoVO, and Amor-MoVO showed the catalytic activity for these reactions. Tet-MoVO was completely inactive for these reactions. Since the difference between Tet-MoVO and the other three catalysts is only the structural arrangement in the cross-section of the rod-shaped crystals, the crystal structure determines the catalytic activity. It is clear during the consideration of the crystal structure of the four catalysts that the heptagonal channel in the crystal structure is solely responsible for the catalytic activity of the reactions.

Orth-MoVO catalysts with the same microporosity but with a different external surface area due to different particle sizes were tested for the selective oxidation of ethane and acrolein in order to clarify the roles of the heptagonal channel in the catalysis. If the catalytic activity is correlated with the external surface area, it can be concluded that the reaction takes place at the heptagonal channel over or near the catalyst surface. On the other hand, if the catalytic activity is observed independent of the external surface area, the result means that the reaction takes place inside the heptagonal channel. Figure 3 shows the ethane or acrolein conversion as a function of the external surface area of Orth-MoVO catalysts. The selectivity to the main product (ethane oxidation, ethene; acrolein oxidation, acrylic acid) for both the reactions was over 90% for all the catalysts. Surprisingly, the ethane conversion was changed only slightly despite the significant change of the external surface area, indicating that ethane is converted inside the heptagonal channel. On the other hand, the acrolein conversion was well correlated with the external surface area, indicating that acrolein is converted at the heptagonal channel over the catalyst surface. This observation should be natural because the size of acrolein is much larger than the size of the pore (0.4 nm), while that of ethane is similar, allowing it to enter into easily. What is striking is that all parts of the channel can be utilized for the catalytic ethane

Table 1. Relationship between microporosity and catalytic activity.

Catalyst	Elemental composition[a] (V/Mo)	Number of 7-membered ring within 100 nm^2	External surface area[b] (m^2 g^{-1})	Micropore volume[b] (10^{-3} cm^3 g^{-1})	Ethane conv.[c]/%	ACR conv.[d]/%
Orth-MoVO	0.38	73	7.2	14.0	42.2	53.8
Tri-MoVO	0.32	68	18.0	4.0	25.2	99.8
Tet-MoVO	0.38	0	2.7	0	<1	<1
Amor-MoVO	0.38	10–30	5.7	2.8	5.9	9.7

[a] Determined by ICP.

[b] Measured by N_2 adsorption at liquid N_2 temperature and estimated by t-plot method.

[c] Reaction condition: catalyst amount, 0.5 g; reaction gas feed, $C_2H_6/O_2/N_2$ = 5/5/40 mL min^{-1}; reaction temperature, 313~319°C.

[d] Reaction condition: catalyst amount, 0.25 g; reaction gas feed, $ACR/O_2/H_2O/N_2/He$ = 2.5/8.0/27.1/39.5/30.5 mL min^{-1}; reaction temperature, 217~218°C. ACR represents acrolein.

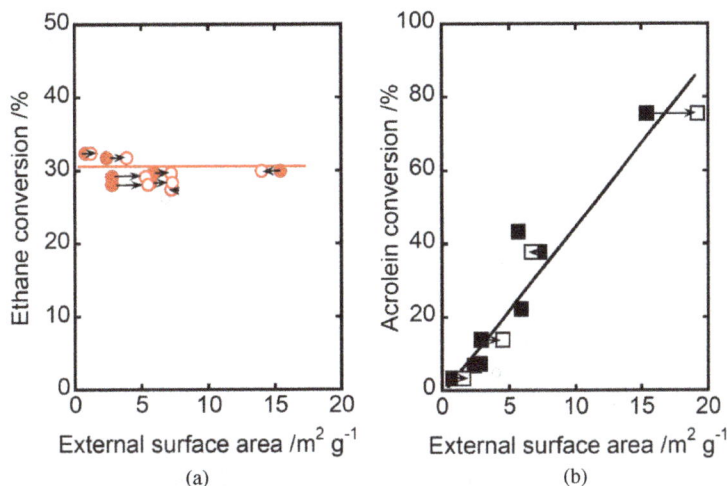

Figure 3. Conversion changes as a function of external surface area in the selective oxidation of (a) ethane and (b) acrolein. Solid symbols, surface area before the reaction; open symbols, surface area after the reaction.

oxidation to ethene, which might explain why Orth-MoVO catalysts are superior to other related catalysts which have poor microporosity.

Local structure around the micropore for selective oxidation

Based on the above-mentioned results, the change of the heptagonal channel size is expected to influence the catalytic performance of ethane oxidation. Since it has already been reported that the size of the heptagonal channel in Orth-MoVO can be continuously and reversibly controlled by appropriate redox treatments,[7,22] the redox treatments of Orth-MoVO may cause local structural changes around the heptagonal channel micropore as well. In fact, the bridging oxygen in the pentamer facing the heptagonal channel in the crystal structure of Orth-MoVO was found to be preferentially removed (MoVO (4.2) in Figure 4) in the early stage of the reduction. The removal of this lattice oxygen then resulted in the expansion of the size of the heptagonal channel, and the resulting heptagonal channel was capable of adsorbing molecules slightly larger than the heptagonal channel (e.g. propane). Then the catalytic activity for the selective oxidation of ethane was drastically increased. Further reduction caused the expansion of the pentagonal $\{Mo_6O_{21}\}^{6-}$ unit which resulted in the shrinkage of the heptagonal channel size (MoVO (6.8) in Figure 4) and at the same time in the disappearance of the microporosity. This MoVO (6.8) catalyst became almost inactive for ethane conversion. This fact again indicates that ethane is converted inside the heptagonal channel. The reoxidation of reduced MoVO restored the size of the pentagonal $\{Mo_6O_{21}\}^{6-}$ unit which again expanded the heptagonal channel size (MoVO (6.8)-AC in Figure 4). Interestingly, the oxygen removed in the early stage of the reduction hardly returns to the structure by this reoxidation. Consequently, the crystal structure of MoVO (6.8)-AC was quite similar to MoVO (4.2), which resulted in the same level of ethane conversion to MoVO (4.2).

Based on the above results, the following reaction scheme in ethane oxidation (Figure 5) is deduced: (1) both ethane and molecular oxygen are captured in the heptagonal channel; (2) molecular oxygen is activated at the oxygen defect site facing the heptagonal channel; (3) activated molecular oxygen is attached to ethane and oxidatively dehydrogenate ethane efficiently to ethene and water, leaving the oxygen defect which is the original state. During this catalytic cycle, the structural framework formed with six $\{Mo_6O_{21}\}^{6-}$ pentagonal units plays a role to support the pentamer unit inside the framework and to maintain the structure under the dynamic redox

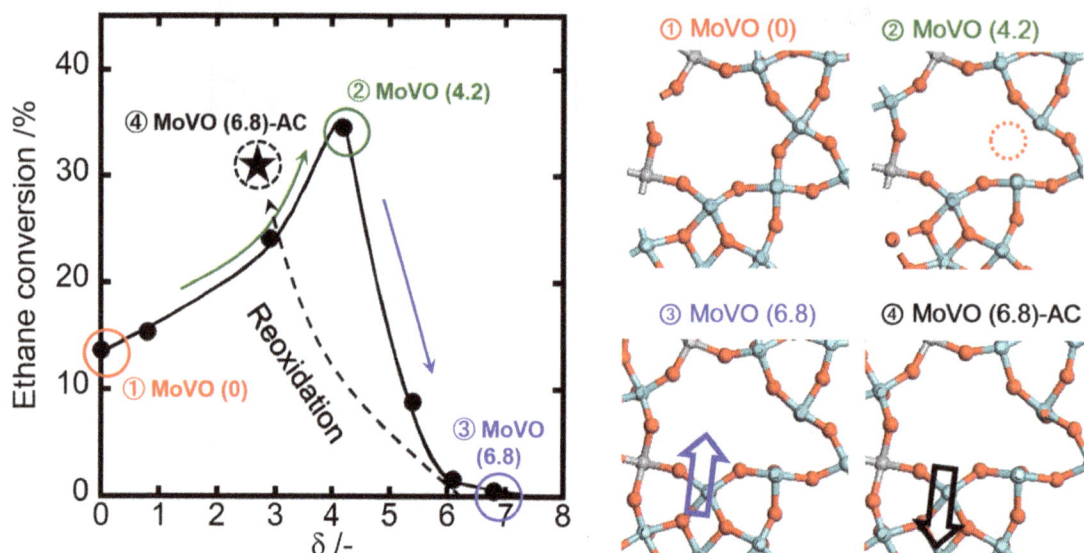

Figure 4.   Ethane conversion change vs the local structure changes of Orth-MoVO. Orth-MoVO after the appropriate reductions is abbreviated as MoVO (δ), where d indicates the number of lattice oxygen removed from the unit cell of Orth-MoVO ($Mo_{29}V_{11}O_{112-\delta}$). MoVO (0) indicates the Orth-MoVO after the air calcination of as-prepared Orth-MoVO at 400°C for 2 h. MoVO (d)-AC means the catalyst prepared by calcining MoVO (δ) under air atmosphere at 400°C for 2 h.

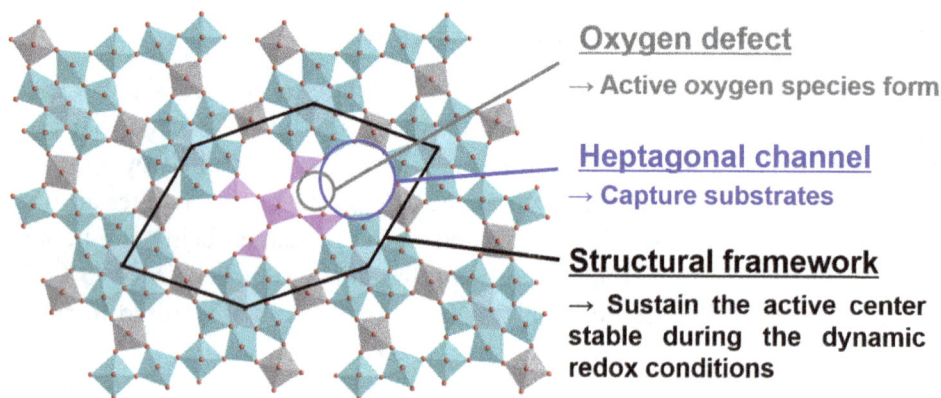

Figure 5.   Role of local crystal structure of Orth-MoVO for the selective oxidation of ethane. Pentamer unit is displayed in purple.

phenomenon (Figure 5). Thanks to the collaboration of each structural part, highly organized crystal structure of Orth-MoVO realizes an outstanding catalytic activity for the selective oxidation of ethane.

Other example similar to microporous Mo_3VO_x oxides

As mentioned above, the creation of a highly organized local crystal structure is crucial for selective oxidations. This fact can also be seen in other selective oxidations. Highly organized W–V–O (HDS-WVO) is one of the examples.[23,24] HDS-WVO is synthesized by the hydrothermal method. The structure of HDS-WVO comprises slabs constituted with the pentagonal $\{W_6O_{21}\}^{6-}$ unit and $\{MO_6\}$ octahedral (M = W, V), and the slabs are stacked together to form a rod-shaped crystal (Figure 6(a)). HDS-WVO showed outstanding catalytic activity for the ammoxidation of 3-picoline (PIC) to 3-cyanopuridine (CP), and the activity was clearly superior to the other W- and V-based catalysts (e.g. tungsten oxide-supported vanadia (VO_x/WO_3), WO_3, V_2O_5)(Figure 6(b)).

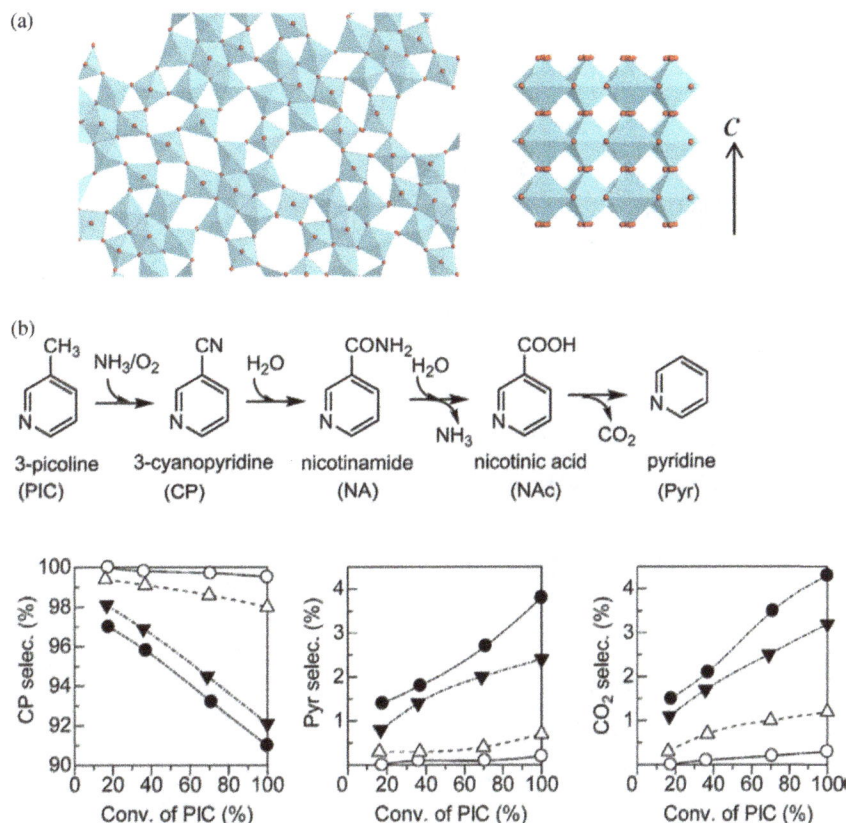

Figure 6. (a) Structural model of HDS-WVO. (b) Selectivity vs PIC conversion for ammoxidation of PIC over (○) HDS-WVO, (△) VO$_x$/WO$_3$, (▲) VO$_x$/TiO$_2$, and (●) V$_2$O$_5$. Reaction conditions: catalyst amount, 0.5 g; reaction temperature, 380°C; gas composition, PIC/H$_2$O/NH$_3$/O$_2$/He = 1/8/6/4.4/19.6 mL min^{-1}.

Epsilon-Keggin-Based Mixed Metal Oxide

Structure of Mo–V–Bi-based oxide

The above-mentioned orthorhombic Mo–V-based oxides have been synthesized by the hydrothermal reaction of Mo and V sources. In the course of our research to introduce an additional element into the orthorhombic Mo–V-based oxide, we found that a new oxide is formed by the addition of Bi sources (Figure 7). This was the first step to find the epsilon-Keggin-based mixed metal oxides (Mo–V–Bi oxide).[25,26]

Hydrothermal reaction of an aqueous solution containing $(NH_4)_6Mo_7O_{24}$, $VOSO_4$, and several Bi sources produces a new crystalline product with cubic crystal system (crystal system: cubic, space group: $Fd3m$, a = ca. 19.66 Å) (Table 2). This oxide has an epsilon-Keggin-type vanadomolybdate, $[\varepsilon\text{-}Mo_{9.4}V_{3.6}O_{40}]^{9.7-}$ where one V^{5+}–O_4 tetrahedron is surrounded 12 M–O$_6$ octahedra with mixvalent Mo and V (M = Mo^{6+}, Mo^{5+},V^{5+} and V^{4+}), which is connected by Bi^{3+} to form a 3D framework structure, $[\varepsilon\text{-}Mo_{9.4}V_{3.6}O_{40}Bi_2]^{3.7-}$ (Figure 8). The epsilon-Keggin-type compound has four planes in a tetrahedral fashion where three oxygens can bind to Bi^{3+}. Bi^{3+} binds two epsilon-Keggin-type vanadomolybate (Figure 8(c)) by binding Bi^{3+} cations with three oxygens of one epsilon-Keggin-type vanadomolybdate on one side and three oxygens of the next vanadomolybdate on the other side. Because Bi^{3+} binders exist in a tetrahedral fashion, the formed 3D framework has a diamond-like structure. The negative charge is compensated by ammonium cations and protons, and the formula is $(NH_4)_{2.8}H_{0.9}[\varepsilon\text{-}Mo_{9.4}V_{3.6}O_{40}Bi_2]$.

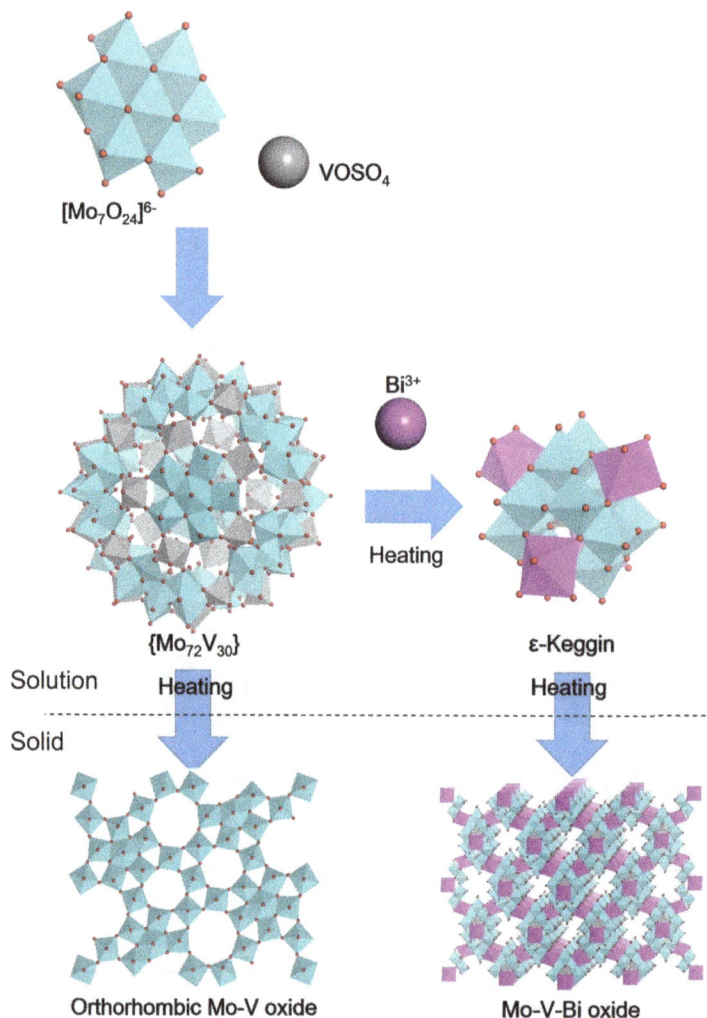

Figure 7.   Synthesis of orthorhombic Mo–V oxide and Mo–V–Bi oxide.

Table 2.   Epsilon-Keggin-based mixed metal oxides and the starting compounds.

	Mo source	Reductant	Other cation source
Mo-V-Bi oxide	$(NH_4)_6Mo_7O_{24}$	$VOSO_4$	$Bi(OH)_3$ Bi_2O_3 $BiOCl$ $Bi_2(SO_4)_3$ $Bi(NO_3)_3$
Mo-Zn oxide	$(NH_4)_6Mo_7O_{24}$ Na_2MoO_4	Mo metal	$ZnCl_2$
Mo-Mn oxide	$(NH_4)_6Mo_7O_{24}$ Na_2MoO_4	Mo metal	MnO
Mo-Fe oxide	$(NH_4)_6Mo_7O_{24}$	Mo metal	Fe_3O_4
Mo-Co oxide	Na_2MoO_4	Mo metal	$CoCl_2$ $CoSO_4$

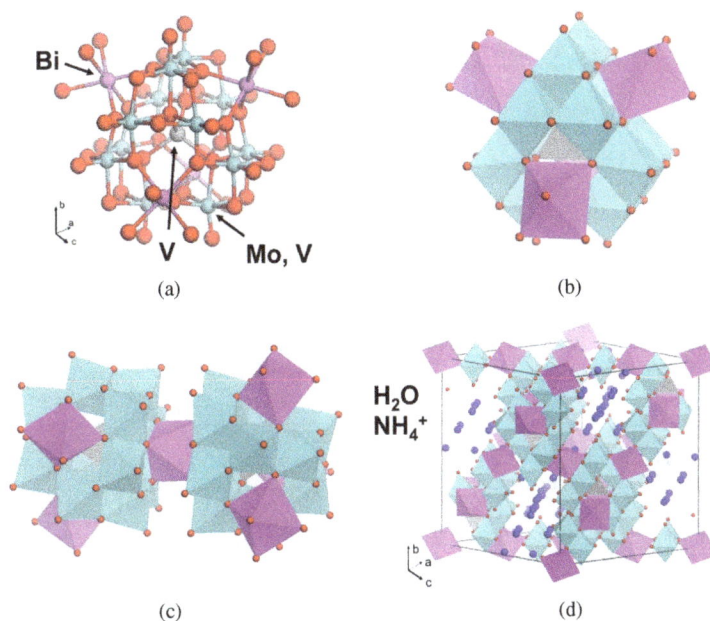

Figure 8.　Structure of Mo–V–Bi oxide. (a) ball-and-stick and (b) polyhedral representation of epsilon-Keggin-type vanadomolybdate with four Bi^{3+}. (c) Polyhedral representation of two epsilon-Keggin-type vanadomolybdate connected by Bi^{3+}. (d) Polyhedral representation of framework of Mo–V–Bi oxide.

Introduction of other elements

Several Bi sources such as $Bi(OH)_3$, Bi_2O_3, $BiOCl$, $Bi_2(SO_4)_3$, and $Bi(NO_3)_3$ can be used to produce the Mo–V–Bi oxide. The Mo in the Mo–V–Bi oxide is partially reduced, and the reducing reagent is $V^{4+}OSO_4$. Vanadyl sulfate is not only the source of V but also a reducing reagent. In order to obtain the epsilon-Keggin-based oxide with other elements, we found that Mo metal is a suitable reducing reagent (Table 2). By combining the Mo source, such as $(NH_4)_6Mo_7O_{24}$ or Na_2MoO_4, Mo metal, and other metal salts, we can prepare epsilon-Keggin-based Mo–Zn oxide,[27,28] Mo–Mn oxide,[27,28] Mo–Fe oxide,[29] and Mo–Co oxide.[30] The newly added Zn, Co, Fe, and Mn are incorporated both in the central M–O_4 tetrahedron in the epsilon-Keggin structure and in the linking site which connect two epsilon-Keggin units. The central M–O_4 tetrahedrons are surrounded by 12 Mo–O_6 octahdedra, and therefore the formula is $[MMo_{12}O_{40}M_2]^{n-}$ (M = Zn, Co, Fe, or Mn). The valence of these additional metals is 2+, and some of the 12 Mo are reduced to Mo^{5+}. The negative charges are compensated by the counter cation such as proton, NH_4^+, or Na^+.

Microporosity

TG-DTA and TPD (temperature-programed desorption)-MS analysis of the generated gases indicate that the ammonium cations and water occupying the micropore can be removed by thermal treatments (Figure 9). The Mo–V–Bi oxide is stable at a temperature of 623 K, and the 3D micropore in the diamond-like framework structure can be opened by heating at 623 K (Figure 10). The N_2 isotherm at liquid N_2 temperature shows sudden N_2 uptake at very low relative pressure indicative of the existence of the micropore (Figure 10(b)). The micropore is so large that CO_2, methane, and ethane can enter.

The counter cation, NH_4^+, of Mo–V–Bi oxide can be exchanged with other cations, such as H^+, Li^+, Na^+, K^+, Rb^+, and Cs^+. K^+ cation blocks micropore, and the N_2 isotherm of the Mo–V–Bi oxide with K^+ does not show any N_2 uptake at low relative pressure (Figure 10(b), red).

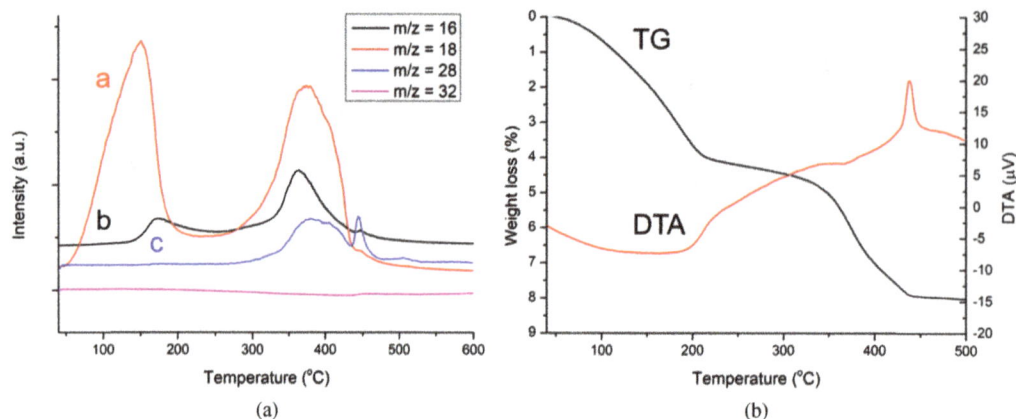

Figure 9.   (A) MS and (B) TG-DTA of Mo–V–Bi oxide. (a) H_2O, (b) NH_3, and (c) N_2.

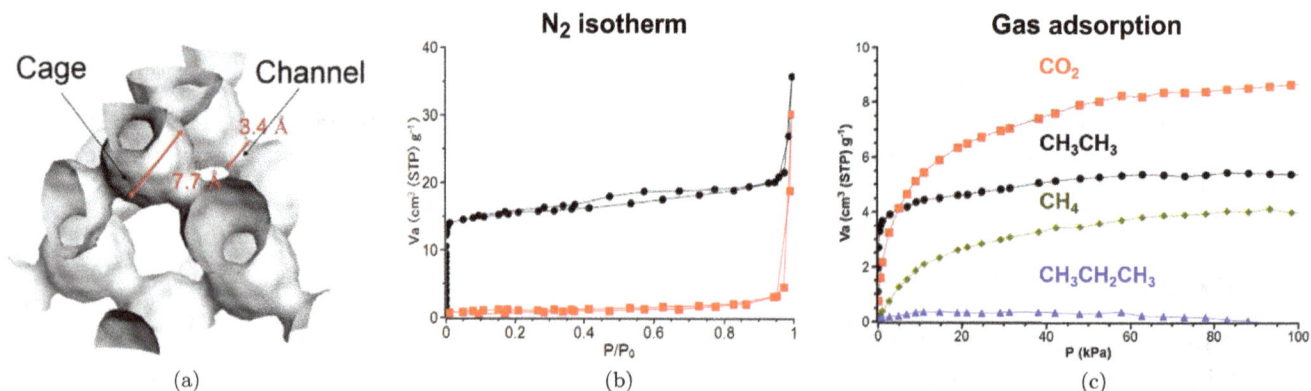

Figure 10.   (a) Micropores in Mo-V-Bi oxide, (b) N_2 isotherm of (black) Mo-V-Bi oxide and (red) K exchanged Mo-V-Bi oxide at 77 K, (c) CO_2, ethane, methane, and propane isotherm at 298 K.

The other oxides such as Mo–Zn oxide, Mo–Mn oxide, Mo–Fe oxide, and Mo–Co oxide have smaller micropore volume. The reasons for this are that (1) these oxides have lower thermal stability and all NH_4^+ cannot be removed, and (2) Na^+ counter cation, which comes from a Mo source, Na_2MoO_4, cannot be removed by thermal treatment, and so blocks micropores.

Na^+-containing Mo–Mn oxide and Mo–Zn oxide have a smaller micropore volume.[32] However, Na^+-containing Mo–Mn oxide and Mo–Zn oxide are more basic than the NH_4^+-containing compound and selectively adsorb CO_2 compare to methane. We found that Na^+-containing Mo–Mn oxide and Mo–Zn oxide can be used for separation of CO_2 and methane.

Ion exchange property

TPD-MS analysis indicates that there are two NH_3 desorption peaks at the top at ca. 440 K and 620 K (Figure 3), indicating that there are two kinds of acidic sites (two kinds of NH_4^+: weakly interacted NH_4^+ and strongly interacted NH_4^+) (Table 3). NH_4^+ in the Mo–V–Bi oxide can be exchanged with other cations such as H^+, Li^+, Na^+, K^+, Rb^+, and Cs^+ by mixing with the corresponding salts in water.[25]

H^+, Li^+, and Na^+ selectively exchange with weakly bound NH_4^+, and K^+, Rb^+, and Cs^+ selectively exchange with strongly bound NH_4^+.

Table 3. Amount of NH_4^+ in Mo–V–Bi oxide before and after ionexchange.

Cation	Formula	Amount of NH_4^+/ε-Keggin	
		NH_4^+ (weak)	NH_4^+ (strong)
NH_4^+ (before ion-exchange)	$(NH_4)_{2.8}H_{0.9}[\varepsilon\text{-VMo}_{9.4}V_{2.6}O_{40}Bi_2]$	0.7	1.4
H^+	$(NH_4)_{2.1}H_{0.7}H_{0.9}[\varepsilon\text{-VMo}_{9.4}V_{2.6}O_{40}Bi_2]$	0.0	1.6
Li^+	$(NH_4)_{2.6}Li_{0.2}H_{0.9}[\varepsilon\text{-VMo}_{9.4}V_{2.6}O_{40}Bi_2]$	0.3	1.4
Na^+	$(NH_4)_{2.2}Na_{0.6}H_{0.9}[\varepsilon\text{-VMo}_{9.4}V_{2.6}O_{40}Bi_2]$	0.2	1.4
K^+	$(NH_4)_{0.9}K_{1.9}H_{0.9}[\varepsilon\text{-VMo}_{9.4}V_{2.6}O_{40}Bi_2]$	0.6	0.0
Rb^+	$(NH_4)_{0.7}Rb_{2.1}H_{0.9}[\varepsilon\text{-VMo}_{9.4}V_{2.6}O_{40}Bi_2]$	0.2	0.0
Cs^+	$(NH_4)_{0.8}Cs_{2.0}H_{0.9}[\varepsilon\text{-VMo}_{9.4}V_{2.6}O_{40}Bi_2]$	0.3	0.0

Table 4. Acid catalyst property of Mo–V–Bi oxide for benzyl alcohol dehydration to produce benzyl ether.

Catalyst	Conv. (%)	Yield (%)	Side product (%)
No catalyst	5	0	
NH_4^+ (before ion exchange)	3	3	
Cal. 473 K	92	91	
Cal. 623 K	95	94	
Cal. 623 K, second use	100	97	
Cal. 673 K	98	91	Benzaldehyde (4%)
H^+	99	95	Benzaldehyde (3%)

Note: Condition — Amount of cat. 20 mg, benzylalcohol (10 mmol), 403 K, 3 h.

Acid catalyst property

The Mo–V–Bi oxide shows acid catalytic activity after the production of a protonic site (Table 4).[25] The as-synthesized Mo–V–Bi does not show catalytic activity. Calcination at 473 K that produces a weak acid site is enough to show the catalytic activity for benzyl alcohol dehydration. The compound is reused and filtration of the solid stops the reaction indicating that the catalyst is a heterogeneous one. Nanosized Mo–V–Bi oxide can be produced by controlling the production condition, and the nanosized Mo–V–Bi oxide shows better activity compare to large-size Mo–V–Bi oxide because the reaction occurs on the surface.[26]

Crystalline Complex Metal Oxides with Molecular Wire Structure

Nanowires have been attracting increasing attention due to their large surface area and quantum mechanical effects, resulting in unique properties for the materials. Among the various types of nanowires, molecular wires are more interesting, and they are composed of a repeating single molecular unit in a certain axis. The most common type of molecular wire is an organic or organometallic polymer,[31–34] which has been widely applied in nanotechnology, semiconductors, electrochemistry, and cell biology. A more interesting material is the molecular wire with an all-inorganic composition. However, the all-inorganic molecular wire is very rarely synthesized.

Herein, a family of transition metal oxide-based molecular wires is introduced. The molecular wire is formed by repeating a hexagonal molecular unit of $[XY_6O_{21}]$ (X = P, Se, Te. Y = W, Mo) along the *c*-axis, denoted as MoSeO,[35] MoTeO,[35] WPO,[36] WSeO,[37] and WTeO,[37] respectively. The molecular wires assemble in a hexagonal fashion with the interaction of water and ammonium cation to form crystals, and the molecular wires are isolable from the crystal. The ultrathin molecular wire-based material acts as an active acid catalyst.

Synthesis and structure of molecular wire

The crystalline transition metal oxides based on molecular wires were synthesized using a hydrothermal method. The aqueous precursor contained Mo/W species and central elements. The ratio of the Mo/W:central elements was mainly 6:1. The detailed reaction conditions are summarized in Table 5. The synthesis conditions, such as concentration of the starting solution, reaction time, reaction temperature, pH value of the precursor, and the starting compounds, affect the resulting materials. The acidic solution was suitable for material formation, and no solid would be recovered when the pH was higher than 5. Concentration, temperature, and time of synthesis would not highly affect the crystal structure of the material. So far, we have found that the crystalline materials based on transition metal oxide molecular wire can be formed using Mo/W as surrounding atoms and P, Se, and Te as central atoms. Finally, five different isostructural materials were obtained.

There were five isostructural materials. The materials showed similar XRD patterns, indicating that the crystal structure of the materials was the same (Figure 11 (A)). FTIR spectra of the materials were

Table 5.   Synthesis conditions for molecular wires3

Material	Mo/W source	Central element	pH	Temperature (°C)	Time (h)
MoSeO	AHM[a]	SeO_2	2.8	175	24
MoTeO	AHM	TeO_2/$VOSO_4$+$Te(OH)_6$	2.8	175	24
WSeO	AMT[b]	SeO_2	1.5	175	24
WTeO	AMT	TeO_2	1.5	175	24
WPO	AMT	H_3PO_3	1.5	175	24

Notes. [a]AHM: $(NH_4)_6Mo_7O_{24}\cdot4H_2O$.
[b]AMT: $(NH_4)_6H_2W_{12}O_{40}\cdot nH_2O$.

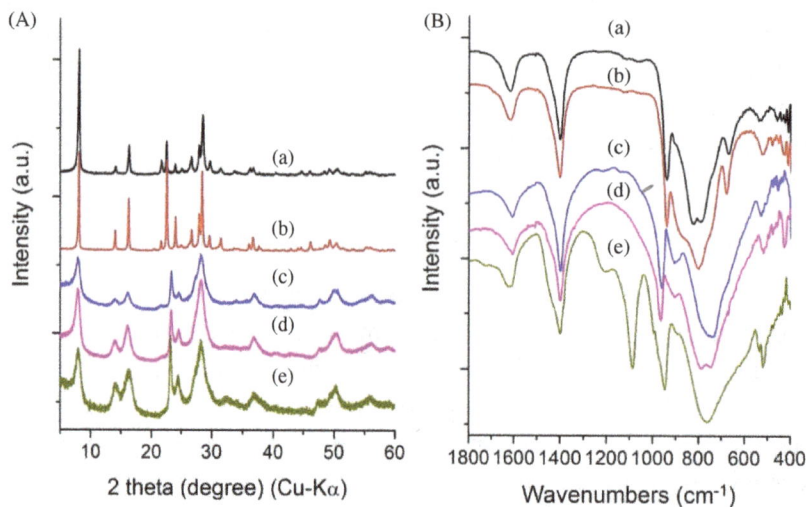

Figure 11.   A) XRD patterns and B) FTIR spectra of a) MoTeO, b) MoSeO, c) WTeO, d) WSeO, and e) WPO.

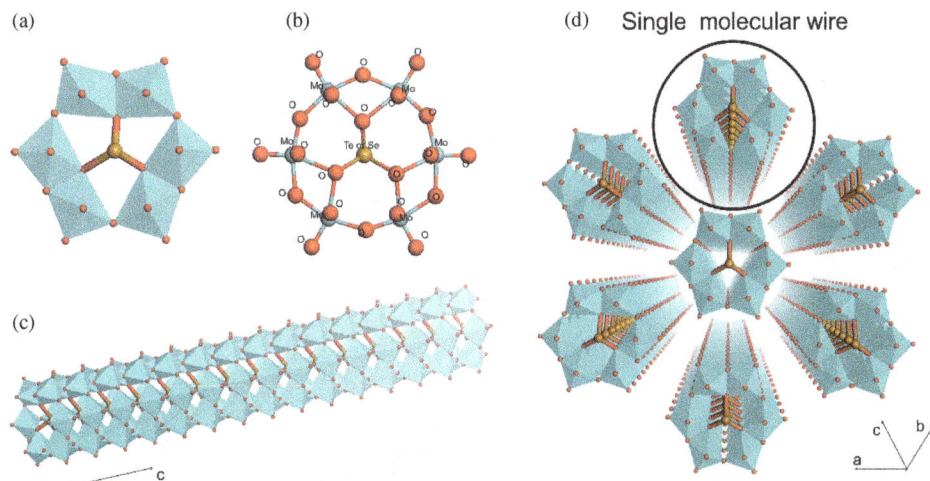

Figure 12. Structural representations. (a) Polyhedral representation of a hexagonal unit of $[Te^{IV}Mo^{VI}_6O_{21}]^{2-}$, (b) ball-and-stick representation of a hexagonal unit of $[Te^{IV}Mo^{VI}_6O_{21}]^{2-}$ with labels, (c) a single molecular wire of MoTeO. (d) Assembly of single molecular wires into crystalline MoTeO. Mo: blue, Te: brown, O: red.

also similar, indicating that the bonding structure of the materials was similar (Figure 11(B)). The materials have the same basic structure with different chemical composition. The Mo-based material was grown to be a well-crystallized sample, which was suitable for single-crystal structure analysis. For MoTeO, single-crystal X-ray structural analysis showed that six MoO_6 octahedra surrounded a central Te in the ab-plane, which constructed the hexagonal unit of $[Te^{IV}Mo^{VI}_6O_{21}]^{2-}$. The MoO_6 octahedra units were connected to each other through two edge-sharing oxygen atoms on one side with equatorial Mo–O bond lengths of 1.97(4)–1.99(6) Å and a corner-sharing oxygen atom on another side with equatorial Mo–O bond lengths of 1.79(6)–1.84(5) Å (Figures 12 (a) and 12(b)). The central Te ion inside the Mo–O cluster is connected to three oxygen atoms with bond lengths of 2.05(3)–2.05(4) Å. The hexagonal units are repeated along the c-axis to form (Figure 12(c)) molecular wires. The molecular wires were further packed parallel in a hexagonal fashion to form the crystalline material with cation species and water (Figure 12(d)). The nanowires were micrometre scale in length and ca. 1.2 nm in width.

For W-based materials, the crystallinity was poor, which was not good for X-ray-based structure analysis. Alternatively, the structure of the material could be visualized by HAADF-STEM. The cross-sectional image of WTeO exhibited hexagonal units corresponding to $[TeW_6O_{21}]^{2-}$ (Figure 13(a)). The hexagonal array with a periodicity of ca. 1.2 nm was also observed (Figures 13(b) and 13(c)). The hexagonal units were stacked with a layer distance of ca. 0.4 nm to form a molecular wire in the side views of WTeO. The atomic positions observed by HAADF-STEM were in good agreement with the structures based on single-crystal analysis (Figure 13 right). A single molecular wire with a diameter of ca. 1.2 nm was observed at the end of a bundle of molecular wires (Figure 13 (d)).

Isolation of the molecular wire

The molecular wire arrays in the crystals could be disassembled to form single molecular wires by a simple process. Based on the materials, there were three different processes for isolation of the molecular wire as summarized in Table 6, which will be discussed in detail.

In the case of W-based materials, the crystallinity of the materials was quite low, and therefore the direct ultrasonication or heating process could isolate the material to form molecular wires. WTeO and WSeO eas-

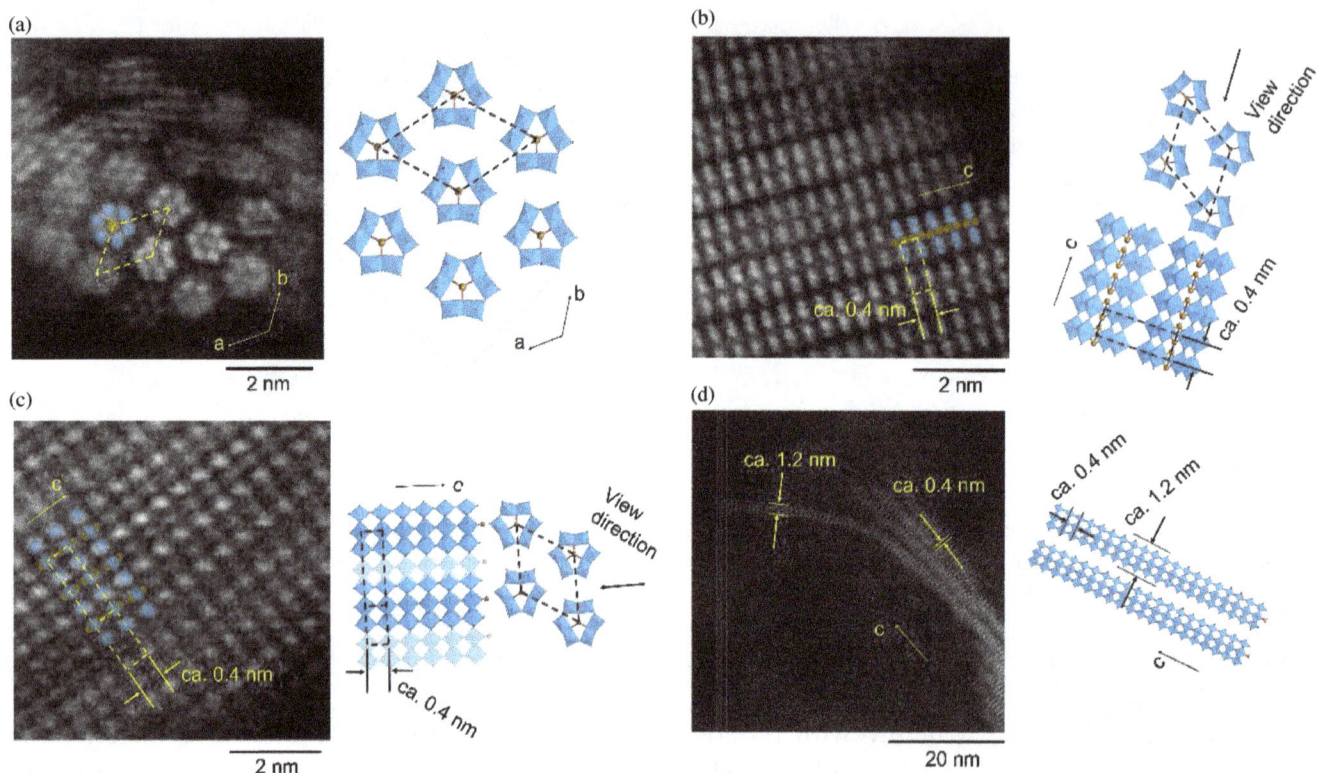

Figure 13.   High resolution HAADF-STEM images (left) and proposed structures (right) of a) WTeO in the a-b plane, b) in the (1 0 0) plane, c) in the (2 -1 0) plane, and d) the observation along the c-axis, dash line indicated the unit cell, W: blue, Te : brown and O: red.

Table 6.   Obtaining ultrafine nanowires from the crystals.

Material	Method	Solvent	Diameter (nm)	Length (nm)
WTeO	Sonication	Water	2–3	40–60
MoTeO	Proton exchange, sonication	Ethanol	2.5	20–80
TBA-WPO	Organic exchange	Acetonitrile	1–2	15–50

ily formed nanowires in aqueous solution. The SEM and TEM images exhibited ultrathin nanowires after dispersal in water (Figure 14). Single molecular wires could be observed, the diameters of which were ca. 1–2 nm. Size distribution based on the TEM observation demonstrated that the diameter of the material was 2–3 nm with a length of 40–60 nm.

The crystallinity of the Mo-based material was better than that of the W-based material. The dispersal of the molecular wire was able to be achieved by two steps. A further ion-exchange process before ultrasonication was necessary to break the crystal. For MoTeO, proton exchange enabled the molecular wire to be isolated easily. Ammonium cations in MoTeO were replaced by protons to form a proton-exchanged material, denoted as HMoTeO. The structure of the material did not change after proton exchange, while elemental analysis demonstrated that half of the ammonium cations were replaced. Proton exchange assisted the cracking of the materials. HMoTeO remained a rod-shaped material after proton exchange. There were several open gaps generated in the crystal surface. The crystals were broken into several small rods (Figure 15(b)). After ion exchange, ultrasonication was applied to the ion-exchanged material, which further separated the

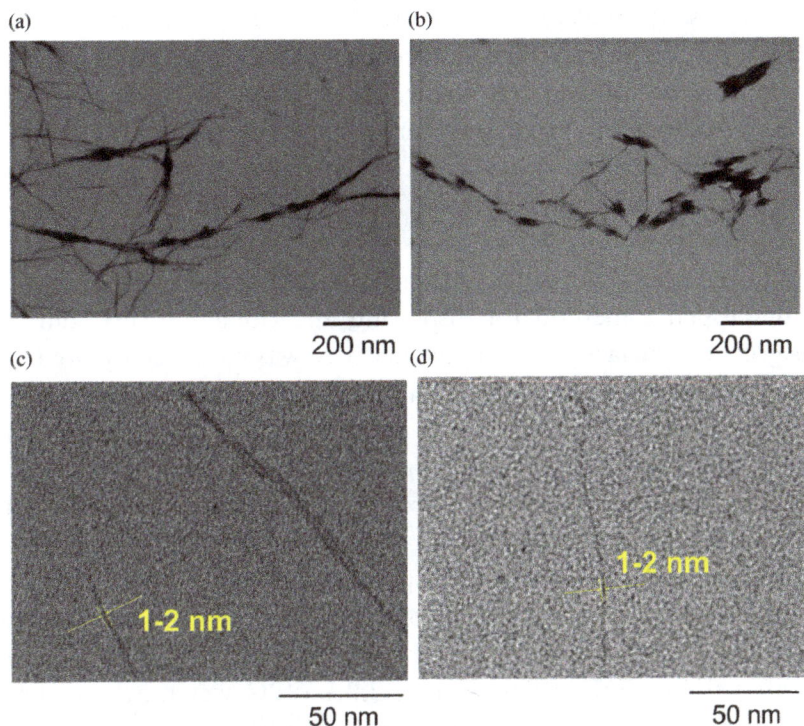

Figure 14. Electron microscopy images of the dispersed materials, TEM image of a) WTeO and b) WSeO, HR-TEM images of c) WTeO and d) WSeO.

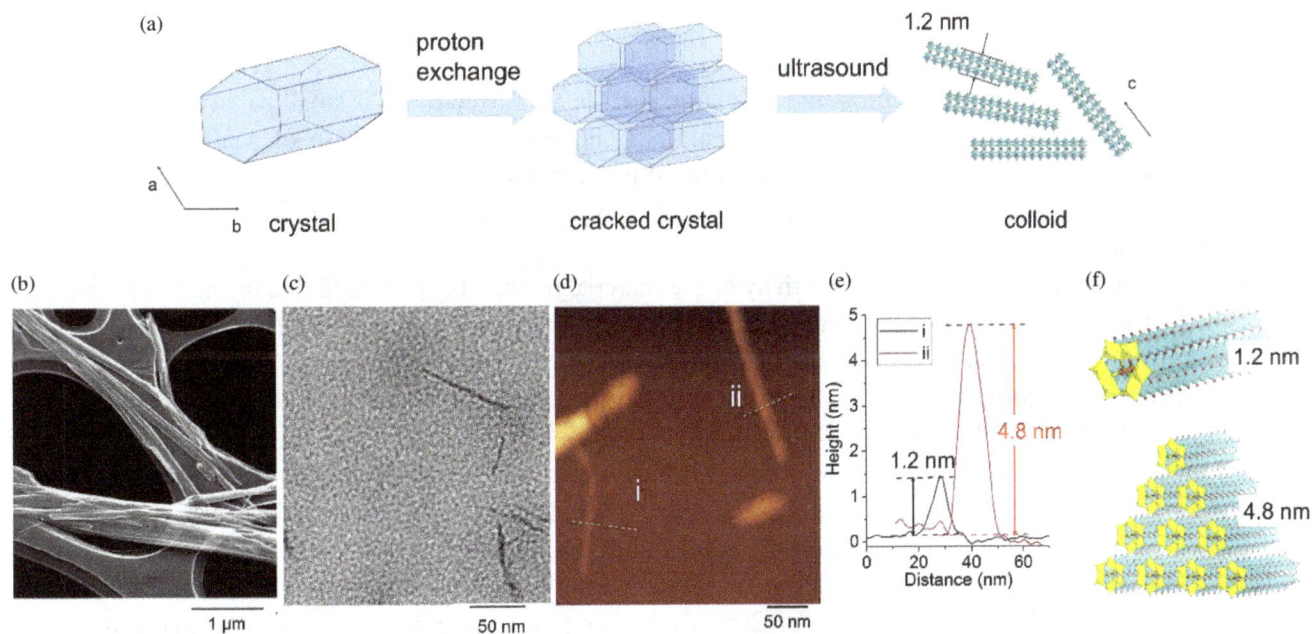

Figure 15. Isolation of crystals into molecular wires. (a) Isolation of HMoTeO to obtain small particles, (b) SEM images of HMoTeO, scale bar = 1 μm (c) HR-TEM images of a colloid sample of HMoTeO, scale bar = 50 nm (d) AFM images of a colloid sample of HMoTeO, scale bar = 50 nm, (e) line profiles from the AFM images, (f) proposed structures from the line profile analysis, with yellow highlighting the a-b plane of the materials.

material into nanowires, even single molecular wires. HR-TEM confirmed that nanowires with smaller sizes were obtained after ultrasound, and some isolated molecular wires with widths of 1.5 nm (Figure 15(c)) were observed. The isolated single molecular wire was further characterized by AFM, which showed that the thicknesses of typical particles were ca. 1.2 nm (particle i) and ca. 4.8 nm (particle ii) (Figures 15(d) and 15(e)), respectively. The thickness of the particle with 1.2 nm was consistent with that of a single HMoTeO molecular wire deduced from the crystallographic data on MoTeO (Figure 15(f)). The size distribution of the HMoTeO nanowires indicated that most of the nanowires had widths less than 10 nm and length between 20 nm and 80 nm, which were much smaller than those of the MoTeO crystals.

Furthermore, we developed a method for dispersing and obtaining ultrathin nanowires in organic media. To achieve this goal, the surface of the molecular wire was modified by organoammonium cations, and the dispersal of the molecular wire in liquid media was tuned. First, the original cation of the material was replaced by the organoammonium cation, such as tetraammonium bromide (TBAB), in an aqueous solution, denoted as TBA-WPO. After ion exchange, the distance between each molecular wire was increased by insertion of a cation species and the hexagonal unit along the c-axis almost did not change, indicating that the structure of the WPO molecular wire was stable. After the surface was modified by TBAB, the material could be readily isolated and dispersed in organic media, and the materials dispersed well in organic polar solvents such as acetonitrile and DMF, generating a transparent solution. The surface-modified material, TBA-WPO, was able to be rapidly dispersed by sonication in acetonitrile. TEM observation exhibited that the typical particle size of TBA-WPO was much smaller than that of WPO (Figures 16 (a)–(e)). Furthermore, the dispersion in acetonitrile generated a uniform particle of the molecular wire based on the distribution of the diameter and the length based on the TEM observation. The diameter of most of the particles was less than 3 nm and the maximum distribution length was 35 nm (Figure 16(f)).

Catalytic applications

Because the materials with nanoarchitecture had large surface areas due to their small particle size, nanomaterials were anticipated to have applications in catalysis (Figure 17). In the case of MoTeO, half of the ammonium cations in MoTeO were able to be replaced by protons, which acted as the acid site for catalysis. HMoTeO catalyzed esterification of ethanol with acetic acid. The activity of the material before isolation had low activity under the present conditions.

After ultrasound treatment (58%), the activity of the material remarkably increased due to the smaller particle size of the former. The catalyst was recyclable and reused three times with only a slight decrease in activity.

The acid site of the molecular wire was able to be created by calcination when the thermal stability of the material was high enough. WTeO was stable in air up to 350°C. The calcined WTeO was denoted as WTeOAC350. Note that 50% of the original ammonia was removed from WTeO, which left acid sites in the material for acid catalysis. WTeOAC350 was used as an acid catalyst for polysaccharide hydrolysis in an aqueous solution. Soluble polysaccharides, such as cellobiose (Table 7, entry 1), sucrose (Table 7, entry 2), and starch (Table 7, entry 3), were hydrolyzed by WTeOAC350 at 130°C for 4 h under hydrothermal conditions, and glucose was obtained as the main product with a high yield. Cellulose is insoluble and difficult to be reacted and was hydrolyzed at 175°C for 2 h (Table 7, entry 4). Hexoses (glucose and mannose) were the primary products, with other minor products such as 5-hydroxymethylfurfural (HMF), levulinic acid, and formic acid. Ball-milling decreased the crystallinity of cellulose, which further activated the cellulose. The activity of the reaction increased while using the ball-milled cellulose (Table 7, entries 5–6). In the absence of the catalyst, no products were detected (Table 7, entry 7). The recovery of the nanoparticle in the dispersed state was a challenging task, and the difficulty was caused by the ultrathin molecular wire. The

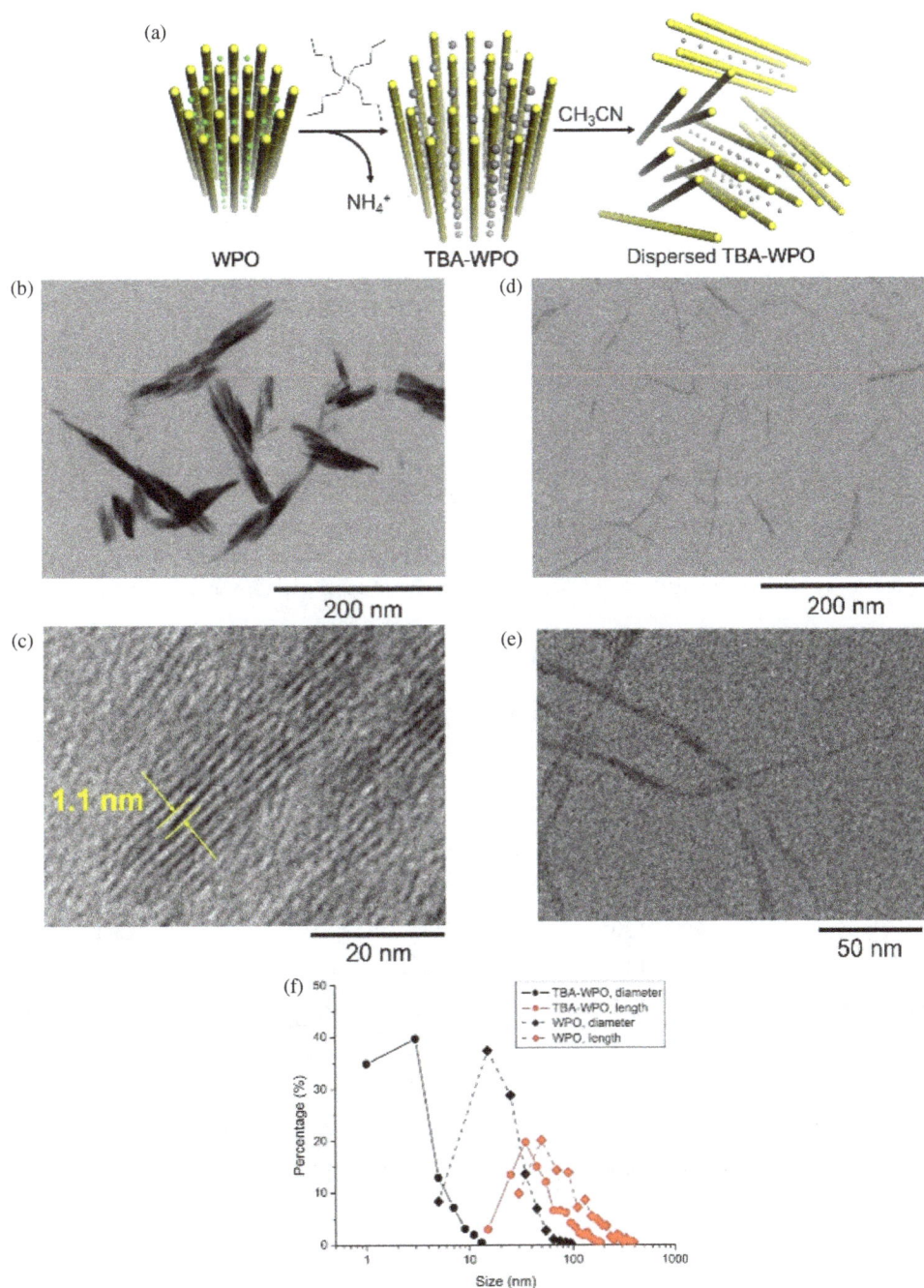

Figure 16. a) Surface modification of **WPO** nanowire and dispersal of **TBA-WPO** in organic media, $\{[HPW_6O_{21}]^{2-}\}_n$ (yellow column), NH_4^+ (green sphere), and **TBA** (grey sphere). b) TEM image and c) HR-TEM image of **WPO** after dispersal in water, d) TEM image and e) HR-TEM image of **TBA-WPO** after dispersal in acetonitrile, and f) diameter and length distributions of the materials.

catalyst could be recovered by a high-speed centrifugation condition and reused. The used catalyst showed the same XRD results, indicating that the structure of the material did not change. A high catalytic activity of the recovered catalyst was achieved.[37]

The materials based on W or Mo showed multielectron transfer property. The material could be used as a catalyst for oxidation. After surface modification and tuning, the surface property of the material

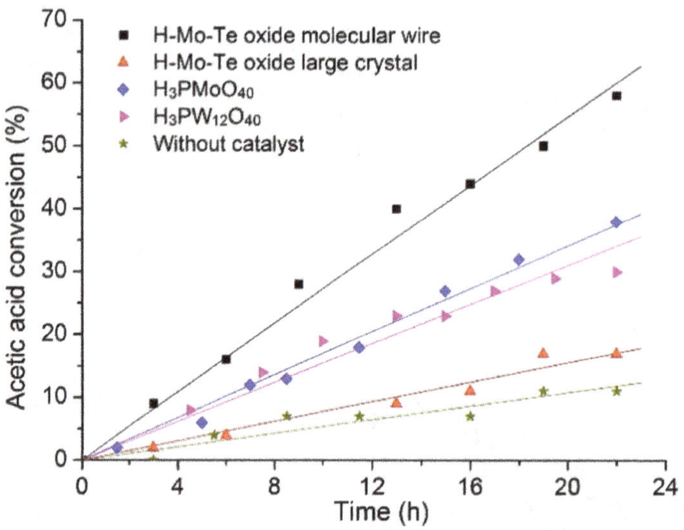

Figure 17. Ethanol esterification with acetic acid. Reaction conditions, catalyst, 0.1 mg; ethanol, 5 mL; acetic acid, 0.1 mL; decane as internal standard, 0.1 mL; reaction temperature, 92 °C. The selectivity of ethyl acetate was 99% and the carbon balance was over 95% in all cases.

Table 7. Biomass hydrolysis by WTeOAC350 under different conditions.[a]

Entry	Biomass	t(h)	T(°C)	Conv.(%)	Yield based on carbon (%) Glucose	Mannose	Formic acid	Levulinic acid	HMF	Total yield of organic (%)
1	Cellobiose	4	130	93.1	90.5	0	0.1	0.1	0.7	91.4
2	Sucrose	4	130	99	43.9	1.9	3.3	15.1	7.8	72.0
3	Starch	4	130	—	78.4	0.2	0.5	1.0	1.2	81.3
4	Macrocrystalline cellulose[b]	2	175	—	8.8	2.9	0.5	1.1	0.7	14.0
5	Ball-milled cellulose	2	175	—	19.0	7.2	0.7	2.5	1.2	30.6
6	Ball-milled cellulose[c]	2	175	—	25.8	7.8	3.0	8.5	2.0	47.1
7	Microcrystalline cellulose[d]	2	175	—	0	0	0	0	0	0

Notes: [a]Reaction condition: biomass: 0.308 mmol based on glucose unit, WTeO AC350: 0.05 g, water 0.5 mL, carbon balance was 86.7% (see detail in supporting information).
[c]Cellulose 0.185 mmol.
[d]Without catalyst.

changed from hydrophilic to hydrophobic. The nanowire of WPO was well dispersed in an organic solution. TBA-WPO was a well-defined catalyst for epoxidation of olefin in an organic solvent, because it dispersed well in organic solvents and because of its nanosize, which made the active site highly accessible to the substrate.[36] Cyclohexene was oxidized to cyclohexene oxide using H_2O_2 as an oxidant catalyzed by TBA-WPO in acetonitrile for 4 h at 60°C. The yield of cyclohexene oxide reached 85% (Table 8, entry 1), with selectivity of 91%. H_2O_2 efficiency was high, and most of the H_2O_2 was used for producing the cyclohexene oxide. Different olefins could be converted to corresponding olefin oxides under the current conditions (Table 8, entries 2–6). The result demonstrated that the material was an effective heterogeneous nanocata-

Table 8. Epoxidation of different olefins catalyzed by TBA-WPO using H_2O_2.[a]

Entry	Substrate	Time (h)	Sel. (%)	Yield (%)[b]
1	Cyclohexene	4	91	85
2	Cyclooctene	4	99	85
3	Cyclopentene	4	91	94
4	1-octene	6	98	14
5	1-octene	8	98	19
6	Styrene	4	34 (66)	24 (44)

Notes. [a] Reaction conditions: TBA-WPO: 0.02 g, acetonitrile: 1 mL, olefin: 2.5 mmol, H_2O_2: 0.5 mmol, decane as internal standard, 60°C, 4 h.
[b] Yield based on H_2O_2.

lyst, which was suitable for a wide range of olefins. In the present research, the dispersed ultrathin molecular wires of TBA-WPO were recovered from the solution using high-speed centrifugation (48,000 G, 24 h) directly. After centrifugation, about 50% of the catalyst was recovered from the solution. XRD and FTIR showed that the structure of the material did not change during the reaction. The activity of the recovered catalyst was tested using the same reaction, and the yield of the epoxide was almost the same compared with the fresh catalyst.

Assembly of molecular wire to a 3D network

We found that the molecular wire was a building block which could be used to constructed new 3D transition metal oxide frameworks. Our strategy for constructing all-inorganic soft framework materials was the connection of molecular wires with metal cations. By adding linker ions in the precursor solution of the molecular wire, a 3D network based on the molecular wire was obtained using a self-assembly approach. Heating the precursor solution, including $(NH_4)_6H_2W_{12}O_{40} \cdot nH_2O$, SeO_2, and $Co(OAc)_2$, at 100°C for 17 h produced CoWSeO.[38]

The structure of CoWSeO was determined by powder XRD. The result demonstrated that the framework of the material formed via connection of the $\{[Se^{IV}W^{VI}_6O_{21}]^{2-}\}_n$ molecular wire with the Co ions in a hexagonal manner (Figures 18(a)–18(c)). The micropore was originally occupied by water or ammonium cation (Figure 18(d)). Atomic resolution HAADF-STEM images showed the cross-sectional image of the (001) plane of CoWSeO, which revealed the hexagonal unit of $[SeW_6O_{21}]^{2-}$. The side view of CoWSeO exhibited that the hexagonal units, $[SeW_6O_{21}]^{2-}$, were stacked with a layer distance of ca. 0.4 nm to form a molecular wire (Figure 19). The material showed unique water adsorption property, which caused the change of Co coordination, increase in the lattice parameter, and eventually alteration of the structure. The unique structural feature of the porous soft framework material resulted in the unique water adsorption property, and a high level of adsorption of water was achieved. The structure of the material changed reversibly with changes to the water content in the structure.

Crystalline Microporous Vanadotungstate

A recent progress on the crystalline microporous transition metal oxide is zeolitic vanadotungstate (Figure 20).[39] The building block of the vanadotung state ($[W_4O_{16}]$) comprised four W–O octahedra in a

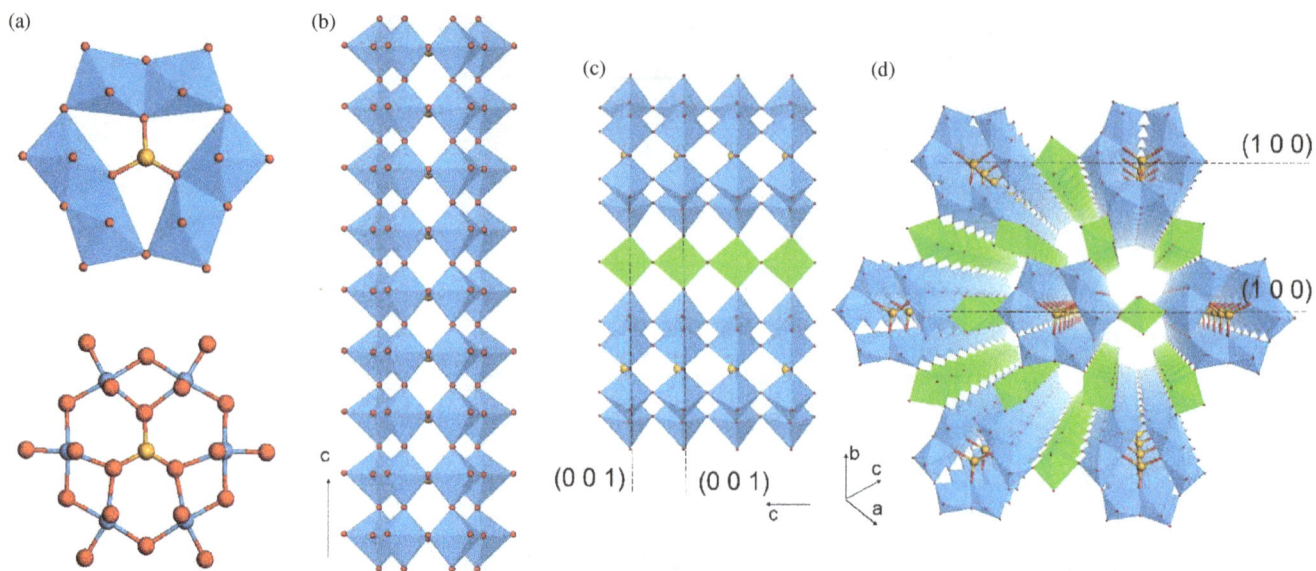

Figure 18.   a) The polyhedral representation and ball-stick representation of the hexagonal building block of $[SeW_6O_{21}]^{2-}$, b) the polyhedral representation of the $\{[SeW_6O_{21}]^{2-}\}_n$ molecular wire, c) the connection of the $\{[SeW_6O_{21}]^{2-}\}_n$ molecular wires with Co ions, d) the 3D structure of **CoWSeO·H$_2$O**, W: light blue, Se: yellow, Co: green, O: red, N: deep blue.

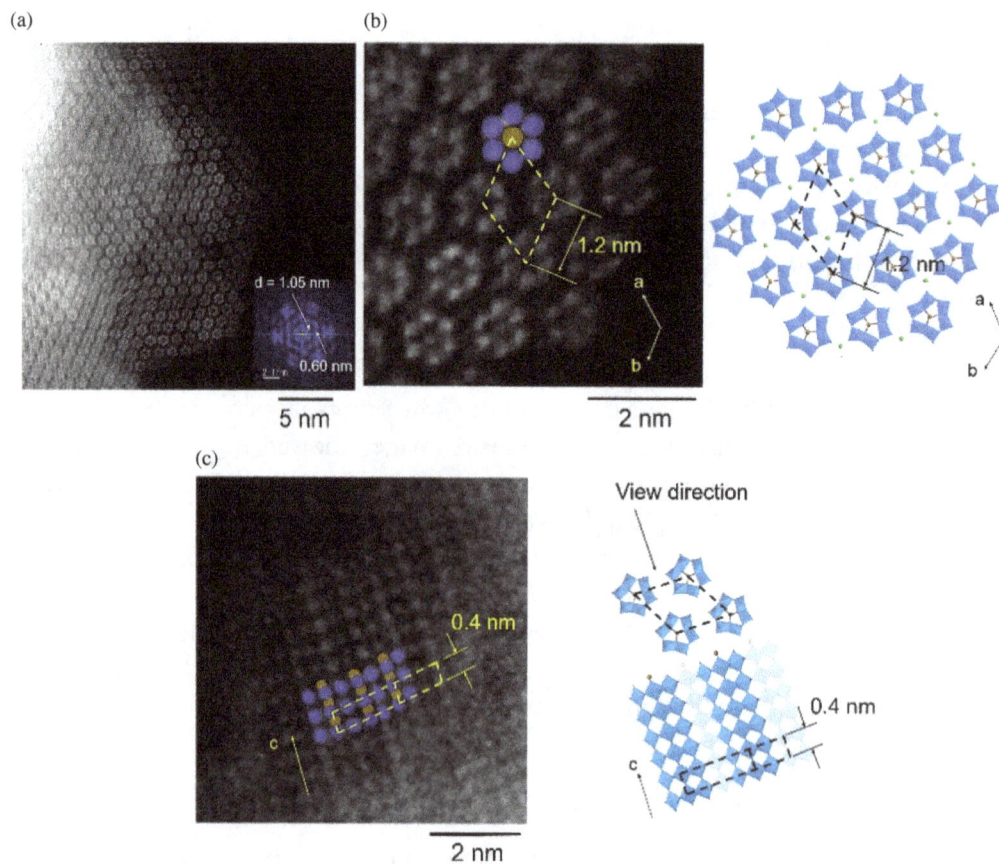

Figure 19.   HAADF-STEM images of a) CoWSeO in the a-b plane, b) enlarged HAADF-STEM image of CoWSeO in the a-b plane (left) with the proposed structure (right), and c) CoWSeO along the c-axis (left) with the proposed structure (right). Dashed line indicates the unit cell of the material; W: blue, Se: yellow, Co: green.

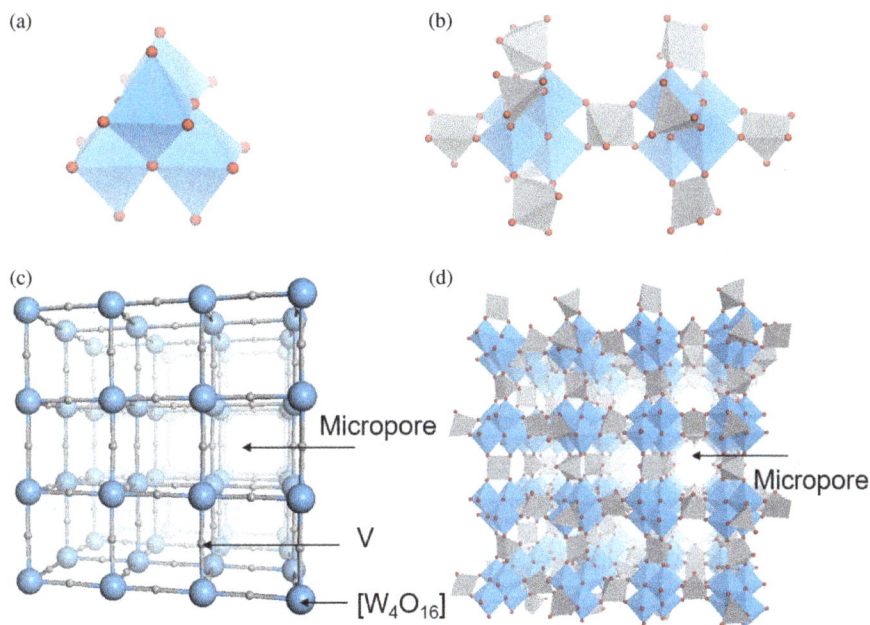

Figure 20. a) Building block of $[W_4O_{16}]$, b) connection of $[W_4O_{16}]$ with V linker, c) schematic representation of the framework, and d) crystal structure of vanadotungstate, W: blue, V: gray, O: red.

tetrahedral manner. The building block of the material was connected by six linked VO_5 in an octahedral fashion. The micropore of the material was sourronded by eight $[W_4O_{16}]$ building blocks with linkers, which was originally occupied by water and K^+. The micropores of the material were accessible to small molecules. The material showed great application potentials in separation and catalysis.

Conclusive Remarks

Solid-state catalysts, particularly complex metal oxide catalysts, for catalytic selective oxidation are quite complex, the dynamic changes of the active sites occurs during the redox reaction conditions. This situation has long prevented the deep understanding of the true catalytically active site at a nanoscale. For overcoming this situation, creation of catalytically active 3D crystalline materials based on complex metal oxides must be one of the most desired research direction among various possible scientific approaches. The materials introduced in this chapter are good examples and clearly encourage catalysis scientists to create revolutionary complex metal oxide catalysts.

References

1. *Selective Oxidation by Heterogeneous Catalysis*. Centi, C.; Cavani, F.; Trifiro, F. (Eds.), Kluwer Academic/Plenum Publishers: New York, 2001.
2. Cheetham, A.; Ferey, G.; Loiscau, T., Open-Framework Inorganic Materials, *Angew. Chem. Int. Ed.*, 1999, 38, 3268–3292.
3. Katou, T.; Vitry, D.; Ueda, W., Hydrothermal Synthesis of a New Mo-V-O Complex Metal Oxide and Its Catalytic Activity for the Oxidation of Propane. *Chem. Lett.* 2003, 32, 1028–1029.
4. Sadakane, M.; Watanabe, N.; Katou, T.; Nodasaka, Y.; Ueda, W., Crystalline Mo_3VO_x Mixed-Metal-Oxide Catalyst with Trigonal Symmetry. *Angew. Chem. Int. Ed.* 2007, 46, 1493–1496.
5. Ishikawa, S.; Ueda, W., Microporous Crystalline Mo-V Mixed Oxides for Selective Oxidations. *Catal. Sci. Technol.* 2016, 6, 617–629.
6. Sadakane, M.; Kodato, K.; Kuranishi, T.; Nodasaka, Y.; Sugawara, K.; Sakaguchi, N.; Nagai, T.; Matsui, Y.; Ueda, W., Molybdenum-Vanadium-Based Molecular Sieves with Microchannels of Seven-Membered Rings of Corner-Sharing Metal Oxide Octahedra. *Angew. Chem. Int. Ed.* 2008, 47, 2493–2496.

7. Sadakane, M.; Kodato, K.; Kuranishi, T.; Nodasaka, Y.; Sugawara, K.; Sakaguchi, N.; Nagai, T.; Matsui, Y.; Ueda, W., Molybdenum-Vanadium-Based Molecular Sieves with Microchannels of Seven-Membered Rings of Corner-Sharing Metal Oxide Octahedra. *Angew. Chem. Int. Ed.* 2008, 47, 2493–2496.

8. Ueda, W., Establishment of Crystalline Complex Mo-V-Oxides as Selective Oxidation Catalysts. *J. Jpn. Petrol. Inst.* 2013, 56, 122–132.

9. Tichy, J., Oxidation of Acrolein to Acrylic acid over Vanadium-Molybdenum Oxide Catalysts. *Appl. Catal. A: Gen.* 1997, 157, 363–385.

10. Kampe, P.; Giebeler, L.; Samuelis, D.; Kunert, J.; Drochner, A.; Haaß, F.; Adams, A. H.; Ott, J.; Endres, S.; Schimanke, G.; Buhrmester, T.; Martin, M.; Fuess, H.; Vogel, H., Heterogeneously Catalysed Partial Oxidation of Acrolein to Acrylic Acid-Structure, Function and Dynamics of the V-Mo-W Mixed Oxides. *Phys. Chem. Chem. Phys.* 2007, 9, 3577–3589.

11. Heid, M.; Knoche, S.; Gorra, N.; Ohlig, D.; Drochner, A.; Etzold, B. J. M.; Vogel, H., Dynamics of Bulk Oxygen in the Selective Oxidation of Acrolein, *ChemCatChem* 2017, 9, 2390–2398.

12. Gärtner, C. A.; Veen, A. C.; Lercher, J. A., Oxidative Dehydrogenation of Ethane: Common Principles and Mechanistic Aspects. *ChemCatChem* 2013, 5, 3196–3217.

13. Zhu, Y.; Sushko, P. V.; Melzer, D.; Jensen, E.; Kovarik, L.; Ophus, C.; Sanchez, M. S.; Lercher, J. A.; Browning, N. D., Formation of Oxygen Radical Sites on MoVNbTeO$_x$ by Cooperative Electron Redistribution. *J. Am. Chem. Soc.* 2017, 139, 12342–12345.

14. Sadakane, M.; Endo, K.; Kodato, K.; Ishikawa, S.; Murayama, T.; Ueda, W., Assembly of a Pentagonal Polyoxomolybdate Building Block, $[Mo_6O_{21}]^{6-}$, into Crystalline MoV Oxides. *Eur. J. Inorg. Chem.* 2013, 10–11, 1731–1736.

15. Konya, T.; Katou, T.; Murayama, T.; Ishikawa, S.; Sadakane, M.; Buttrey, D. J.; Ueda, W., An Orthorhombic Mo$_3$VO$_x$ Catalyst Most Active for Oxidative Dehydrogenation of Ethane among Related Complex Metal Oxides. *Catal. Sci. Technol.* 2013, 3, 380–387.

16. Pyrz, W. D.; Blom, D. A.; Sadakane, M.; Kodato, K.; Ueda, W.; Vogt, T.; Buttrey, D. J., Atomic-Level Imaging of Mo-V-O Complex Oxide Phase Intergrowth, Grain Boundaries, and Defects using HAADF-STEM. *Proc. Natl. Acad. Sci. USA* 2010, 107, 6152–6157.

17. Pyrz, W. D.; Blom, D. A.; Sadakane, M.; Kodato, K.; Ueda, W.; Vogt, T., Buttrey, D. J., Atomic-Scale Investigation of Two-Component MoVO Complex Oxide Catalysts Using Aberration-Corrected High-Angle Annular Dark-Field Imaging. *Chem. Mater.* 2010, 22(6), 2033–2040.

18. Andrushkevich, T. V., Heterogeneous Catalytic Oxidation of Acrolein to Acrylic Acid: Mechanism and Catalysts. *Catal. Rev. Sci. Eng.* 1993, 35, 213–259.

19. Thorsteinsn, E. M.; Wilson, T. P.; Young, F. G.; Kasai, P. H., The Oxidative Dehydrogenation of Ethane over Catalysts Containing Mixed Oxides of Molybdenum and Vanadium. *J. Catal.* 1978, 52, 116–132.

20. Ishikawa, S.; Yi, X.; Murayama, T.; Ueda, W., Heptagonal Channel Micropore of Orthorhombic Mo$_3$VO$_x$ as Catalysis Field for the Selective Oxidation of Ethane. *Appl. Catal. A: Gen.* 2014, 474, 10–17.

21. Ishikawa, S.; Yi, X.; Murayama, T.; Ueda, W., Catalysis Field in Orthorhombic Mo$_3$VO$_x$ Oxide Catalyst for the Selective Oxidation of Ethane, Propane and Acrolein. *Catal. Today* 2014, 238, 35–40.

22. Ishikawa, S.; Kobayashi, D.; Konya, T.; Ohmura, S.; Murayama, T.; Yasuda, N.; Sadakane, M.; Ueda, W., Redox Treatment of Orthorhombic Mo$_{29}$V$_{11}$O$_{112}$ and Relationships between Crystal Structure, Microporosity and Catalytic Performance for Selective Oxidation of Ethane. *J. Phys. Chem. C* 2015, 119, 7195–7206.

23. Goto, Y.; Shimizu, K.; Murayama, T.; Ueda, W., Hydrothermal Synthesis of Microporous W-V-O as an Efficient Catalyst for Ammoxidation of 3-picoline. *Appl. Catal. A: Gen.* 2016, 509, 118–122.

24. Goto, Y.; Shimizu, K.; Kon, K.; Toyao, T.; Murayama, T.; Ueda, W., NH$_3$-efficient ammoxidation of toluene by hydrothermally synthesized layered tungsten-vanadium complex metal oxides. *J. Catal.* 2016, 344, 346–353.

25. Zhang, Z.; Sadakane, M.; Murayama, T.; Izumi, S.; Yasuda, N.; Sakaguchi, N.; Ueda, W., Tetrahedral connection of ε-Keggin-type polyoxometalates to form an all-inorganic octahedral molecular sieve with an intrinsic 3D pore system. *Inorg. Chem.* 2014, 53, 903–911.

26. Zhang, Z.; Sadakane, M.; Murayama, T.; Ueda, W., Investigation of the formation process of zeolite-like 3D frameworks constructed with ε-Keggin-type polyoxovanadomolybdates with binding bismuth ions and preparation of a nano-crystal. *Dalton Trans.* 2014, 43, 13548–13590.

27. Zhang, Z.; Sadakane, M.; Murayama, T.; Sakaguchi, N.; Ueda, W., Preparation, Structural Characterization, and Ion-exchange Properties of two New Zeolite-like 3D Frameworks Constructed by ε-Keggin-type Polyoxometalates with Binding Metal Ions, $H_{11.4}[ZnMo_{12}O_{40}Zn_2]^{1.5-}$ and $H_{7.5}[Mn_{0.2}Mo_{12}O_{40}Mn_2]^{2.1-}$. *Inorg. Chem.* 2014, 53, 7309–7318.

28. Zhang, Z.; Sadakane, M.; Noro, S.I.; Murayama, Kamachi, T.; Yoshizawa, K.; Ueda, W., Selective Carbon Dioxide Adsorption of ε-Keggin-type Zincomolybdate-based Purely-Inorganic 3D Frameworks. *J. Mater. Chem. A* 2015, 3, 746–755.

29. Zhang, Z.; Ishikawa, S.; Tsuboi, Y.; Sadakane, M.; Murayama, T.; Ueda, W., New Crystalline Complex Metal Oxides Created by Unit-Synthesis and Their Catalysis Based on Porous and Redox Properties. *Faraday Discuss* 2016, 188, 81–98.

30. Igarashi, T.; Zhang, Z.; Haioka, T.; Iseki, N.; Hiyoshi, N.; Sakaguchi, N.; Kato, C.; Nishihara, S.; Inoue, K.; Yamamoto, A.; Yoshida, H.; Tsunoji, N.; Ueda, W.; Sano, T.; Sadakane, M., Synthesis of ε-Keggin-type Cobaltomolybdate-based 3D Framework Material and Characterization Using Atomic-scale HAADF-STEM and XANES. *Inorg. Chem.* 2017, 56, 2042–2049.

31. Gothard, C. M.; Rao, N. A.; Nowick, J. S., Nanometer-Sized Amino Acids for the Synthesis of Nanometer-Scale Water-Soluble Molecular Rods of Precise Length. *J. Am. Chem. Soc.* 2007, 129, 7272–7273.

32. Sakaguchi, H.; Matsumura, H.; Gong, H., Electrochemical Epitaxial Polymerization of Single-Molecular Wires. *Nat. Mater.* 2004, 3, 551–557.

33. Davis, W. B.; Svec, W. A.; Ratner, M. A.; Wasielewski, M. R., Molecular-Wire Behaviour in p-Phenylenevinylene Oligomers. *Nature* 1998, 396, 60–63.

34. Mahato, R. N.; Lülf, H.; Siekman, M. H.; Kersten, S. P.; Bobbert, P. A.; de Jong, M. P.; De Cola, L.; van der Wiel, W. G., Ultrahigh Magnetoresistance at Room Temperature in Molecular Wires. *Science* 2013, 341, 257–260.

35. Zhang, Z.; Murayama, T.; Sadakane, M.; Ariga, H.; Yasuda, N.; Sakaguchi, N.; Asakura, K.; Ueda, W., Ultrathin Inorganic Molecular Nanowire Based on Polyoxometalates. *Nat. Commun.* 2015, 6, 7731.

36. Zhang, Z.; Sadakane, M.; Hara, M.; Ueda, W., Ultrathin Anionic Tungstophosphite Molecular Wire with Tunable Hydrophilicity and Catalytic Activity for Selective Epoxidation in Organic Media. *Chem. Eur. J.* 2017, 23, 17497–17503.

37. Zhang, Z.; Sadakane, M.; Hiyoshi, N.; Yoshida, A.; Hara, M.; Ueda, W., Acidic Ultrafine Tungsten Oxide Molecular Wires for Cellulosic Biomass Conversion. *Angew. Chem. Int. Ed.* 2016, 55, 10234–10238.

38. Zhang, Z.; Sadakane, M.; Noro, S.; Hiyoshi, N.; Yoshida, A.; Hara, M.; Ueda, W., The Assembly of an All-Inorganic Porous Soft Framework from Metal Oxide Molecular Nanowires. *Chem. Eur. J.* 2017, 23, 1972–1980.

39. Zhang, Z.; Zhu, Q.; Sadakane, M.; Murayama, T.; Hiyoshi, N.; Yamamoto, A.; Hata, S.; Yoshida, H.; Ishikawa, S.; Hara, M.; Ueda, W.; A zeolitic vanadotungstate family with structural diversity and ultrahigh porosity for catalysis. *Nat. Commun.* 2018, 9, 3789.

Index

Material Index